Liquid Crystals and Ordered Fluids

Liquid Crystals and Ordered Fluids

Proceedings of an American Chemical Society Symposium on Ordered Fluids and Liquid Crystals, held in New York City, September 10-12, 1969

Edited by
JULIAN F. JOHNSON
Department of Chemistry and
Institute of Materials Science
University of Connecticut
Storrs, Connecticut

and

ROGER S. PORTER
Chairman, Polymer Science and Engineering Program
University of Massachusetts
Amherst, Masachusetts

PLENUM PRESS • NEW YORK–LONDON • 1970

Library of Congress Catalog Card Number 76-110760

ISBN 978-1-4684-8216-4   ISBN 978-1-4684-8214-0 (eBook)
DOI 10.1007/978-1-4684-8214-0

Softcover reprint of the hardcover 1st edition 1990

A Division of Plenum Publishing Corporation
227 West 17th Street, New York, N.Y. 10011

United Kingdom edition published by Plenum Press, London
A Division of Plenum Publishing Company, Ltd.
Donington House, 30 Norfolk Street, London W.C.2, England

PREFACE

This volume contains papers presented at the Second Symposium on Ordered Fluids and Liquid Crystals held at the 158th National Meeting of the American Chemical Society, New York, September, 1969. The Symposium was sponsored by the Division of Colloid and Surface Chemistry.

The proceedings for the first symposium on this subject were published by the American Chemical Society in the Advances in Chemistry Series. In the preface to the volume for the first meeting held four years ago, we noted that research on liquid crystals had gone through tremendous fluctuations, with peaks of activity around 1900 and again in the early 1930's. The present period of high activity which started about 1960 has continued to exhibit acceleration. The reason for the persistent growth in the field is due to the increasing recognition of the important role played by liquid crystals in both biological systems and in items of commerce as diverse as detergents and electronic components. Additionally, more powerful and sophisticated instrumentation is providing a basis for understanding the properties of the liquid crystalline state as well as yielding incisive tests for the theories of mesophase structure which are only now reaching a state of maturity.

Julian F. Johnson

Roger S. Porter

CONTENTS

Thermal Phase Transitions in Biomembranes

Joseph M. Steim

Chemistry Department, Brown University

Providence, Rhode Island

Many years ago Davson and Danielli (1) suggested that the lipids in biological membranes are organized in a two-dimensional array, similar to the liquid-crystalline lamellar phase which phospholipids exhibit in excess water. Much of the membrane protein was assumed to be bound on each side to the polar ends of the lipid molecules. Although this bilayer concept, with minor modifications, is the most commonly accepted model for membrane structure, recently many investigators have suggested alternative arrangements which emphasize non-polar association of lipid hydrocarbon chains with proteins (2). Some of the current speculation and divergence of opinion can be attributed to a lack of definitive experimental approaches.

If in fact lipids exist in membranes in a conformation similar to the lamellar phase which free lipids assume in water, some unique property shared by both systems might be detected by a direct physical technique. Such a property is the reversible thermotropic gel-liquid crystal phase transition observed in phospholipid myelin forms in water. It has been studied by differential scanning calorimetry, differential thermal analysis, nuclear magnetic resonance spectroscopy, X-ray diffraction, and light microscopy (3,4,5). Unlike transitions between liquid-crystalline phospholipid mesophases (6), the transition does not necessarily result in a molecular rearrangement but arises from melting of the hydrocarbon interiors of lipid bilayers. The lipids can exist in the lamellar conformation both above and below the transition temperature.

Calorimetric observations of such transitions in both membranes and in dispersions of membrane lipids in water have been reported for Mycoplasma laidlawii (7,8). This report extends the observation of transitions to both gram negative and gram positive bacteria, by the use of differential scanning calorimetry, infrared spectroscopy, and proton nuclear magnetic resonance spectroscopy.

Experimental

The growth media were tryptose for M. laidlawaii (Strain B, PG9), phosphate-buffered salts with glucose for E. coli (Strain B650), and nutrient broth for M. lysodeikticus. All cultures were grown to stationary phase. The membranes of M. laidlawii were prepared by lysis in deionized water, those of M. lysodeikticus by treatment with lysozyme, and those of E. coli by lysis of spheroplasts prepared in lysozyme with ethylenediamine tetraacetic acid.

Samples of 90-100 mg of sedimented membranes (10-15% dry weight) were sealed in specially constructed stainless steel sample pans (8,9) and scanned in a Perkin-Elmer DSC-1B differential scanning calorimeter at 1 mcal/sec full-scale sensitivity and a heating rate of 5 degrees per minute. The calorimeter cell was mounted in a brass cylinder immersed in a refrigerated ethanol bath. The suspending medium for the samples was 0.1 M NaCl buffered at pH 8.0 with 0.01 M Tris. Lipids were extracted from membranes with chloroform-methanol 2:1 v/v, repeatedly dried in a rotary evaporator, and filtered to remove any proteolipid protein. Samples were dried over P_2O_5 in a vacuum, then suspended in the buffer used for membranes and allowed to equilibrate for several hours before examination in the scanning calorimeter.

E. coli membranes and lipids were prepared in the same way for nuclear magnetic resonance studies as for calorimetry except that the final suspending medium was 0.1 M NaCl in D_2O. Free induction decay (90°pulse) experiments were performed at Bell Telephone Laboratories through the courtesy of Dr. W. P. Slichter and Mr. D. D. Davis. The spectrometer operated at 30 MHz with diode detection. Infrared spectroscopy was carried out using air-dried films on KRS-5 (thallium bromide-iodide) plates in a Perkin-Elmer 257 spectrophotometer.

Results

Figure 1 shows representative thermograms of M. laidlawii, E. coli, and M. lysodeikticus membranes, and the lipids of M. laidlawii and E. coli. The peaks are broader than those of synthetic phospholipids (7), and occur

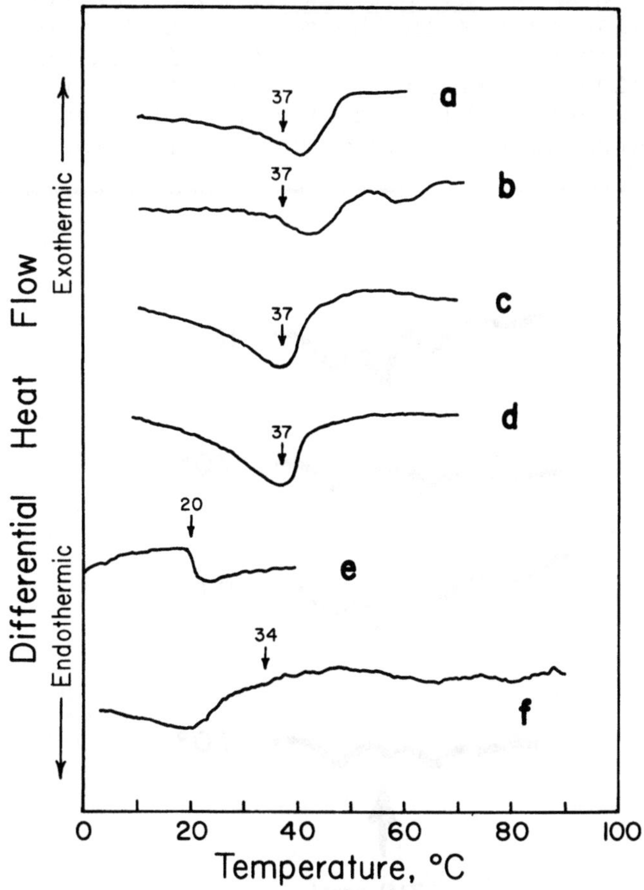

Figure 1. Calorimeter scans of intact membranes and extracted membrane lipids dispersed in buffer: (a) lipids extracted from M. laidlawii membranes; (b) intact M. laidlawii membranes; (c) lipids extracted from E. coli membranes; (d) and (e) intact E. coli membranes; (f) intact M. lysodeikticus membranes. In (e) the trace is inverted because the scan was made from high to low temperature. Protein denaturation can be seen in (b) and (f). Growth temperatures are indicated.

at the same temperature in both membranes and lipids extracted from them. We have previously shown that if the fatty acid composition of M. laidlawii membranes is altered by varying the growth medium, the transition temperature of both membranes and lipids can be made to change over an 80 degree range (8). A similar effect occurs with E. coli, in which the fatty acids become increasingly unsaturated as the growth temperature is decreased (10). Protein denaturation can be easily seen on the thermogram of M. laidlawii; such irreversible denaturation also occurs in the membranes from other organisms, but is often less apparent. As in the protein-free dispersions of lipids, the lipid transitions in the membranes are reversible and appear to be unaffected by thermal denaturation of the membrane proteins.

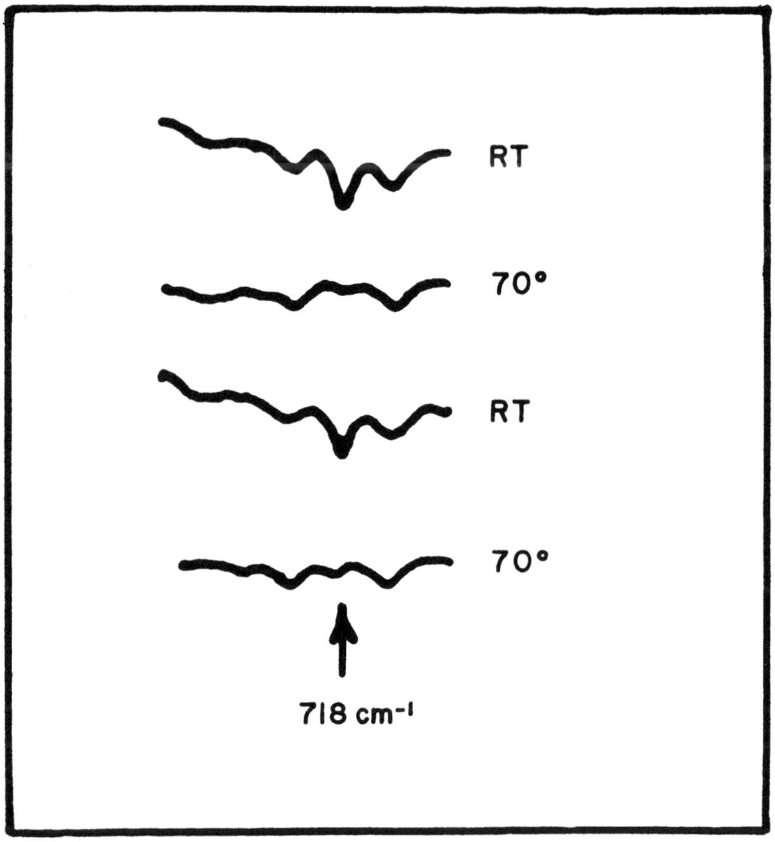

Figure 2. Infrared spectra of a dry film of E. coli membranes at room temperature and 70°C, showing reversibility of the change in conformation of fatty acid chains.

That the phase change involves a change in state of fatty acid hydrocarbon chains is suggested by the infrared spectra shown in Figure 2. A discrete peak for phospholipids in the neighborhood of 720 cm^{-1} has been reported by other workers (11), and is assigned to C-H rocking in polymethylene chains packed in the all trans conformation. A melt of hydrocarbon chains results in a decrease in absorbance of this band (12). Although calorimetric and magnetic resonance studies were carried out with wet samples, the infrared spectra were obtained with dry films and must be viewed cautiously. However, in the scanning calorimeter dried membranes show a phase transition similar to that observed in hydrated samples but with the transition temperature somewhat elevated. The results obtained with dried films, then, appear to have some relevance to the hydrated case.

An alternative method for detecting transitions is proton nuclear magnetic resonance spectroscopy. High-resolution NMR of intact membranes does not yield acceptable spectra (7,13) but broad-line or pulsed techniques are promising. The transition in E. coli membranes was followed by observing, as a function of temperature, the free induction decay signal following a 90 degree pulse (14). The components of the free induction decay, which is the Fourier transform of the steady-state NMR signal, were resolved from semi-logarithmic plots of the signal amplitude against time. The slope of the plots gives the transverse relaxation time, T_2. If more than one T_2 is present the log plot is the sum of several straight lines. The extrapolated intercept of each line on the ordinate is proportional to the number of protons which contribute to that particular relaxation time. The procedure is equivalent to resolving a broad-line spectrum into separate peaks and measuring the area under each peak, but has the advantage that modulation effects do not distort the lines. A representative plot, taken at 36°, is shown in Figure 3. The curve apparently can be resolved into three components, but the behavior of the first component, possessing the longest T_2 and hence arising from protons in molecules having the greatest molecular motion, is of particular interest. In Figure 4 its y intercept has been plotted against temperature. The curve, which is proportional to the number of mobile protons, indicates a phase change in the same temperature region as the transition detected by scanning calorimetry.

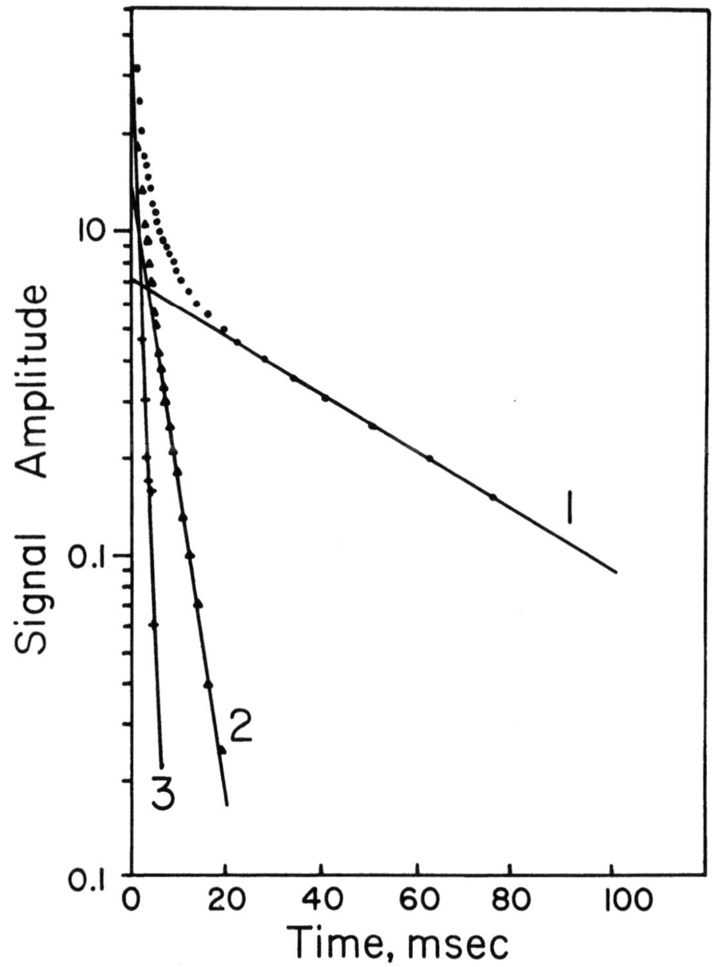

Figure 3. Resolution of the free induction decay signal for E. coli membranes into its components. Component 1, with a T_2 *of 45 msec, is of particular interest.*

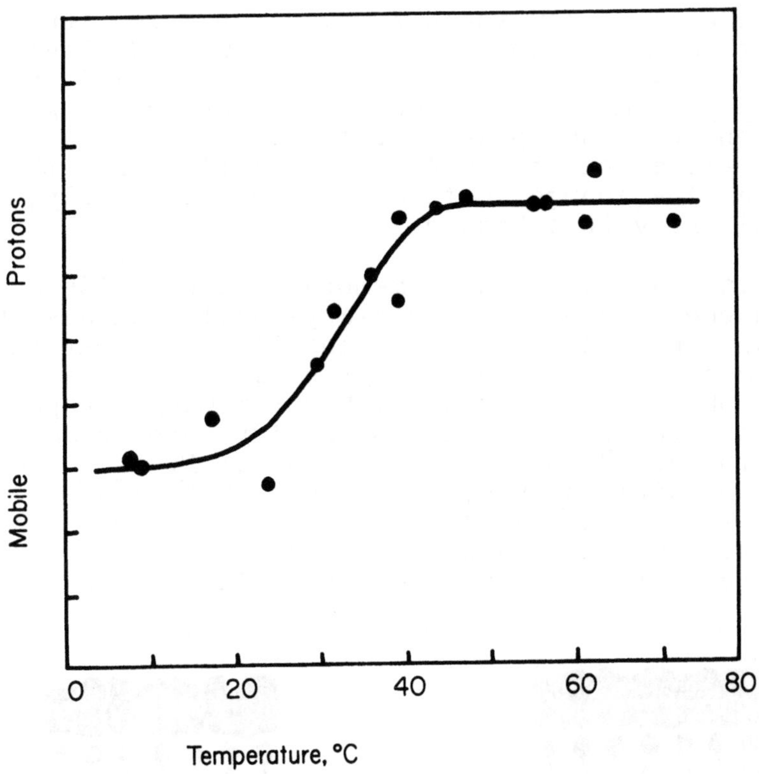

Figure 4. Detection of the phase change in E. coli membranes by pulsed NMR. The y intercept of component 1 (Figure 3) has been plotted as a function of temperature.

Discussion

There is little doubt that the transitions observed in membranes are associated with lipid and not protein, since they always occur at the same temperature in both membranes and dispersions of membrane lipid but can be varied over a wide temperature range by changing the fatty acid composition (7,8). The transition temperatures are unaffected either by thermal denaturation of protein or treatment with proteolytic enzymes (8). In view of the great similarity of these phase changes in natural systems to those observed in synthetic phospholipids, which are also dependent upon fatty acid composition, it is reasonable to conclude that the effect observed in membranes is an order-disorder transition among hydrocarbon chains. Infrared spectra of dry films, as well as low-angle X-ray diffraction studies, support this mechanism. Working at King's College, London, Dr. D. M. Engelman has shown that the "melting" of M. laidlawii membranes is accompanied by the progressive replacement of a sharp X-ray reflection at 4.2 Å with a diffuse reflection at 4.6 Å, an indication that the fatty acid chains change conformation from a closely packed hexagonal array to a liquid-like state (15).

It is possible that in membranes the lipids could exist in one of several liquid-crystalline phases, but identical lipid and membrane transition temperatures suggest a bilayer as the most probable structure. The lamellar phase is well established for many phospholipids in excess water [including phosphatidyl ethanolamine, which comprises 80-90% of the lipids of E. coli (16)]. Similar transitions in synthetic phospholipids are known to occur within the lamellar phase. It is suggested, therefore, that the transitions observed in membranes result from a melt of hydrocarbon chains within bilayers.

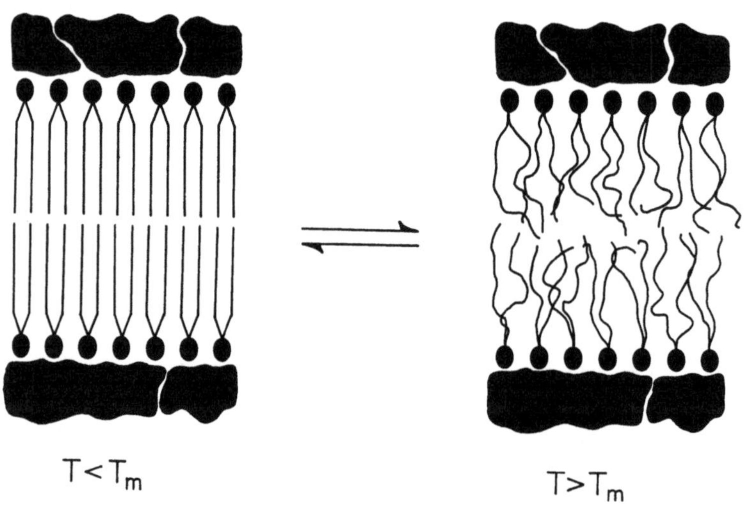

Figure 5. Schematic representation of the phase change observed in biomembranes. Below the transition temperature the hydrocarbon chains of the fatty acids are in an ordered crystalline-like state; above the transition temperature they are in a disordered liquid-like state. Protein is shown covering the polar ends of the phospholipid molecules.

The nature of the organisms exhibiting phase transitions should be noted. Although M. laidlawii is a unicellular free-living organism, it is not a bacteria but belongs to the group of pleuropneumonia-like organisms, order Mycoplasmatales. On the other hand E. coli and M. lysodeikticus are true bacteria, the former gram negative and the latter gram positive. Most bacteria can be roughly divided into these two groups, the gram negatives and the gram positives, and it is known that the envelopes of organisms within each group share common chemical and morphological features (17). With respect to the properties of their boundary layers, then, these two organisms can be taken as representative of most bacteria. It is likely that in the membranes of all the Mycoplasmateles, and in the membranes of most or all the bacteria, lipids are present in the bilayer conformation or at least are in a liquid-crystalline state.

The evidence cited here indicates that an appreciable portion of the lipids in membranes must be in a liquid-crystalline state. However, it is possible that some of the lipids in the membranes do not participate in the transition and are bound in some other way to proteins (e.g. by apolar association of fatty acid chains to apolar amine acid residues). Some technique is necessary to determine quantitatively the extent of the bilayer phase. At least two approaches are possible. One is to compare the enthalpy of transition of membranes with that of the equivalent amount of membrane lipids in water. Another is to treat quantitatively the results from NMR, as described earlier. Pulsed spin-echo sequences appear to be the most promising, and are being investigated in our laboratory at the present time. The quantitative calorimetric approach has been carried out with M. laidlawii membranes. The transitions are broad and accurate determinations of areas are difficult, but measurements on five separate runs indicate that the heat of transition of membranes is at least 75% of that observed for the equivalent amount of membrane lipids in water. Similar experiments have not been carried out with the bacteria.

It is likely that phase transitions will be directly observed in other organisms, especially the bacteria, but they cannot be expected to occur universally. In particular, mammalian plasma membranes must be ruled out because of the presence of cholesterol. Mixed with phospholipids, cholesterol in sufficient concentration (about 1/3 molar) behaves as a plasticizer and maintains the fatty acid chains in a liquid-like state so that no discrete transitions occur (5,18). Since order-disorder transitions cannot be observed calorimetrically in either the lipids extracted from membranes rich in cholesterol or in the native membranes, the calorimetric evidence for structure is inconclusive. Even without interference from cholesterol, phospholipids extracted from some membranes may not exhibit phase transitions which are sufficiently well defined to be detectable. We have found, for example, that phospholipids extracted from beef heart mitochondria show a weak, very broad transition that is quite difficult to detect calorimetrically. No calorimetric response was obtained from the lipid extracts of spinach chloroplasts, H. cutirubrum, or an L form of Proteus vulgaris. These systems may indeed possess a bilayer structure, but a different approach will be needed to demonstrate its existence.

Some physiological implications are apparent. In M. laidlawii growth ceases when the membrane transition temperature is above the temperature of the growth medium (8). This correlation between growth and the physical state of the lipids suggests that for proper physiological conditions to be maintained, the hydrocarbon chains must be in a liquid-like state. Although the transition temperature cannot rise above the growth temperature without producing damage to the cell, evidence obtained thus far indicates that the phase change can occur far below the growth temperature without producing aberrations. In the stationary phase E. coli may appear to be hovering on the brink of a phase transition, but M. lysodeikticus is not, and the transition temperature of M. laidlawii can be varied at will to as much as 60 degrees below the growth temperature. To a first-order approximation cells seem to require only that the membranes be in a liquid state. Because of the plasticizing effect of cholesterol mentioned earlier, speculations concerning the role of such phase transitions in most mammalian membranes are not valid.

This work was supported by United States Public Health Service grants GM 14696 and GM 10906.

References

1. Davson, H. A., and J. Danielli, J. Cell. Comp. Physiol 5, 495 (1935).
2. Korn, E. D., Ann. Rev. Biochem. 38, 263 (1969).

3. Chapman, D., P. Byrne, and G. G. Shipley, Proc. Roy. Soc. (London) A290, 115 (1966).
4. Chapman, D., R. M. Williams, and B. D. Ladbrooke, Chem. Phys. Lipids 1, 445 (1967).
5. Ladbrooke, B. D., R. M. Williams, and D. Chapman, Biochim. Biophys. Acta 150, 333 (1966).
6. Bangham, A. D., Prog. Biophys. Molec. Biol. 18,29 (1968).
7. Steim, J. M., Adv. Chem. Series (ACS) 84, 259 (1968).
8. Steim, J. M., M. E. Tourtellotte, J. C. Reinert, R. N. McElhaney, and R. L. Rader, Proc. Natl. Acad. Sci. 63, 104 (1969).
9. Steim, J. M., Perkin-Elmer Inst. News 19, 12 (1968).
10. Marr, A. G., and J. L. Ingraham, J. Bact. 84, 1260 (1962).
11. Chapman, D., V. B. Kamat, and R. J. Levene, Science 160, 314 (1968).
12. Chapman, D., The Structure of Lipids, Methuen and Co. Ltd., London (1965).
13. Chapman, D., V. B. Kamat, J. de Gier, and S. A. Penkett, J. Molec. Biol. 31, 101 (1968).
14. Kaufman, S., J. H. Gibbs, and J. M. Steim, unpublished data.
15. Engelman, D. M., J. Molec. Biol., in press.
16. Kanfer, J. N., and E. P. Kennedy, J. Biol. Chem. 238, 2919 (1963).
17. Osborn, M. J., Ann. Rev. Biochem. 38, 501 (1969).
18. Shah, D. O., and J. H. Schulman, Adv. Chem. Series (ACS) 84, 189 (1968).

CONDITIONS OF STABILITY FOR LIQUID-CRYSTALLINE PHOSPHOLIPID MEMBRANES

Demetrios Papahadjopoulos and Shinpei Ohki

Roswell Park Memorial Institute, Buffalo, New York and

State University of New York at Buffalo, New York

SUMMARY

Four different classes of phospholipids, extracted and purified from natural sources, have been studied as model membranes in the form of 1) multilamellar, liquid-crystalline particles; 2) unilamellar vesicles produced by sonication; and 3) thin films of approximately bimolecular thickness (black films). The study centers around the importance of the polar groups and the aqueous environment on the stability and permeability properties of the resulting membranes. All the phospholipids, studied at physiological ionic strength (0.1M) and ambient temperature (22°-30° C), produce stable membranes, exhibiting high electrical resistance and low permeability to ions. However, differences in the chemical composition of the head-groups of phospholipids, ionic strength, ion composition or pH of the aqueous phase, produce substantial differences in ion self-diffusion rates and electrical resistance. These observations suggest two types of dynamic conformational changes of the orientation of the phospholipid molecules. One: a micelle-bilayer transformation as a result of charge-charge interaction on the same plane, and second: an "inversion" of the orientation of molecules or clusters of molecules resulting from asymmetricity of the surface energy between the two planes of a bilayer.

Submitted to the A.C.S. as an invited participation to the "Second Symposium on Ordered Fluids and Liquid Crystals" to be held at the meeting of the American Chemical Society, New York, September 7-12, 1969.

INTRODUCTION

It is well recognized that phospholipids, similarly to other amphipathic compounds, undergo a series of phase transitions following variations of the temperature and the amount of water present in the system (26). Solid phases are present predominantly at low temperatures and low water content, while the liquid phase or isotopic solutions are to be found at high temperature-low water content systems (19, 7, 13). Between these extremes there exist a remarkable variety of liquid-crystalline structures exhibiting short-range disorder associated with a long range order (27). This polymorphism is certainly a property that makes phospholipids uniquely interesting as structural components of biological membranes.

However, when phospholipids are dispersed in excess water (more than 30-50 percent w/w) they tend to assume only a lamellar phase specifically defined as smectic mesophase (28, 8). Furthermore, when the temperature of the system is above a certain critical point (which depends on the fatty acid chain-length and degree of unsaturation) the lamellae tend to form closed systems which exhibit characteristic permeability properties (11, 14). This is the temperature at which the fatty acid chains melt, forming a liquid state within the two dimensions of each lamella, while the translamellar long-range forces keep the system crystalline in the third dimension (15). The liquid state is also apparently required for the formation of thin lipid films from an organic solvent-phospholipid solution (black films); it has been observed that stable black films can form only with phospholipids above their transition temperature (22, 24). An important consideration for biological membranes is that most of the naturally occurring phospholipids at ambient temperatures are above the transition point and consequently assume a liquid-crystalline structure.

Most of the published work on phospholipids as model membranes has been performed either with neutral amphipathic compounds (such as lecithins, cephalins, sphingomyelins, diglycerides, monoglycerides, etc.) or with complex, random mixtures of naturally occurring lipids (8, 50). However, recent evidence indicates that each head-group represented in the various classes of phospholipids (PE, PS, PI, PA, PG) confers characteristic properties to the rest of the molecule (25, 38, 43). The head-groups carrying formal charges are of particular importance to the interaction with cations or anions and in general to the permeability properties of the phospholipid membranes (40). In this paper, we describe the properties of membranes composed of a variety of naturally occurring phospholipid classes in purified form. The subject under investigation is the effect of the environmental salt solution, ionic strength, pH and bivalent metals, on the stability, permeability and electrical properties of different phospholipids. Thus a comparison has been made between the behaviour of neutral and acidic phospholipids and the different head-groups. In making such a comparison, we neglect the differences in the individual fatty acids that may be present in each class of phospholipids. However, the variety of fatty acids within each species tends to minimise individual differences (39). The phospholipid membranes described here are either the unilamellar vesicles (39) formed after sonication of the multilamellar liquid-crystalline particles (11) or the "black films" formed from a phospholipid solution in decane on a plastic frame (30) and separating two aqueous phases.

MATERIALS AND METHODS

Phospholipids were prepared as described earlier (39). Phosphatidylcholine (PC) was isolated from egg yolk; phosphatidic acid (PA) prepared by enzymatic hydrolysis of PC using cabbage phospholipase D; phosphatidyl glycerol (PG) was prepared by enzymatic glycerolysis of PC using the same enzyme (17); and finally phosphatidylserine (PS) was isolated from beef brain. The fatty acid chains present in each family of phospholipids was as determined before (39).

All the above phospholipids were kept in chloroform solutions in concentrations of approximately 20 mM/ml at low temperature (-30° C) in sealed ampules under nitrogen. Each ampule contained 10-20 μ moles total and a new ampule was opened for each experiment. All these precautions were taken in order to minimise air oxidation or hydrolysis and to reduce discrepancies between different experiments.

Phospholipid Vesicles

Preparation of the phospholipid vesicles was performed as described earlier (43) except that the whole operation was completed under a nitrogen atmosphere. The procedure involves 5 minutes mechanical agitation of the dry phospholipid with the salt solution, and 30 minutes sonication in a bath type sonicator. One hour equilibration at room temperature, is followed by passage through sephadex. Finally 0.5 ml aliquots of the fine suspension of washed vesicles is dialysed against 10 ml of aqueous salt solution changed at regular time intervals (usually one hour).

The salt solution used throughout the experiments contained NaCl (140 mM), Histidine (2mM), TES (2mM) and EDTA (0.1 mM) adjusted to pH 7.4 with NaOH. Tris-HCl has also been used as a buffer, but the combination of histidine-TES is preferable as it covers a much wider pH range.

The diffusion rates were expressed as percent of the total amount incorporated. The amount of phospholipid in each bag was 0.5 μ mole measured as phosphate.

Bilayer Membranes

The apparatus and method of formation of the black films was described earlier (31, 34). The phospholipid solutions were evaporated to dryness and dissolved in decane containing a small amount of methanol (0.5% v/v). The salt solution used, contained NaCl (100 mM), Tris-Cl (0.2 mM), and EDTA (0.1 mM). All measurements were performed at room temperature.

RESULTS AND DISCUSSION

Permeability and Electrical Properties of Phospholipid Membranes

Phospholipid model membranes in the form of either "black" films (bilayers) or vesicles are characterized by very low permeability to ions. Table I summarizes date from different laboratories for a useful comparison. The observed d.c. resistance for Egg PC bilayers varies from 2.0 x 10^5 to 7.7 x 10^8 Ω cm^2. It appears that when such membranes are formed from a solution containing chloroform the resistance is usually around the lower figure, approximately 10^6 Ω cm^2 (22, 36). When the same membranes are made from a decane solution the resistance is considerably higher, approximately 10^8 Ω cm^2 (20, 31). In view of the recently reported effect of chloroform and other general anaesthetics in increasing the permeability of PC membranes (10), it seems that the lower resistance could be the result of such anaesthetic action. However, the possibility of "border leaks" has also been shown as an important factor in producing lower resistance (20).

Further examination of the data contained in Table I reveals the following interesting points.

1. The d.c. resistance of membranes composed of purified acidic phospholipids (25, 31) is very high (10^7 -10^8 Ω cm^2), similar to the resistance of membranes made of PC, a neutral molecule. Thus it appears that one charge per molecule (1 charge/60 $Å^2$) does not produce instability in the integrity of the bilayer structure.

2. The observed d.c. resistance of bilayers (20, 24, 31) is in very good agreement with the resistance of phospholipid vesicles of same composition calculated from isotopically determined ionic fluxes (43). This is an important comparison indicating that both systems of model membranes (bilayers and vesicles) are similar in

TABLE I

Summary of data on electrical d.c. resistance and isotopic flux for K^+, Na^+, and Cl^-. The numbers in parenthesis given in the first column represent the reference number of the publication in which the data appeared.

Phospholipid	Observed d.c. resistance (Ω cm^2)	Ionic Flux, M ($\mu\mu$ moles/cm^2/sec)		Calculated d.c. resistance Ω cm^2		$E_{(Cl)}$ $\frac{KCal}{Mole}$	$\frac{M_{Cl}}{M_{Na}}$
		K or Na	Cl	K or Na	Cl		
PC bilayers (20)	$7.7x10^8$	-	-	-	-		
PC bilayers (34)	$1-5x10^7$	-	-	-	-		
PC bilayers (22)	$0.2-4.0x10^5$	-	-	-	-		
PC bilayers (36)	$2.0-7.0x10^5$	0.39	90.2	$6.7x10^5$	$3.0x10^4$	10.7	230
PC vesicles (43)	-	$0.4x10^{-3}$	$24x10^{-3}$	$5.8x10^8$	$1.1x10^7$	9.5	40
PS bilayers (31)	$0.5-1.0x10^8$	-	-	-	-		
PS vesicles (43)	-	$2.8x10^{-3}$	$28x10^{-3}$	$1.0x10^8$	$1.0x10^7$		
PI bilayers (25)	$0.5x5.0x10^8$	-	-	-	-		
PI vesicles (43)	-	$0.7x10^{-3}$	-	$3.7x10^8$	-		

their basic properties. Moreover, the resistance calculated from the efflux of $^{42}K^+$ through the small, sonicated vesicles of PC (43) is comparable to the highest reported values for PC bilayers (20).

3. The relative diffusion rates of $^{36}Cl^-$ and $^{42}K^+$ through vesicles (11,43) indicate a much higher rate for Cl^- ($^MCl/^MNa$=40), and similar diffusion rates for $^{42}K^+$ and $^{22}Na^+$. This appears to be in contradiction to the transference numbers calculated from the transmembrane potentials of bilayers composed of red blood cell lipids (6). A recent study with big, spherical bilayers (36) seems to have resolved the apparent contradiction. This latter work indicated that although the isotopic diffusion rate of Cl^- is much higher than that of Na^+($^MCl/^MNa$=230), the transference numbers calculated from the same membranes favor K^+ over Cl^- (tm^+=0.83, tm^-=0.17). It has been concluded (36) that the chloride flux contains a large component which does not contribute to a steady-state electric current. Consequently it may be stated that the ($^MCl/^MNa$) flux ratio is rather similar in both model membrane systems (36, 43).

4. Finally, the activation energies for Cl^- diffusion have been calculated using both systems. Pagano and Thompson(36) observed a value of 10 KCal/mole using spherical bilayers between 10° and 30° C. Papahadjopoulos and Watkins (43) reported a value of 4-5.6 KCal/mole with vesicles between 22° and 50° C. A recent further study of the problem (41) revealed a change of slope for the Arhenius plot at approximately 30° C. Thus the activation energy between 10° - 30° C was found to be 9.5 KCal/mole while at higher temperatures 30° - 50° C was considerably lower (5.3 KCal/mole).

Effect of Ionic Strength and pH

Calculations on the stability of phospholipid bilayers indicate that the bilayer structure is the most stable form for phospholipids carrying 0 to 1.0 charge per molecule and a bulk phase containing 0.1 N monovalent salts (16, 33). However, with more than one charge per molecule or at lower ionic strength of the salt solution the bilayer would be in equilibrium with "inverted cylindrical micelles" (33). Such structural arrangement, representing a statistical aqueous pore through the bilayer, would be expected to drastically increase the conductivity of the membranes. Figure 1 illustrates the results obtained with two phospholipids at different ionic strength of the salt solution. The acidic phospholipid PS shows a great dependence on ionic strength while the neutral PC considerably smaller. Thus for PS the d.c. resistance changes by a factor of 5 $\sim$ 10 for a 10-fold decrease in salt concentration and does not form a stable membrane at very low salt concentration. In contrast, PC forms stable membranes with high resistance even in pure water and the d.c. resistance does not

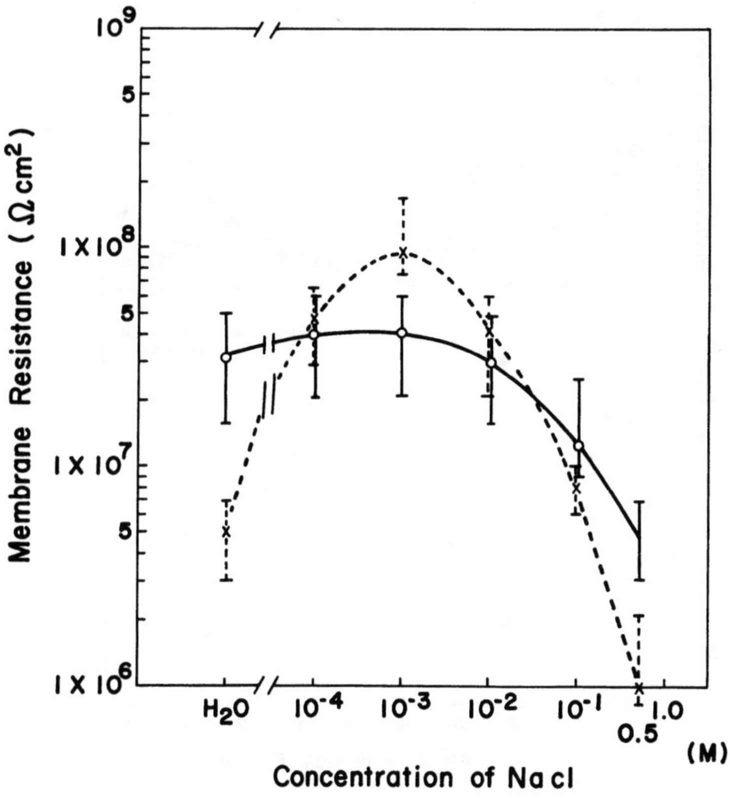

Fig. 1. Electrical resistance of PC and PS membranes (black films) in aqueous solution containing NaCl (100 mM) and Tris-Cl pH 7.4 (0.2 mM).

change so much with various salt concentrations. These experiments were performed at pH 7.4 stabilized with a small concentration of Tris-Cl (0.2 mM).

The effect of pH on the conductivity and capacitance of the same phospholipid membranes in 0.1M NaCl solutions is shown in Figure 2. Here, both PS and PC show maxima in d.c. resistance at slightly acidic solutions (31, 32). The maximum at pH 3.0 for PS probably coincides with the isoelectric point of this phospholipid (37). No stable membranes can be formed above pH 8.5 which is close to the beginning of the de-protonation of the amino-groups (37). The decrease of the PC resistance between pH 5 and 8, however, cannot be easily reconciled with titrations of PC monolayers which show no change in surface potential even at pH 11.0 (37). Trace contaminants of free fatty acids or oxidation products could be responsible for such a decrease. Changes in capacitance, also shown in Figure 2 generally follow the resistance curves with

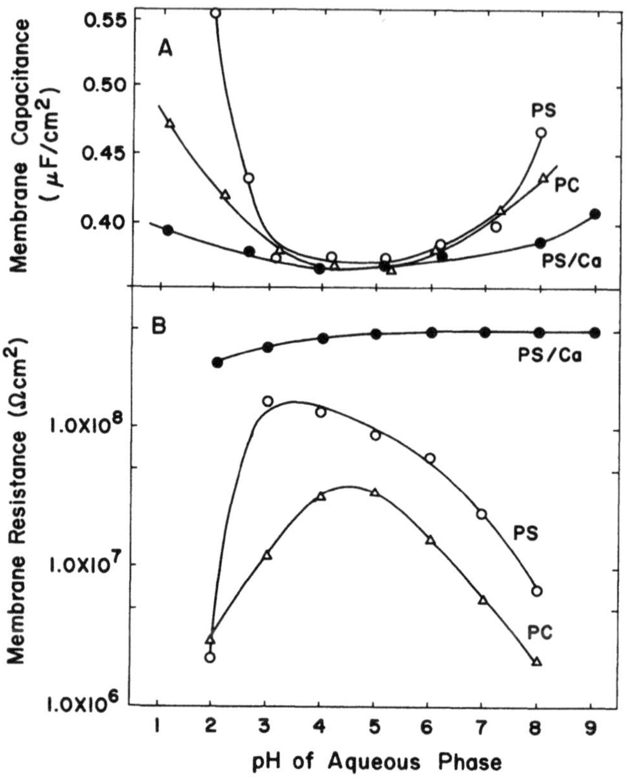

Fig. 2. Electrical resistance and capacitance of PC and PS membranes (black films) in aqueous solution of NaCl (100 mM) at different pH.

minima between pH 4.0 and 6.0) (32). The inclusion of $CaCl_2$ to the bulk phase has a pronounced effect on the behaviour of PS and negligible effect on PC. Thus, PS in the presence of 1 mM $CaCl_2$, and 100 mM NaCl has a much higher resistance which shows very little change with pH (Figure 2). The curve for PC in the presence of 1 mM $CaCl_2$ is identical to that in the absence of Ca^{2+}. Higher amounts of Ca^{2+}; however, do show an increase in PC resistance at high pH (31). The stability of PS-Ca membranes over a wide range of pH is remarkably similar to the titration of PS monolayers in the presence of the same amounts of Ca^{2+} (37). This result is in contrast to the effect of adding small amounts of Ca^{2+} to one side of the PS membranes (40,42), a phenomenon discussed in the following section. It should also be noted that the PS curve in Figure 2B was obtained without the addition of EDTA. When this compound is present (in 0.1 mM) the resistance is usually lower and closer to that of PC. It thus appears that even small amounts of Ca^{2+} (or other bivalent metals) either introduced as an impurity in the bulk phase or extracted along with PS from natural sources, contribute appreciably to the properties of PS membranes.

Effect of Ca^{2+} on Permeability of Phosphatidylserine

One of the most interesting observations on the permeability properties of phospholipid liquid crystals is the effect of divalent metals (Ca^{2+} and Mg^{2+}) on phosphatidylserine vesicles (38, 43). It was observed that small amounts of Ca^{2+} (10^{-3}M) when added after the formation of PS vesicles in NaCl or KCl (10^{-1}M) had a pronounced effect in increasing the diffusion rate of $^{42}K^{+}$ and $^{22}Na^{+}$ out of the vesicles into the outside bulk phase. Figure 3 presents some additional evidence on the relative effects of Ba^{2+}, Ca^{2+}, Sr^{2+}, Mg^{2+} using basically the same system of PS vesicles.

The mechanism for this phenomenon is not well understood. However, recent observations on the effect of Ca^{2+} upon the stability of PS bilayers (black films) have indicated that PS membranes become unstable only when Ca^{2+} is added to one side of the membrane (40). The instability is manifested as a drop in d.c. resistance and rupture of the membrane when Ca^{2+} concentration reaches a certain level characteristic of each pH, and it has been attributed to difference in surface energy between the two surfaces of the membrane (40, 42).

It has been proposed upon two entirely independent lines of evidence (4, 29), that Ca^{2+} favors the formation of "inverted

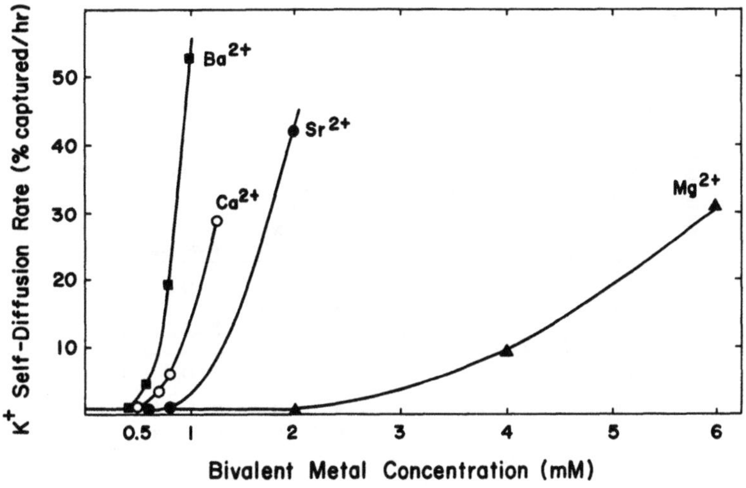

Fig. 3. K^{+} self-diffusion rate through PS vesicles at different bivalent metal concentration in the aqueous salt solution. The PS vesicles were formed by sonication in 130 mM NaCl, 15 mM Tris-Cl aqueous solution at pH 7.4. The bivalent metals were added outside of the vesicles.

micelles", structural arrangements which would result in increased permeability when compared with that of a bilayer.

However, X-ray diffraction data on PS liquid crystals in the presence of $CaCl_2$ and 145 mM KCl indicated a very well ordered lamellar arrangement with a spacing of 52 Å (39). Since PS dispersed in KCl alone gave a spacing of 75 Å (with considerable spreading) it was apparent that Ca^{2+} produced a "shrinking" of the inter-lamellar water space within each particle as well as aggregation of the particles as evidenced by precipitation. A similar phenomenon of "shrinking" produced by Ca^{2+} on brain "cephalins" was observed many years ago by Bear, Palmer and Schmitt (12).

It seems reasonable to conclude that Ca^{2+} penetrates the PS vesicles, and by binding to PS molecules in each lamella with a

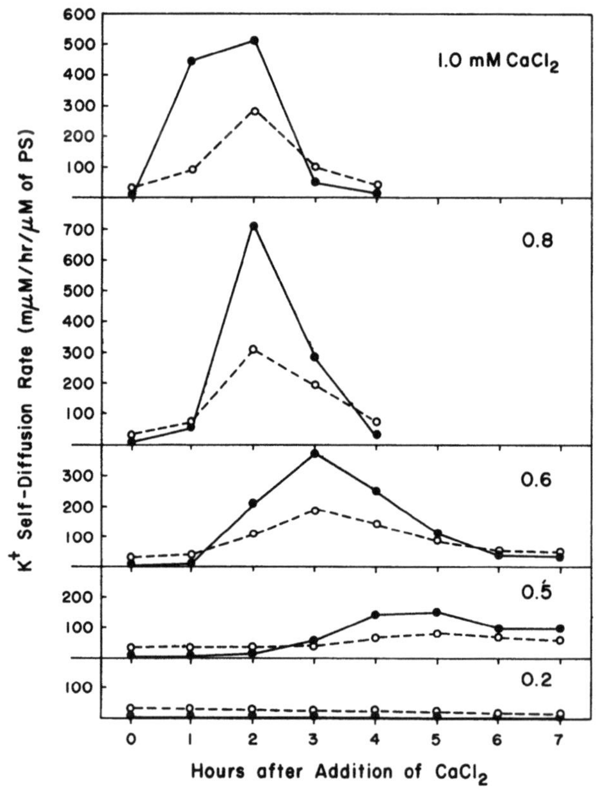

Fig. 4. K^+ and Cl^- self-diffusion rates through PS vesicles, at different Ca^{2+} concentrations in the external aqueous salt solution.

Diffusion rate of K^+, ●—●; diffusion rate of Cl, o--o.

ratio of 1 Ca^{2+} per two molecules (9), it reduces the surface charge density. This brings a reduction of the equilibrium distance between each lamella, with water and salt extruded out of the vesicles as a consequence of the shrinkage and collapse of the interlamellar aqueous space.

Figure 4 describes an experiment designed to show the extrusion of salt (both K^+ and Cl^-) from the PS vesicles as a result of increasing the external $CaCl_2$ concentration from 0 to 0.2 or 0.5,0.6, 0.8 and 1.0 mM. It is apparent that although with no Ca^{2+} the ratio of Cl^- to K^+ is 10/1 (43), at the peak of the diffusion rate curves out, this ratio is reduced to approximately 0.5 which is the ratio of concentrations of the two ions inside the particles. This would indicate unrestricted diffusion through fairly large aqueous pores. Water is presumably extruded at the same time since the final water space inside the particles is much smaller than the initial (39). Figure 4 also illustrates qualitatively the kinetics of the Ca^{2+} induced permeability changes. The time required for a measurable increase in ^{42}K efflux, increases from minutes with 1 mM Ca^{2+} to hours with 0.5 mM.

It has already been shown (38) that there is an appreciable difference on the amounts of K^+ and Na^+ diffusing out of PS vesicles at the same concentration of Ca^{2+}. Nevertheless, the difference was based on a comparison of different experiments containing either K^+ or Na^+ as monovalent cations. Because of the physiological importance of Na^+-K^+ specificity, the experiments were repeated with both K^+ and Na^+ were included within the same PS vesicles. These experiments are illustrated in Figure 5. For this experiment, the phospholipid vesicles were formed in a medium containing equal amounts of KCl and NaCl (65 mM each) and with $^{42}K^+$ and ^{22} Na^+ as tracers. The medium also included Tris-Cl (15 mM) pH 7.4. Finally, after the elimination of nonincorporated ions by passage through sephadex in NaCl/KCl/Tris solution, aliquots were dialysed against solutions containing either KCl or NaCl (130 mM, with 15 mM Tris-Cl pH 7.4) and Ca^{2+} or Mg^{2+} at varying concentrations. The amounts of ^{42}K and ^{22}Na present in each dialysate were determined independently.

The results presented in Figure 5 indicate that significant differences were obtained between the diffusion of Na^+ and K^+, confirming earlier observations (38). The amounts of K^+ diffusing out of the PS vesicles were generally higher than the amounts of Na^+ at the same Ca^{2+} concentration. Significantly, the composition of the external medium also seems to influence the results. The presence of K^+ outside the vesicles results in increased permeability to both Na^+ and K^+ as compared to the situation with only Na^+ outside. Essentially the same results were obtained with Mg^{2+} (also shown in Figure 5) with higher diffusion rate for K^+ than Na^+ in the presence of either K^+ or Na^+ outside. It seems reasonable to conclude that the competition between monovalent and bivalent

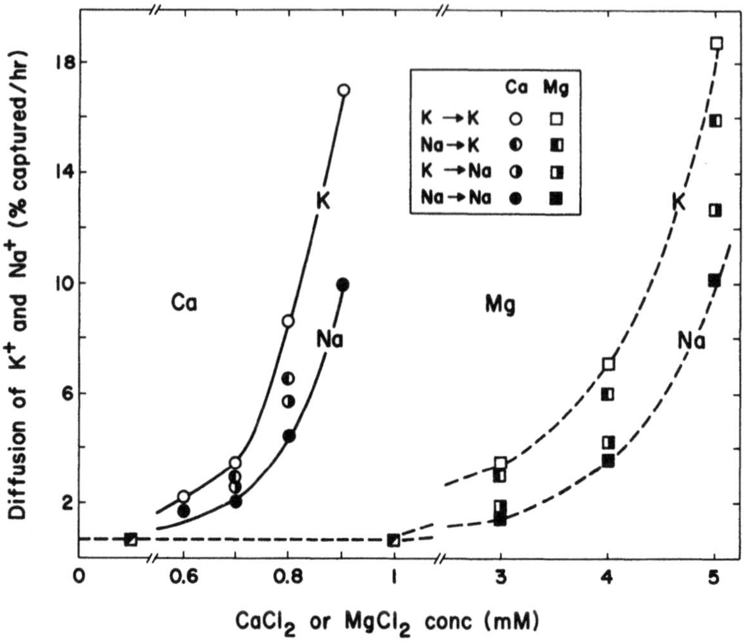

Fig. 5. K^+ and Na^+ diffusion rates through PS vesicles at different Ca^{2+} (Mg^{2+}) concentrations in the external aqueous salt solution. Diffusion rates in the presence of Ca^{2+}: diffusion of $^{42}K^+$ with K^+ outside, o; and with Na^+ outside, ◐; diffusion of $^{22}Na^+$ with Na^+ outside, ●; and with K^+ outside, ◑. Diffusion rates in the presence of Mg^{2+}: diffusion of $^{42}K^+$ with K^+ outside, □ ; and with Na^+ outside ◧ ; diffusion of $^{22}Na^+$ outside, ■ ; and with K^+ outside ◨.

metals for head-groups of the phospholipid molecules (PS) is an important factor in determining the permeability properties. When the same experiment was repeated with vesicles composed of PA, no appreciable difference was noted between the efflux of ^{42}K or ^{22}Na. Thus PS appears to have some specificity toward monovalent cations which is not the case for the other phospholipid tested namely PA.

Effect of Ca^{2+} on Different Phospholipids

Although most phospholipids have been shown to bind divalent metals (1,2,9,23,37,46,47), the extent of binding varies considerably. For example, no appreciable binding of Ca^{2+} to egg PC can be demonstrated in the presence of 0.1M concentrations of either Na^+ or K^+ (23,37,46). The ability of Ca^{2+} to induce permeability changes also varies considerably with different phospholipids. Figure 6 illustrates the results obtained with four different phospholipids. It is clear that a concentration of Ca^{2+} up to 20 mM

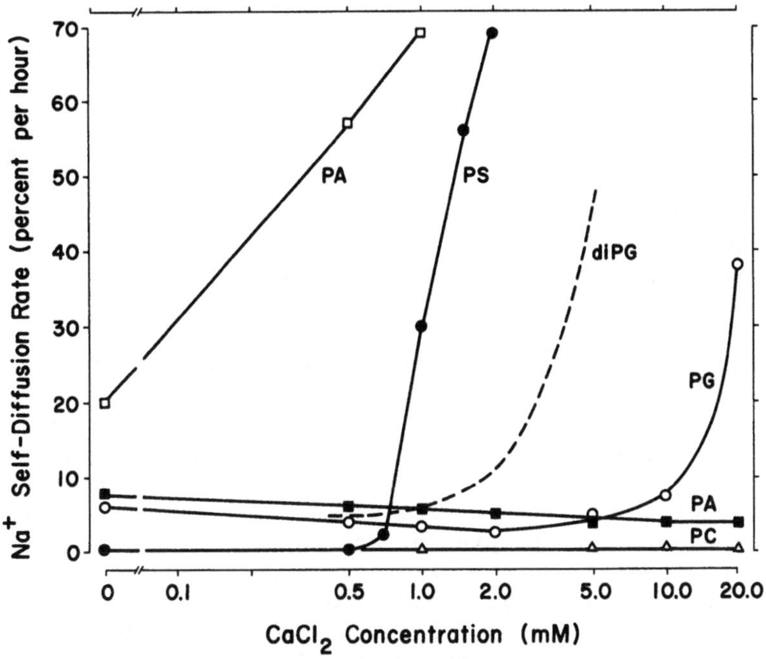

Fig. 6. Na^+ self-diffusion rates through vesicles composed of different phospholipids, at various Ca^{2+} concentrations in the external aqueous salt solution. The salt solution contained 140 mM NaCl and 4 mM Tris-Cl or 2 mM Histidine and 2 mM TES. PA at pH 9.0, □ ; PA at pH 6.0, ■ ; PS at pH 7.4, ● ; diPG at pH 7.4, ---; PG at pH 7.4, o; PC at pH 7.4, Δ.

has no effect on PC. This is easily explained on the basis of the internally neutralised charge-distribution of the PC molecule. Studies with PC monolayers have indicated that PC does not attract monovalent cations from the bulk phase (5). Recent calculations by Parsegian (45) depict the quaternary ammonium ion as a counter-ion with a considerable degree of freedom to interact with the neighboring phosphate groups.

In the case of the acidic phospholipids the situation seems to be more complex. Thus, PS, PG and PA (at pH 6.0), all carrying one extra negative charge per molecule, show considerable differences in the concentration of Ca^{2+} needed for a certain increase in permeability (Figure 6).

It has been suggested that the strong interaction of PS with Ca^{2+} is the result of chelation complexes between the metal and four different phospholipid molecules forming a linear polymeric arrangement (37). Such complexes would tend to increase the binding constant and thus would explain the obtained results. PA at high pH would also be able to form similar complexes.

Cardiolipid, which itself is a dimer of phosphatidic acid exterified with glycerol (18), tends to support this argument. Recent studies employing this phospholipid (48) indicate that the Ca^{2+} curve falls closer to PS than to PG. Although in this case, all phosphates are diesterified, the presence of two phosphates per molecule would tend to favor a linear polymeric arrangement.

Interaction of Ca^{2+} with Mixtures of Phospholipids

All the previously described experiments were performed with purified phospholipids. The properties of well defined mixtures of phospholipids might well be very important although not necessarily characteristic of the individual components. It has been shown recently that the amino-groups present at the surface of PE and PS vesicles are not easily identifiable by reagents reacting with amino-groups in solution (44). However when PE is mixed with PC in molar proportions of 1 to 9, the amino groups become reactive toward the same reagents. It was concluded that in the case of pure PE, the amino-phosphate interactions between neighboring molecules diminish the reactivity of these groups. However when PE is dispersed in the presence of large amounts of PC, each PE molecule, being statistically distributed, is always surrounded by PC molecules, and thus it has a limited probability of interacting with phosphates from other PE molecules (44). The Ca^{2+} induced permeability changes are also drastically influenced by mixing of different classes of phospholipids. Figure 7 illustrates this point with mixtures of PS, PA and PC.

When PS or PA are mixed with PC, the ability of Ca^{2+} to increase $^{22}Na^{+}$ efflux is generally inhibited. PS/PC 2.4/1 and PS/cholesterol 1/1 (mole/mole) mixtures show a decreased response to Ca^{2+} as would be expected from a similar dilution of the surface charge by either neutral molecule (the area per molecule of cholesterol is approximately half that of phospholipids). However, when the ratio of PS to PC is 1.2/1 no effect is observed even up to 10 mM $CaCl_2$ concentration. The Ca^{2+} obviously still binds to these latter mixed vesicles as evidenced by aggregation, although no appreciable increase in permeability is observed. Similar results were obtained with PA/PC mixtures although in this case the 1.1/1 mixture showed some increase in permeability. We interpret these results as evidence for cooperative phenomena in the Ca^{2+} binding to the head groups of phospholipids. Thus it would appear that Ca^{2+} can be chelated to more than one neighbouring molecule. The random statistical distribution of molecules expected for mixtures would thus tend to favor simple ion-molecule interactions rather

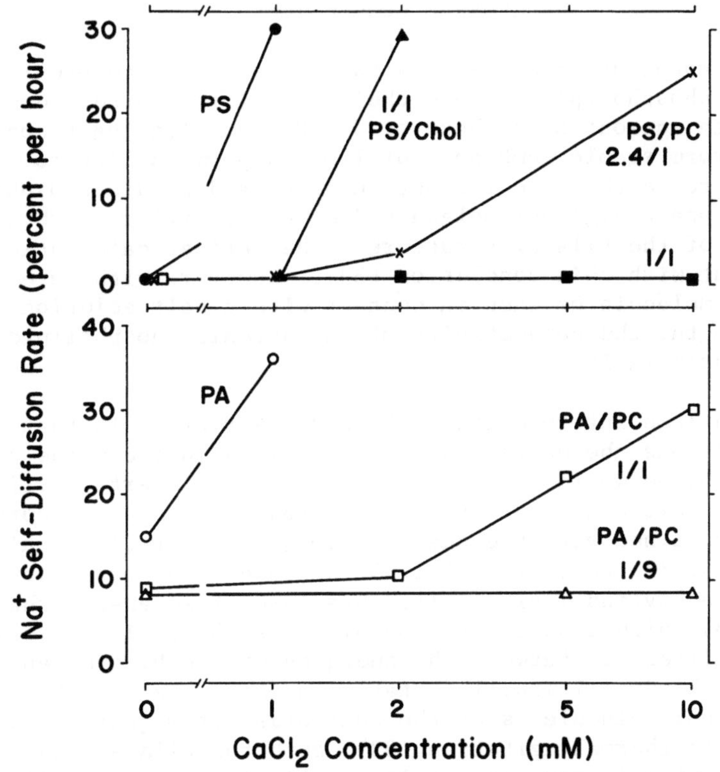

Fig. 7. Effect of Ca^{2+} on the permeability properties of vesicles composed of mixtures of phospholipids. PS alone, ●; PS/cholesterol 1/1, ▲; PS/PC 2.4/1 x; PS/PC 1.1 ■; PA alone, ○; PA/PC 1/1, □; PA/PC 1/9, △; Aqueous salt solution contained NaCl (140 mM) Histidine (2 mM), TES (2 mM), EDTA (0.1 mM), at pH 7.4. Molar ratios were calculated on the basis of phosphate determination.

than polymeric assemblies (37) and consequently tend to minimize the permeability changes induced by Ca^{2+}.

The implications of these interactions to biological membranes are of some importance. The extracted lipids from natural sources are complex mixtures but such mixtures are probably artificial. This stems from the possibility that lipids are segregated topologically at different areas of biological membranes or at different membranes, and are mixed homogeneusly only during the extraction with organic solvents. There is no direct evidence at the present time for such segregation in biological membranes, but it seems a reasonable supposition.

GENERAL CONCLUSIONS

It is apparent from the results discussed in the previous sections that phospholipids, above their transition temperature and in contact with aqueous bulk phase of 0.1 M monovalent salt concentration, can form stable membranes of limited permeability to ions. This seems to be the case for both neutral and acidic phospholipids. Apparently one charge per molecule does not interfere with the cohesiveness of the bilayer structure. Theoretical calculations are in agreement with this conclusion (33). However, with more than one charge per molecule or when in contact with a salt solution of low ionic strength, the conductivity of the anionic phospholipids increases considerably.

A plausible interpretation of this observation is that under these conditions the charge-charge repulsions in the plane of the bilayer membrane would favor the formation of "inverted cylindrical micelles". This could be best illustrated as an aqueous pore through the bilayer, lined with the charged groups. Theoretical calculations indicate that in the region of 1.25 to 2.0 charges per phospholipid molecule the "cylindrical micelle" has lower free energy than the bilayer (33) which consequently becomes unstable. In any case, since the difference between the energies of the bilayer and the "inverted cylindrical micelle" states in the range of 1.0 to 1.5 net charge per molecule is of the same order of magnitude as the energy due to thermal agitation (33), both the bilayer and the cylindrical micelle may coexist in equilibrium. A gradually increasing number of holes (cylindrical micelles) would be produced in the membrane, until the bilayer form disappears. It should be noted here that the same calculations indicate that even at a charge of 2.0 per molecule, the energy required for the formation of spherical micelles (the classical micelle formed by soaps) is still too high to favor their formation.

The other type of instability discussed in this paper is the result of the introduction of Ca^{2+} or H^+ to the bulk aqueous phase on one side of a bilayer composed of anionic phospholipids. In this case, the instability would be the effect of a difference in the surface energy between the two surfaces. In a recent attempt to explain this phenomenon (40) it was suggested that the condition of asymmetry would result in re-orientation (inversion) of molecules or clusters of molecules between the two sides in order to equilibrate the difference in surface energy. During such a re-orientation, it would be reasonable to expect a substantial change in the permeability properties of the bilayer. The experimental evidence presented here concerning the effect of Ca^{2+} on the permeability of phospholipid vesicles, and the effects of pH and Ca^{2+} on PS bilayers (40, 42) tend to support this hypothesis. Furthermore, calculations based on a difference of one charge per phospholipid molecule between the two sides of the bilayer indicate such membranes would be unstable (35).

The biological implications of such instability due to asymmetric distribution of charges or divalent cations are considerable. It is well known that the electrical properties of nerve membranes are influenced profoundly by changes of pH and of divalent cation concentrations (21,49). It is also recognized that phospholipids are an integral part of the axon membrane and necessary for the manifestation of the action potential (3). It therefore seems possible that the changes in permeability due to differences in the ionic environment of acidic phospholipids described here could be intimately connected with the electrical properties of excitable membranes.

ACKNOWLEDGEMENT

This work was supported in part by the United States National Institutes of Health, Grant GM-16106-01 (D.P.) and the Damon Runyon Memorial Fund, Grant DRG-1001 (S.O.)

REFERENCES

1. ABRAMSON, M.B., KATZMAN, R. and GREGOR, H.P., J. Biol. Chem., 239, 70 (1964).

2. ABRAMSON, M.B., KATZMAN, R., GREGOR, H.P. and CURCI, R., Biochemistry, 5, 2207 (1966).

3. ALBUQUERQUE, E.X. and THESLEFF, S., J. of Physiol., 190, 123 (1967).

4. ALLEN, B.T., CHAPMAN, D. and SALSBURY, N.J., Nature, 212, 282 (1966).

5. ANDERSON, P.J. and PETHICA, B.A., in Biochemical Problems of Lipids, (Eds. G. Popjak and E. le Breton). Butterworth, London 1956, p. 24.

6. ANDREOLI, T.E., BANGHAM, J.A., and TOSTESON, D.C., J. General Physiol., 50, 1729 (1967).

7. BANGHAM, A.D., Advances in Lipid Research (Eds. D. Kritchevsky, R. Paoletti) Acad. Press, 1, 65 (1963).

8. BANGHAM, A.D., in Progress in Biophys. & Molecular Biology (Eds. J.A.V. Butler, D. Noble) Pergamon Press, Oxford, p. 29 (1968)

9. BANGHAM, A.D. and PAPAHADJOPOULOS, D., Biochim. Biophys. Acta, 126, 181 (1966).

10. BANGHAM, A.D., STANDISH, M.M., and MILLER, N., Nature, 208, 1295 (1965).

11. BANGHAM, A.D., STANDISH, and WATKINS, J.C., J. Mol. Biol., 13, 238 (1965).

12. BEAR, R.S., PALMER, K.J., SCHMITT, F.O., J. Cellular Comp. Physiol., 17, 355 (1941).

13. CHAPMAN, D., The Structure of Lipids, Methuen and Co., Ltd., London (1965).

14. CHAPMAN, D. and FLUCK, D.J., J. Cell. Biol., 30, 1 (1966).

15. CHAPMAN, D., WILLIAMS, R.M. and LANDBROOKE, B.D., Chem. Phys. Lipids, 1, 445 (1967).

16. DANIELLI, J.F., in Molecular Associations in Biology, Academic Press, N.Y., p. 529 (1968).

17. DAWSON, R.M.C., Biochem. J., 102, 205 (1967).

18. DE HAAS, G.H., BONSEN, P.P.M., and VAN DEENEN, L.L.M., Biochim. Biophys. Acta, 116, 114 (1966).

19. DERVICHIAN, D.G., in Progress in Biophysics (Eds. J.A.V. Butler, H.E. Huxley) Pergamon Press, 14, 263 (1964).

20. HANAI, T., HAYDON, D.A., TAYLOR, J., J. Theor. Biol., 9, 433 (1965).

21. HILLE, B., J. General Physiol. 51, 221 (1968).

22. HUANG, C., WHEELDON, L., and THOMPSON, T., J. Mol. Biol., 8, 148 (1964).

23. KIMIZUKA, H., NAKAHARA, T., VEJO, H., and YAMAUCHI, A., Biochim. Biophys. Acta, 137, 549 (1967).

24. LAUGER, P., LESSLAUER, W., MARTI, E., and RICHTER, J., J. Biochim. Biophys. Acta, 135, 20 (1967).

25. LESSLAUER, W., RICHTER, J., and LAUGER, P., Nature, 213, 1224 (1967)

26. LUZZATI, V., in Biological Membranes (Edit. by D. Chapman) Academic Press, p. 71 (1968).

27. LUZZATI, V., GULIK-KRZYWICKI, T. and TARDIEU, A., Nature, 218, 1031 (1968).

28. LUZZATI, V., and HUSSON, F., J. Cell. Biol., 12, 207 (1962).

29. MAAS, J.W., and COLBURN, R.S., Nature, 208, 41 (1965).

30. MUELLER,P., RUDIN, D.O., TIEN, H.T., and WESCOTT, W.C., Nature, 194, 979 (1962).

31. OHKI, S., J. Coll. Interf. Science, 30, 413 (1969).

32. OHKI, S., Biophysical J. (in press).

33. OHKI, S., J. Coll. Interf. Science (in press).

34. OHKI, S. and GOLDUP, A., Nature, 217, 458 (1968).

35. OHKI, S., in D. Kursunoglu and A. Perlmutter, Eds: Physical Principles of Biological Membranes, Gordon & Breach, Science Publ. Inc. (in press).

36. PAGANO, R. and THOMPSON, T.E., J. Mol. Biol., 38, 41 (1968).

37. PAPAHADJOPOULOS, D., Biochim. Biophys. Acta, 163, 240 (1968).

38. PAPAHADJOPOULOS, D. and BANGHAM, A.D., Biochim. Biophys. Acta, 126, 185 (1966).

39. PAPAHADJOPOULOS, D. and MILLER, N., Biochim. Biophys. Acta, 135, 624 (1967).

40. PAPAHADJOPOULOS, D. and OHKI, S., Science, 164, 1075 (1969).

41. PAPAHADJOPOULOS, D. and OHKI, S., J. Amer. Oil Chem. Soc. Symposium on Model Membranes, San Francisco, April 1969, (in press).

42. PAPAHADJOPOULOS, D. and OHKI, S., Symposium on Surface Chemistry of Biological Systems, A.C.S. Meeting, Sept., 1969. M. Blank, Ed., in Advances in Experimental Medicine and Biology, Plenum Press (in press).

43. PAPAHADJOPOULOS, D. and WATKINS, J.C., Biochim. Biophys. Acta, 135, 639 (1967).

44. PAPAHADJOPOULOS, D., and WEISS, L., Biochim. Biophys. Acta, (in press).

45. PARSEGIAN, V.A., Science, 156, 939 (1967).

46. ROJAS, E. and TOBIAS, J.M., Biochim. Biophys. Acta, 94, 394 (1965).

47. SHAH, D.O., SCHULMAN, J.H., J. Lipid Res., 8, 227 (1967).

48. SHAH, J., PAPAHADJOPOULOS, D. and WENNER, C., (manuscript in preparation).

49. TASAKI, I., WATANABE, A. and LERMAN, L., Amer. J. Physiol., 213, 1465 (1967).

50. TIEN, H.T. and DIANA, A.L., Chem. Phys. Lipids, 2, 55 (1968).

51. TIEN, H.T. and DIANA, A.L., J. Coll. Interf. Science, 24, 287 (1967).

BIOPOLYMERIZATION OF PEPTIDE ANTIBIOTICS

G. T. Stewart,* B. T. Butcher,* S. S. Wagle+
and M. K. Stanfield*

*Schools of Medicine and Public Health, Tulane University, New Orleans, La.
+Research Department, Kremers-Urban Company, Milwaukee, Wisconsin

In the course of research on penicillin allergy (1, 11, 12), it was found that natural penicillin and its chemical parent nucleus, 6-aminopenicillanic acid (6-APA) contained traces of a macromolecular proteinaceous residue with striking allergenic properties in human subjects. Efforts were then made in the laboratory to remove these residues, but were only partly successful, because of the formation by penicillin in solution of a second residue with the characteristics of a polymer (1, 12, 13, 15). This finding, since confirmed by other workers (5, 17) led to the examination of a range of penicillins, of the related β-lactam antibiotics cephalosporins and of bacitracin, which is also a cyclic oligopeptide antibiotic. Spontaneous polymerization occurred in all the antibiotics examined (16).

The polymers thus formed differed in certain chemical and biological properties from their parent antibiotic monomers. The spontaneous formation of ordered macromolecules from simple precursors is however a necessary feature of many biochemical processes, especially in situations where high specificity and structural fit are necessary for reactions to proceed (14). Such considerations are highly relevant to the highly specific antibacterial, immunogenic and allergenic potency of penicillin and its analogues. The results of chemical investigation of these, hitherto undescribed polymers are therefore reported here.

EXPERIMENTAL

Materials

Antibiotics used in this study were in all instances obtained either, in small amounts, by purchase of standard therapeutic preparations from hospital pharmacies or, in bulk, by donation from various major manufacturers. All therapeutic preparations studied conformed to the standard requirements of the U.S.P., B.P. and F.D.A. Non-therapeutic preparations were supplied as gratis donations by the research laboratories of manufacturers to whom the purpose of the study was explained.

Methods.

The chemical, immunological and clinical methods used in the present report have been described in detail elsewhere (2, 3, 12, 13, 16). Essentially, they consisted of the following:

Fractionation of Antibiotics

(a) Column chromatography. Antibiotics in solution were introduced into the matrix of a Sephadex G25 coarse column (4 x 100 cm) under slight positive nitrogen pressure. The eluant from the column was collected in fractions of 10 ml. which were read over the range 340-200 mμ using a Beckman DB ultraviolet spectrophotometer. The fractions were pooled according to their photometric homogeneity, by physical characteristics, such as smell and color, and were lyophilized. The lyophilized pools were weighed and stored for further testing in nitrogen-flushed containers.

(b) Ultrafiltration. Samples of antibiotic in suitable solution were placed in an Amicon Diaflo filter cell containing a 1000 mol. wt. retention filter and subjected to 100 lbs/sq. in. pressure of nitrogen. The resulting ultrafiltrate was lyophilized and stored as above.

(c) Dialysis. Samples were dialyzed to exhaustion in Visking tubing against a continuous flow of distilled water. The final contents of the tubes (retentates) were lyophilized and stored as above. Dialysis was used essentially to free residues with mol. wt. exceeding 5000 from other soluble fractions.

Spectrophotometry of Starting Material and Fractions

(a) Ultraviolet. Column effluent samples were examined in a Beckman DB model spectrophotometer over the wavelength 340-200 mμ for pooling and lyophilization. For characterization, more accurate spectrophotometry was carried out in a Hitachi model 124 or Cary model 14.

(b) Infrared. Dried samples of about 80 mg. of a 1 mg. sample in 200 mg. dry spectroscopic grade KBr were incorporated into clear discs in a press at 22, 000 lbs/sq. in. pressure. The discs were examined in a Perkin Elmer model 257 or 337 spectrophotometer.

(c) Nuclear Magnetic Resonance. Samples were examined in dimethyl sulfoxide and deuterium oxide as solvents at 60 and 100 megacycles over the range 0-1000 Hz.

Chemical Tests

(a) Potentiometric titrations. Potentiometric titrations were conducted in a Beckman Expandomatic pH meter with magnetic stirrer, Coleman recorder, and a continuous infusion pump. 0. 1 N HCl or 0. 1 N NaOH was pumped at a rate of 1. 0 ml/min. into the reaction mixture containing 5 ml. of 20 mg/ml of the sample. No correction was made in obtaining pK's of the substance under study.

(b) Mercuric Chloride reaction. The reaction was conducted on solutions containing 1 mg. of sample in 5 ml. of 0. 01 M phosphate buffer at pH 7. 6. A 3 ml. portion was pipetted into a clean cell, titrated with $HgCl_2$ (2. 58 x 10^{-4}M) and increase in 282 mμ absorption was recorded. Titration was continued until no further change at 282 mμ was observed. The solutions were diluted with phosphate buffer when necessary. The Molar Extinction Coefficient was then calculated.

(c) Internal reactions. Internal reactions in solutions of benzylpenicillin and its derivatives were studied in phosphate buffers (pH 5.2 and 7.0) and in carbonate buffers (pH 9.2 and 10.2) as well as in water by observing physical, spectrophotometric and chromatographic changes over periods extending to 1000 hours at various temperatures. Formation of penicillenic acid was detected by absorbance at 322 mμ. Penicilloic acid was detected by the mercuric chloride reaction, described above. Standard chemical tests were performed to detect sulfhydryl and disulfide groups and for nitration of the phenol ring.

(d) Chromatography. Samples were chromatographed before and after acid hydrolysis in various solvents (usually butanol:ethanol:phosphate in the ratio 2:1:2.5) by ascending run for 18 hours or on silica gel for 2-4 hours, dried and developed with Ninhydrin. Selected samples were examined further in an amino acid autoanalyzer, and by elementary analysis.

Antimicrobial Activity

Serial 1:10 dilutions of samples, starting at a concentration of 5 mg. per ml. were added to wells bored in trypticase soy agar plates seeded with S. lutea using a standard dropper. After incubation for 18 hours at 37°C the zones of inhibition of growth were measured in two diameters and the average compared with starting material and other samples of the same antibiotic. The activity of the starting material was regarded as 100%.

Immunology

Animal tests were performed by standard immunological techniques which detect passive cutaneous anaphylaxis in guinea pigs passively sensitized to an antigen (or group of antigens) by transfer of serum from a rabbit or guinea pig previously immunized with that antigen. All animals used were fed antibiotic-free foods and water. Guinea pigs and rabbits were immunized with 1 mg/Kg. body weight of antibiotics or antibiotic fractions in complete Freund's adjuvant followed by booster injections of the same material in physiological saline at weekly intervals starting at day 14. Animals were bled at day 49 and subsequently. With these sera, fresh guinea pigs were passively

Table 1: Photometric Properties of Fractions from Benzylpenicillin

Fraction (Sephadex G 25)	K_{av}	Estimated mol. wt.	Infra-red absorbance μ				UV absorbance mμ	
			N 6 2.95	lactam CO 5.65	salt 6.25	OH 2.95	penamaldate 282	penicillenic 322
55-64	0.04	> 5000	+	-	-	-	+	variable
65-84	0.26	3500	weak	weak	weak	+	+	+
85-99	0.51	2500	+	+	+	+	-	+
100-139	0.92	1500	+	+	+	+	-	+
140-154	1.25	< 1000	++	+	+	+	-	+
155-179	1.51	< 1000	++	+	+	+	-	trace

sensitized by intravenous or intracardial injection. After 18 hours to allow fixation of the transferred antibodies to tissues, the animals were challenged by intradermal injections of 200 μg of the antibiotic fractions. Evans blue dye (1% in saline) was injected intravenously or intracardially to facilitate the interpretation of reactions. The animals were examined over a 72 hour period although skin reactions at sites of challenge usually developed in less than 30 minutes (2, 3, 13).

Results

Fractionation of concentrated solutions in water of the natural penicillins and cephalosporins yielded several components with different average molecular weights. (Tables 1 and 2). The components with molecular weights exceeding 5000 were found in general only in the natural forms of the antibiotics (i. e. benzylpenicillin, phenoxymethyl penicillin, 6-aminopenicillanic acid [6-APA], cephalosporin C, 7-amino-cephalosporanic acid [7-ACA]) and had a composition resembling a protein in that it showed 7 or 8 amino-acids chromatographically after acid hydrolysis, a N:S ratio of 4 or more, a U. V. spectrum with maximal absorbance at about 280 mμ and an IR absorbance profile almost identical with that of an artificial polymer formed by covalent linkage of the penicillenic carbonyl to the epsilon-amino group of lysine. This fraction also possessed strong antigenic potency with immunochemical specificity in sensitized guinea pigs for penicilloyl and cephalosporyl groups (2). The origin and structure of this proteinaceous component is not yet certain but it has been regularly found, in varying amount, in batches of β-lactam antibiotics prepared by standard methods by various manufacturers; in the case of benzyl- and phenoxymethyl penicillins, the batches examined were U. S. P. and F. D. A. approved; in the case of 6-APA, 7-ACA, and cephalosporin C, the batches were standard products used for preparation of acylated derivatives for therapy.

The various fractions of intermediate average molecular weights (i. e. 1000-5000) showed paracrystallinity and yielded after acid hydrolysis those amino-acids which would be expected from degradation of penicillins or cephalosporins i. e. glycine, valine and dimethyl cysteine or methionine respectively, together with amino-acids characteristic of the side chain, free ammonia

Table 2: Characteristics of Macromolecular Fractions from β -lactam Antibiotics

Antibiotic	Mol. Wt. (monomer)	Average mol. wt. of residues		Subunits identified in polymer
		>5.000	5.000 - 1000	
6.- APA	216	Proteinaceous*	Polymer+	amide
Benzylpenicillin	333	Proteinaceous and polymer	Polymer	penicillenic and penicil-loic acids.
a-aminobenzyl penicillin	349	Polymer	Polymer	phenylglycine
dimethoxybenzyl penicillin	380	NIL	Polymer	penicillenic acid
a-phenoxymethyl penicillin	350	Proteinaceous	Polymer	penicillenic acid
7-ACA	272	Proteinaceous	Polymer	amide
7-thioenyl-3pyridino-ACA	384	Polymer (a)	Polymer (b)	(a) with 3-7 substituents (b) with 7-substituent and cephalosporoic acid
7-thioenyl-ACA	364	NIL	Polymer	thioenylcephalosporoic acid
Cephalosporin C	415	Proteinaceous	Polymer	cephalosporoic acid

* Yielding unexpected amino-acids after hydrolysis and showing stron immunogenic activity
\+ Yielding expected amino-acids after hydrolysis.

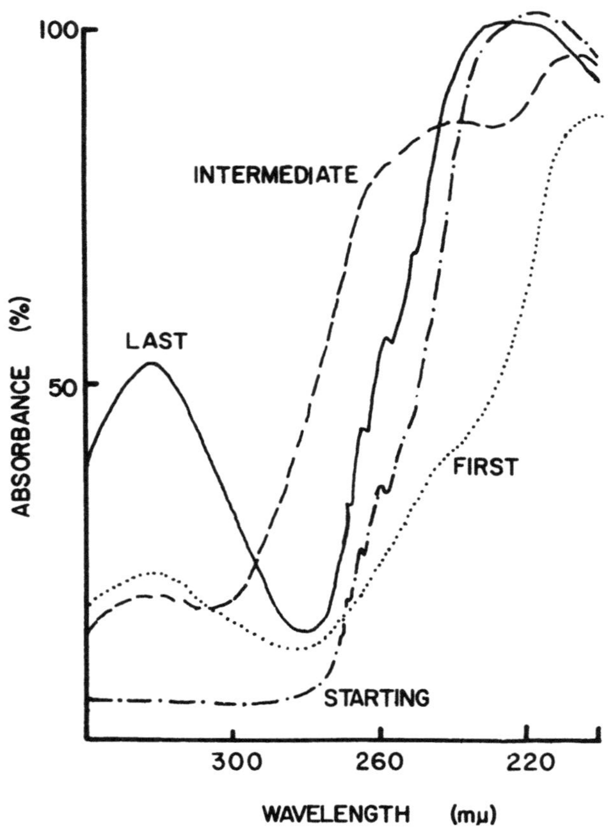

Fig 1 : Column Chromatography of Fractions from Benzylpenicillin.

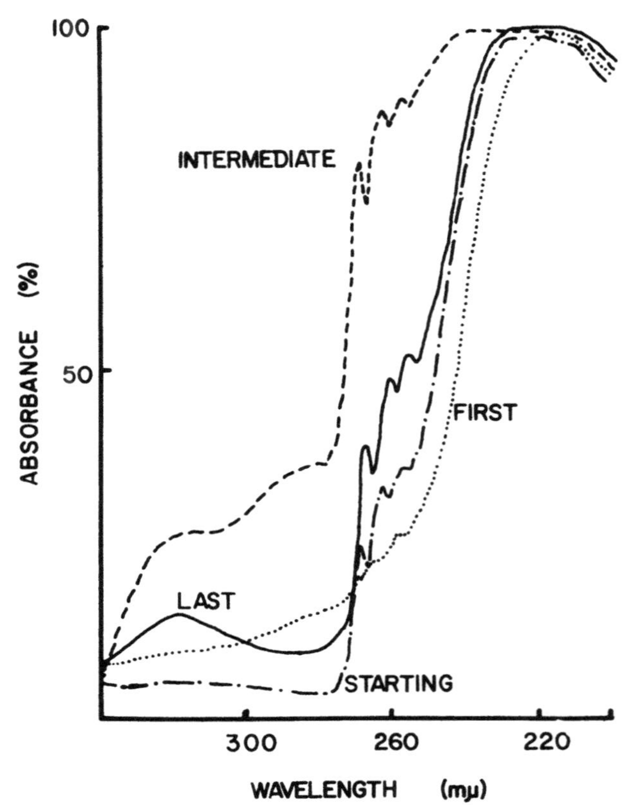

Fig 2: Column Chromatography of Fractions from Ampicillin.

and other ninhydrin-positive products resulting from breakdown of the molecule. These fractions, which invariably had lower antibacterial activity than the original antibiotic, were found in semi-synthetic as well as in natural penicillins and cephalosporins, and were found also in the natural forms of the antibiotics after removal of the fractions with higher molecular weight. The N:S ratios were of the order of 1 or less (Table 3) while spectrometric readings in the UV or IR indicated varying degrees of similarity to the known absorbances of the intact, parent antibiotics (Tables 1 and 4). Fractions with average mol. wts. 1500-3500 showed weakening of the IR signal by the N_4 and N_6 groupings suggestive of substitution at these sites. All fractions with mol. wt. > 2500 showed absorbance of UV light at 282-285 mμ, indicative of penamaldic acid. Fractions with mol. wt. 1000-1500 absorbed strongly at 322 mμ (penicillenic acid). It appeared therefore that these fractions were polymers formed by internal reactions of the antibiotic molecules in solution. The rate and nature of polymerization varied, as did physical appearance and properties, according to the derivative of 6-APA or 7-ACA employed in the fractionation (Table 5). Typical results for benzylpenicillin and α-amino benzylpenicillin (ampicillin) are shown in Figures 1 and 2. These polymers were in general much less reactive as antigens and in immunochemical tests than the proteinaceous components of higher molecular weight (2).

Examination of the penicillin fractions of low and intermediate molecular weight in deuterium oxide for nuclear magnetic resonance (Table 5) showed preservation of the signals from the phenyl, carboxylic and dimethyl protons though the splintering indicative of asymmetry in the dimethyl groups was more pronounced in the low molecular weight fraction 2 consisting of monomers or dimers than in the fraction 1 of intermediate molecular weight or in the starting material (Figure 3). Signals assigned to the carboxylic and amide protons were lost in the (pooled) fraction of intermediate molecular weight. Separation of the low molecular weight fraction yielded a tracing with sharper peaks, due possibly to absence of substitution or linkage in the benzyl, amide, and carbonyl groups. An unexpected finding (Table 5) was the increased integral value of the amide in the low molecular weight fraction, suggestive of 4 hydrogen atoms in this part of the molecule. Some other differences were also noted in these tracings but, since there is little on record about the NMR spectra of penicillin, we are refraining from

Table 3 Elementary analysis of benzylpenicillin before and after fractionation (excluding ash).

	Commercial benzylpenicillin	Fraction with mol. wt. 1000-50000
C	52.07%	50.37%
H	4.74	5.67
N	7.32	7.32
O	16.54	18.63
S	9.67	8.70
N:S	.76	.84

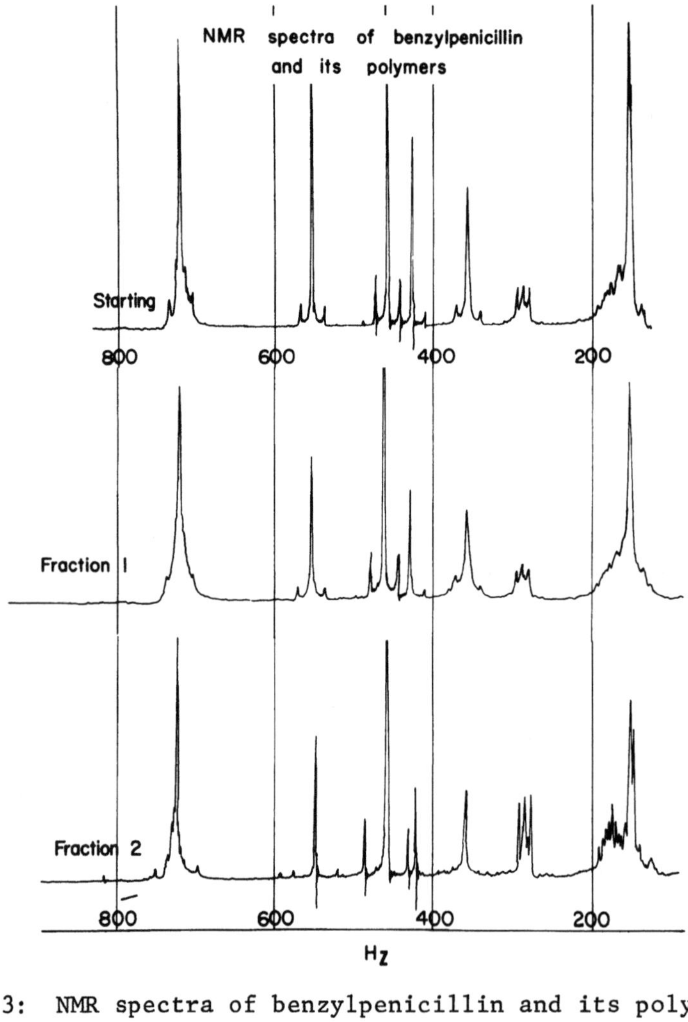

Fig. 3: NMR spectra of benzylpenicillin and its polymers.

further comment at this stage in our work.

In all the antibiotics tested, degradation and polymerization were influenced by temperature and pH, being generally accelerated by rise in temperature from 4° to 37°C but variably affected by pH according to the derivative tested. In the case of benzylpenicillin, spontaneous polymerization was always accompanied by the development of a yellow color and sulfurous odor, with progressive absorbance at 322 mμ, attributable to benzylpenicillenic acid; this change was minimized by buffering solutions in the pH range 7-9.2 (Figure 4) but accelerated by UV light and by half-saturation with ammonium sulfate which produced a plastic, keratin-fixing polymer with lowered aqueous solubility and strongly positive tests for sulfhydryl and disulfide groups. Other derivatives of 6-APA and 7-ACA varied greatly in rate and type of change: thus phenoxymethyl penicillin, α-amino benzylpenicillin and cephaloridine were relatively stable at low pH though polymerization readily occurred during fractionation in Sephadex columns; dimethoxy benzylpenicillin (methicillin) was highly unstable at pH below 5.8 but yielded one polymer with minimal change in UV absorbance at pH 6-7.

Presence or absence of antibacterial activity in these fractions was closely linked to the retention of an intact β-lactam ring, as evidenced by the stretching vibrations of the strained carbonyl bond at a frequency of about 5.65 μ. Immunochemical activity and antigenicity, on the other hand, depended upon molecular weight and heterogeneity of amino-acid structure, being of a low order or altogether absent in the smaller molecular groupings except when conjugated with proteinaceous or large polymer molecules. The results reported above relate mainly to experiments with benzylpenicillin, which is the prototype and the best-known member of the β-lactam antibiotic series. Fractions with molecular weights exceeding those of the appropriate monomer or dimer were obtained from all antibiotics, in common clinical or laboratory use, belonging to the β-lactam group, and from bacitracin, which is also an oligo-peptide antibiotic. The properties of these fractions (Table 2) varied according to whether the cyclic structure was thiazolidine (penicillins: derivatives of 6-APA) or dihydrothiazine (cephalosporins: derivatives of 7-ACA), and according to the nature of the side chain in the 6- or 7- position. Conditions affecting the stability of the monomer, such as pH, exposure to

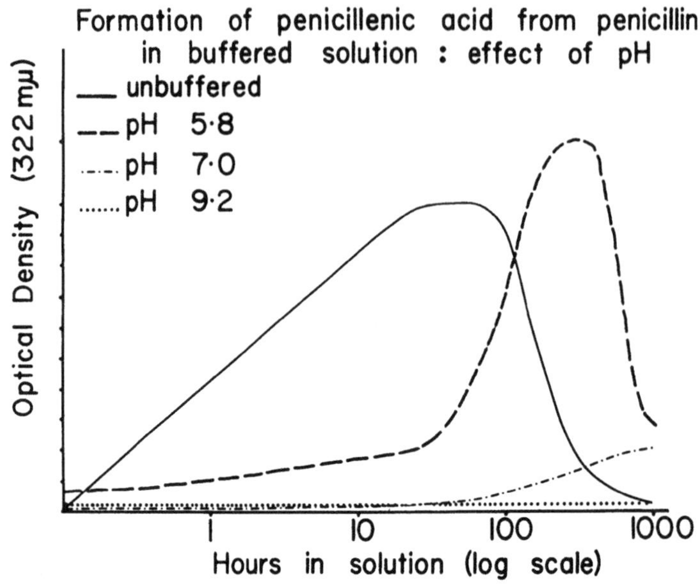

Fig. 4: Formation of penicillenic acid from penicillin in buffered solution: effect of pH.

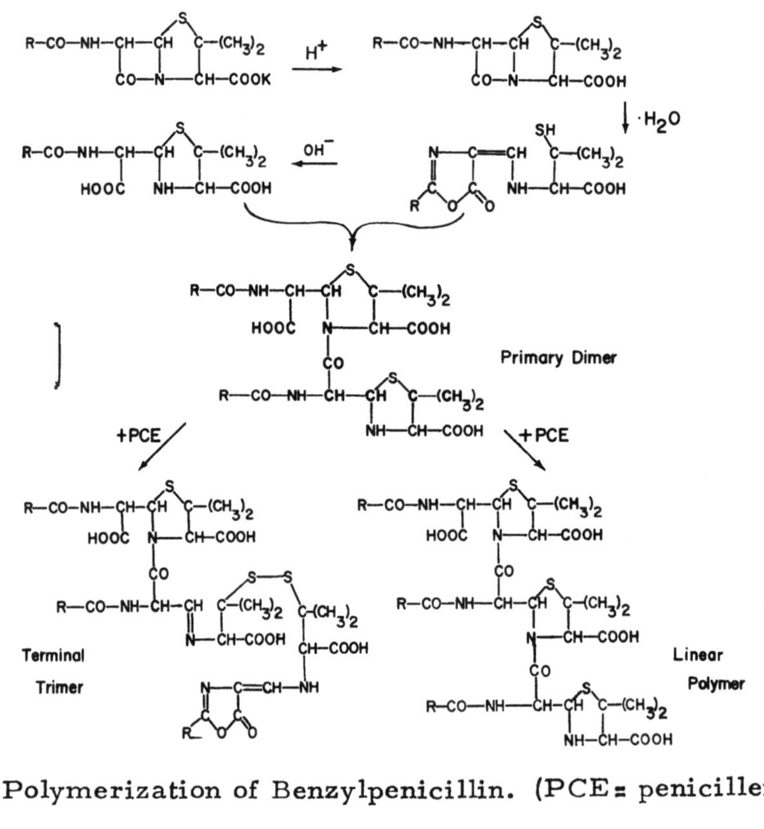

Fig 5: Polymerization of Benzylpenicillin. (PCE = penicillenic acid).

Table 4: Reactions of Products of Benzylpenicillin

Product	Potentiometric Titration		$HgCl_2$ (penamaldate) reaction	Spectrometry		
	α-carboxyl	β-carboxyl		IR	UV (mμ)	NMR
Benzylpenicillin (ultrafiltered)	in β-lactam structure	present	—	β-lactam-amide profile	268, 264, 258, 254	Aromatic and dimethyl protons
Benzylpenicillin polymer mol. wt. 3500	detectable in small amounts	present	Absorbtion at 282 mμ EM = 2400	Pen amide profile β-lactam absent	322, 280-285	Aromatic and dimethyl protons Blurring of intermediate peaks
Penicilloyl amide	absent	absent	Strong abs. at 282 mμ EM = 23, 500		..	not done
Penicilloic acid	present	present	Strong abs. at 282 mμ	..	end absorption	not done

UV light, oxygen and temperature, also influenced the pathway of degradation and of polymerization.

For practical purposes, removal of components with average molecular weight exceeding 1000 was most easily accomplished by ultrafiltration of solutions under positive nitrogen pressure in the dark with immediate lyophilization of the filtrates. Spectrometric tests (Figures 3-5, Tables 1, 4, and 5), together with theoretical reasoning, presented below, indicated that dimers and trimers with molecular weights in the range 600-999 could reform rapidly in filtrates and that preservation of solutions in monomeric form was a problem of great technical difficulty.

Discussion

The chemistry as well as the therapeutic and allergenic properties of penicillin and its derivatives have been intensively studied for nearly thirty years but, until the results which are part of this communication were reported in 1967, there was nothing on record to suggest the presence of a macromolecule, ordered or otherwise, except the identification by Grant et al. (6) of a polyamide formed by 6-APA. It was however well-known that penicillin could degrade in solution to penicillenic, penicilloic and penamaldic acids (4, 8) and to penicillamine, and that these reactions were pH dependent. The present report shows that formation of benzylpenicillenic acid in unbuffered solutions is followed by polymerization, which can be minimized or arrested by buffering. The resulting polymer however has characteristics of penicilloic as well as penicillenic acid, shows some preservation of the benzyl- and thiazolidine structures and appears to be bonded only at the carbonyl and sulfhydryl groups through amide and disulfide linkages. This would suggest that the primary subunit of the polymer is a dimer formed by interaction of penicilloic acid and penicillenic with linkage at the N_4 group of penicilloic acid (Figure 5) causing recyclicization of penicillenic to penicilloic acid. A similar reaction would then proceed with another molecule of penicillenic acid and the amino group of the thiazolidine ring to form a trimer and so on to small linear polymers of repeating units.

This model is consistent with the experimental

Table 5: N. M. R. spectrometry of benzylpenicillin before and after fractionation in Sephadex G. 25

Tentative assignment	Signal	Unfractionated starting material		Mol. wt. 1000-5000		Mol. wt. < 1000	
		τ	H's	τ	H's	τ	H's
- $C(CH_3)2$	doublet peaks downfield	8.45	8.21	no doublet 8.45 + downfield peak	6.7	8.45 + downfield peaks	6.06
Amide - NH	triplet	7.15	1.47	7.15	1.28	7.15	4.15
- CH_2	singlet	6.45	2.08	6.42 (broadened)	2.26	6.4	2.75
-CH CO_2 H	singlet	5.72	1.04	5.7	0.855	5.78	0.805
β-lactam region	singlet	4.44	1.91	4.45	1.28	4.51	1.48
Phenyl.	singlet	2.8	5	2.8	5	2.75	5

observations. The free α-carboxyl group at the growing end would exert only a slight buffering action but the β-carboxyl would exert a strong buffering action and limit linear growth. Only terminal penicilloyl residue would be expected to react with mercuric chloride as N_4 substituted penicilloyl derivatives are resistant to this reagent (4). Chemical tests on the polymers however showed that while some fractions gave reactions for substituted penicilloic acid, the more characteristic reaction, even in small polymers, indicated the presence of disulfide groups with absorbtion at 322 mμ, stability of the N:S ratio (Table 3), absence or reduction of the β-lactam structure, and an absorbtion profile in the infra-red almost identical with that of penicilloyl amide (Tables 1 and 4). This indicates disulfide linkage between the two chains or with penicillenic acid (Figure 5) to form a trimer or a complex polymer of higher molecular weight (2500-5000) with two sets of ordered sequences.

In the case of the natural penicillins and cephalosporins, a third macromolecular component was identified in the form of a proteinaceous substance composed of various amino acids (lysine, threonine, glutamic acid) besides valine and dimethyl cysteine, or valine and methionine, which are the primary units of penicillin and cephalosporin monomer, respectively. This type of macromolecule has chemical and immunochemical properties indicative of amide linkage at its carbonyl group (12, 13) presumably via the epsilon-amino group of lysine; hence it may become bonded with, and greatly enlarge, the small polymers to form complexes with molecular weights exceeding 10, 000. Such complexes are highly immunogenic (2) and are now known to account for some of the allergenicity of penicillin (12, 13, 16).

The structure proposed for dimers and polymers of benzylpenicillin does not necessarily hold for other derivatives of 6-APA or for 7-ACA and its derivatives, since penicillenic acid is not necessarily formed by these compounds. There are however general similarities, since penicilloic or cephalosporoic acids, disulfide and amide, carbonyl or carboxylic linkages can be formed to give repeating subunits varying in size and biologic reactivity.

The experiments described here were performed in solution in concentrations not too different from those used therapeutically; likewise, the residual concentrations of antibiotics with average molecular weights of 2000 or less and below the exclusion capacity of the cross-linked sephadex or silica gel matrices used for fractionation were comparable to those which enter the cross-linked mucopeptide or mucolipopolysaccharide gels of bacterial cell walls under the usual conditions of therapy or microbiological assay. Reference has already been made to the immunochemical activity of the larger macromolecules, in particular to the high degree of penicilloyl and cephalosporoyl activity observed in human subjects and experimental animals (2, 3, 12). For these reasons, it can be assumed that the findings reported here are highly relevant to the biological activity of the β-lactam antibiotics.

Oligo-peptides, including these antibiotics, are well-known for their biological activity which is at times of an extraordinarily high order. This seems to be explained in part at least by the "Signature" principle, as explained by Quastell (10), under which specific activity can be conferred upon a larger molecule by very short chemical sequences or reactive groups. In the case of penicillins, specificity of biological activity is well-illustrated by their potency as antigens and allergens, as well as antibacterial agents. According to Landsteiner (7), antigenic specificity of small peptides is determined by the amino-acid carrying the free carboxyl group. If penicillins and cephalosporins are linked with other molecules via the functional groups and mechanisms described here, they will retain their free carboxyl groups and will therefore be main determinants of antigenic specificity; this is consistent with Landsteiner's haptenic hypothesis, and also with a subsidiary claim made by him that "Pronounced specificity may be connected with the polarity of the CH_2 CONH groups" which are the backbone of the linear polymers described here. Since not all antibodies have allergenic capacity, it can be further postulated that the normal IgG and IgM antibodies formed in response to administration of the antibiotics in solution as dipeptides, dimers, or polymers will exactly fit and neutralize a limited range of thus ordered subunit sequences but that when larger molecules of varied amino acid composition and irregularly spaced penicilloyl haptens are encountered, different antibodies and abnormal responses, such as allergy, might occur. Penicillins are much more prone to cause allergy than cephalosporins; on the oligopeptide polymer hypothesis offered here, this would be explained (9) by the greater reactivity of cysteine - S in penicillins than the methionine - S in cephalosporin dipeptides.

References

1. Batchelor, F.R., Dewdney, Janet M., Feinberg, J. G., and Weston, R.D. Lancet i, 1175 (1967).

2. Butcher, B. T. and Stewart, G. T. Antimicrobial Agents and Chemotherapy, (1968) (In press).

3. Butcher, B. T. and Stewart, G. T. In "Penicillin Allergy" G. T. Stewart and J.P. McGovern, (Eds.), Pub. C.C. Thomas, Springfield, Illinois (to be published).

4. Clarke, H. T., Johnson, J.R. and Robinson, R. (Eds.), "The Chemistry of Penicillin," Princeton Univ. Press, Princeton, New Jersey, (1949).

5. Dursch, F., Lancet ii: 1005, (1968).

6. Grant, N. H., Clark, D. E., and Alburn, H. E., J. Am. Chem. Soc. 84: 876, (1962).

7. Landsteiner, K. "The Specificity of Serological Reactions," pp. 172-183, Dover Publications Inc., New York, New York, (1936).

8. Narasimhachari, N. and Rao, G. R., Current Sci. (India), 28: 488, (1959).

9. Perutz, M. F., "Proteins and Nucleic Acids," pp. 7, American Elsevier Pub. Co., New York, New York, (1962).

10. Quastell, H., "Emergence of Biological Organization," Chapter 2., Yale Univ. Press, New Haven, Conn., (1964).

11. Stewart, G. T., Lancet i: 509, (1962).

12. Stewart, G. T., ibid. i: 1177, (1967).

13. Stewart, G. T., Antimicrobial Agents and Chemotherapy, 543, (1967).

14. Stewart, G. T., Advances in Chem. 63: 141, (1967).

15. Stewart, G. T., Lancet i: 1088, (1968).

16. Stewart, G. T., Antimicrobial Agents and Chemotherapy, (1968), (in press).

17. deWeck, A. L., Schneider, D. H., and Gutersohn, J., Int. Arch. Allergy, 33: 535, (1968).

THE USE OF SLIGHTLY SOLUBLE, NON-POLAR SOLUTES AS PROBES FOR OBTAINING EVIDENCE OF WATER STRUCTURE

Forrest W. Getzen

Department of Chemistry, North Carolina State University, Raleigh, North Carolina 27607

The solubilities of non-polar molecules in water have been examined extensively in recent years in attempts to explain their abnormal behavior as compared to solubility behavior for ideal and for regular solutions. The explanations for the abnormal solubility behavior have involved considerations of enthalpy and entropy changes which have been attributed to iceberg formation, the mixing of molecules, and intermolecular interactions. From their analysis, Frank and Evans[1] concluded that water forms iceberg-like structures around the solute molecules, and such behavior explains the large negative entropies of solution and the small or negative enthalpies of solution. Shinoda and Fujihira[2] have extended the regular solution theory of Hildebrand and Scott[3] to account for the solution behavior in terms of a large positive enthalpy of mixing due to the intermolecular interactions plus a large negative enthalpy of iceberg formation to give the net small or negative enthalpy of solution and the corresponding entropy of solution. They conclude that the iceberg formation of water promotes the solubility of non-polar solutes (at room temperature relative to temperatures above 120°C) and that the small solubilities are not due to entropy effects.

Despite some disagreement in the interpretation of the solubility behavior of non-polar molecules in water there is a general agreement that water exhibits definite structural characteristics as it surrounds the non-polar molecules in solution. Furthermore, from the changes in solubility with temperature, there is considerable evidence that the structural characteristics of water about the solute molecules decrease with increasing temperature. In this paper, a model is developed for the solution of non-polar molecules in water which depends upon the inherent structure of pure water at room temperature. The model explains the temperature dependence of the solubilities of non-polar molecules in water in terms of the decrease in water structure with increasing temperature. Thus, such solubilities may be used not only to obtain evidence of water structure but also to obtain the change in this structure with temperature. The data presented also suggest that the model can be used to obtain enthalpy of fusion values from solubility measurements. The changes in heat capacity differences ΔCp with temperature and the specific volume effects have been neglected in this treatment so that some aspects of the model are obscured; however, this does not affect the overall usefulness of the model as an explanation for the solution behavior of non-polar molecules in water.

DERIVATION OF EQUATIONS

The analysis which is presented here is limited to those solute molecules which follow a Henry's law behavior up to their saturation value in water. The non-polar compounds generally fit into this class of substances at room temperature and at temperatures considerably above room temperature. The chemical potential of the solute in a solution is

$$\mu_2(\text{liq}) = \mu_2^o(\text{liq}) + RT \ln a_2, \text{ with } a_2 = \gamma X_2 \qquad (1)$$

where a_2 is the activity of the solute in a solution of concentration X_2 and γ is the activity coefficient for the solute at this concentration. Likewise, for the gas in equilibrium with the solute in solution

$$\mu_2(\text{gas}) = \mu_2^o(\text{gas}) + RT \ln P_2 \qquad (2)$$

where P_2 is the equilibrium vapor pressure of the solute above the solution. Thus, under equilibrium conditions:

$$\mu_2^o(\text{gas}) = \mu_2^o(\text{liq}) + RT \ln(X_2/P_2) + RT \ln\gamma. \qquad (3)$$

Now, for solutions obeying Henry's law, the ratio (X_2/P_2) is independent of the solute concentration as are also the two standard state chemical potential values. Therefore, where Henry's law holds, the activity coefficient is constant and equal to the value for the infinitely dilute solution. Thus, in the Henry's law region (which is assumed to hold up to saturation values)

$$\mu_2^o(\text{gas}) = \mu_2^o(h) - RT \ln P(h) \qquad (4)$$

where

$$\mu_2^o(h) = \mu_2^o(\text{liq}) + RT \ln \gamma_{i.d.} \neq f(X_2) \qquad (5)$$

and

$$P(h) = (P_2/X_2) = K_H. \qquad (6)$$

The quantity P(h), equal to the Henry's law constant K_H, is the equilibrium vapor pressure of the pure solute in a hypothetical state so that its solution behavior in water up to the saturation value is that of an ideal solute. In summary, the relations which can be used to describe solution behavior up to saturation are, from (1) and (5):

$$\mu_2(\text{liq}) = \mu_2^o(h) + RT \ln X_2 \qquad (7)$$

and

$$P_2 = P(h)\, X_2. \qquad (8)$$

Equation (7) gives the chemical potential for the solute in an ideal solution of mole fraction X_2 and equation (8) shows the Raoult's law behavior for the solute. However, in order to effect such an ideal solution behavior the pure solute must be in the hypothetical state having the equilibrium vapor pressure P(h). The solution process, up to saturation, can be considered as a two step process:
(1) The change in state of the pure solute from a real state having a vapor pressure P_2^o at T to the hypothetical state having a vapor pressure P(h) at T.
(2) The solution of the solute from the hypothetical pure state at T following an ideal solution behavior.

The thermodynamic description of ideal solution behavior is well understood so that the second process furnishes the basis for the observations that the change in enthalpy is zero and the change in entropy is that for random mixing for this process. A second observation is that, from the standpoint of structure, the solute in the hypothetical state must reflect the structure of the solvent (water) at each solution temperature. Therefore, a description of the structural characteristics of water and the changes in these structural characteristics with temperature can be obtained from the thermodynamic behavior of the solute in the hypothetical state as a function of temperature. A third observation is that the resulting structural features apply to the infinitely dilute solution which is pure water. Therefore, the structural characteristics which are revealed by the behavior of the solute in its hypothetical state should be more or less independent of the solute and should depend primarily on the solvent (water).

A thermodynamic description of the solute in its hypothetical pure state and an understanding of the first step of the two step solution process stated above can be obtained from equation (4) which gives the change in chemical potential for the pure solute from the hypothetical standard state to the gas in its standard state as

$$- \Delta\mu^{o} = - (\mu_{g}^{o} - \mu^{o}) = RT \ln P. \quad (9)$$

Here, subscripts have been dropped and the notation has been simplified since only pure solute is involved. Figure 1 shows the general behavior of R lnP versus (1/T) and RT lnP versus T along with the corresponding behavior of the same parameters for the pure liquid and pure solid solute. Two characteristics of these plots should be noted. First, the vapor pressure curve for the hypothetical state

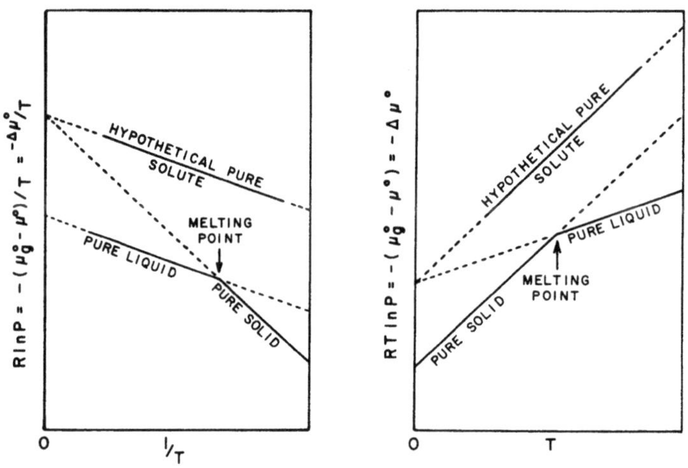

Fig. 1. RlnP versus (1/T) and RTlnP versus T for the pure solute in the hypothetical state, in the liquid state, and in the solid state.

runs parallel to the vapor pressure curve for the pure liquid state in the plot of RlnP versus (1/T). Second, for this same plot, the intercept of the curve for the pure hypothetical state is the same as the intercept for the pure solid state at (1/T)=0. For the plots of RTlnP versus T, the slope of the hypothetical state curve is that for the pure solid state and the intercept for the hypothetical state curve is that for the pure liquid state at T=0.

It should be emphasized that the plots in Fig. 1 are idealized; however, they do show the general behavior of the parameters of interest, namely, the slopes and intercepts of the curves. The behavior shown in Fig. 1 suggests that, of the slopes and intercepts available from the pure solute behavior, the two most appropriate for the description of the hypothetical state behavior are, in the

case of the RlnP versus (1/T) plots, the slope of the pure liquid curve and the intercept of the pure solid curve at (1/T)=0. The corresponding parameters of interest from the RTlnP versus T plots are the slope of pure solid curve and the intercept of the pure liquid curve at T=0. From thermodynamic considerations, these two parameters of interest are:
(1) slope of RlnP versus (1/T) = intercept of RTlnP versus T = $-\Delta H^o_v$ and
(2) intercept of RlnP versus (1/T) = slope of RTlnP versus T = $+\Delta S^o_s$ where ΔH^o_v is the molar enthalpy of vaporization of the liquid and ΔS^o_s is the molar entropy of sublimation of the solid.

The behavior of the curves for the hypothetical standard state in terms of these two thermodynamic parameters is given by:

$$R\ln P = \Delta S^o_s - \Delta H^o_v(1/T) \text{ and } RT\ln P = T\Delta S^o_s - \Delta H^o_v. \qquad (10)$$

These equations describe an approximate behavior for the hypothetical state in terms of the enthalpy of vaporization of the pure liquid and the entropy of sublimation of the pure solid. The actual behavior of the curves for the hypothetical state can be described by the equations (10) plus an excess free energy term. Thus, for the behavior of the hypothetical state for real solutes,

$$R\ln P = \Delta S^o - \Delta H^o(1/T) = \Delta S^o_s - \Delta H^o_v(1/T) - \Delta\mu^o_e/T \qquad (11)$$

and

$$RT\ln P = T\Delta S^o - \Delta H^o = T\Delta S^o_s - \Delta H^o_v - \Delta\mu^o_e \qquad (12)$$

The magnitude and behavior of the excess free energy parameter of equation (12) reflects the degree to which the equations (10) may be relied upon to describe the hypothetical state behavior for real systems. A combination of the terms in equation (12) with the corresponding parameters in the equations (11) leads to a statement for the excess enthalpy change and the excess entropy change:

$$\Delta H^o_e = \Delta H^o - \Delta H^o_v = H^o(\text{liq}) - H^o(\text{h})$$

and (13)

$$\Delta S^o_e = \Delta S^o - \Delta S^o_s = S^o(\text{solid}) - S^o(\text{h}).$$

Thus, the enthalpy term of the excess free energy change is the enthalpy change from the hypothetical state to pure liquid, and the entropy term of the excess free energy change is the entropy change from the hypothetical state to pure solid.

The determination of the excess free energy change is obtained from the temperature, the molar entropy of sublimation of the pure solid, the enthalpy of vaporization of the pure liquid, and the measured equilibrium value for P. The value of P for solids and for liquids is given by equation (6). For gases, the value of P is given by:

$$P = (1/X_2) \tag{14}$$

where X_2 is the mole fraction saturation value of the solute in the solution with its equilibrium gas pressure above the solution equal to one atmosphere.

In the case of solids and liquids, advantage can be taken of the fact that, because of the very low solubilities, the equilibrium vapor pressure of the gaseous solute above the solution in equilibrium with the solid or liquid is nearly equal to the equilibrium vapor pressure of the pure solid or liquid. Under these conditions, the equations (11) can be simplified as follows:

(for solids) $R\ln P^o = \Delta S_s^o - \Delta H_s^o(1/T)$

so

$$R\ln P - R\ln P^o = -R\ln X_2 = \Delta H_f^o(1/T) - \Delta\mu_e^o/T$$

or

$$[RT\ln X_2 + \Delta H_f^o] = \Delta\mu_e^o \tag{15}$$

and (for liquids) $R\ln P^o = \Delta S_v^o - \Delta H_v^o(1/T)$

so

$$R\ln P - R\ln P^o = -R\ln X_2 = \Delta S_f^o - \Delta\mu_e^o/T$$

or

$$T\,[R\ln X_2 + \Delta S_f^o] = \Delta\mu_e^o \tag{16}$$

where X_2 is the equilibrium mole fraction saturation value for the solute having an equilibrium vapor pressure P^o at T. In these equations, ΔH_f^o is the molar enthalpy of fusion and ΔS_f^o is the molar entropy of fusion. Where variations of heat capacity differences with temperature can be neglected, values for ΔH_f^o and ΔS_f^o can be taken as the freezing point or triple point values.

Equations (12), (15), and (16) can be used to test the arguments proposed for the solubility behavior of non-polar solutes in water, particularly with reference to the effect of water structure and changes of water structure with temperature upon such solubilities. Some general observations concerning the excess free energy term are summarized as follows:
(1) The term should be small compared to the free energies derived from equation (10) in order to confirm the highly structured character of water and small interaction of water with the non-polar solutes.

(2) The temperature dependence of the excess free energy term should reflect changes in water structure with temperature and changes in solute-water interactions with temperature.
(3) For low or zero solute-water interaction, the temperature dependence of the excess free energy term should be practically independent of the solute. Thus, this term would not reflect phase transitions of a solute, but it should reflect such transitions for the solvent water.

COMPARISON WITH EXPERIMENTAL RESULTS

The explanation for solubility behavior in terms of water structure and changes in water structure and the use of slightly soluble, non-polar solutes as probes for obtaining evidence of water structure can be tested through the application of equations (12), (15), and (16) to experimental data. The data of Claussen and Polglase[4] for gases and of Bohon and Claussen[5] for liquids and solids are used. Enthalpy and entropy changes are from vapor pressure and calorimetric data and the values used are in Tables I and II.

Table I contains the enthalpy and entropy changes which are used in equations (12), (15), and (16) along with the measured solubilities to obtain the excess free energy term.

TABLE I

THERMODYNAMIC PARAMETERS

Compound	ΔH^o_v(t.p.) (Kcal/mole)	ΔS^o_s(t.p.) (cal/mole-°Abs)	t.p. (°Abs)
Methane[a]	2.102	21.50	90.680
Ethane[b]	4.586	35.28	89.89
	ΔH^o_f(t.p.) (Kcal/mole)	ΔS^o_f(m.p.) (cal/mole-°Abs)	m.p. (°Abs)
Benzene[c]	2.350	8.434	278.683
Toluene[c]	1.582	8.879	178.159
p-Xylene[c]	4.090	14.281	286.413
m-Xylene[c]	2.764	12.271	225.278
Ethylbenzene[c]	2.190	12.292	178.175
Naphthalene[c]	4.315	12.200	353.70
Biphenyl[d]	5.18	13.42	342.65

t.p. = triple point m.p. = melting point
[a]Parameters obtained from Antoine equation data, reference (6).
[b]From Antoine equation and enthalpy of fusion data, reference (7).
[c]From reference (8). [d]From vapor pressure data, reference (9).

TABLE II

VAPOR PRESSURE PARAMETERS

Compound	ΔH°_v (Kcal/mole)	ΔS°_v (e.u./mole)	ΔH°_s (Kcal/mole)	ΔS°_s (e.u./mole)	Temp. range
Benzene[(a)]	8.217	23.217	10.564	31.866	m.p.
p-Xylene[(a)]	10.17	25.27	14.26	39.55	m.p.
Naphthalene[(b)]	-	-	17.31	39.88	6-21°C
Biphenyl[(b)]	12.93	24.49	-	-	70-255°C
Biphenyl[(b)]	-	-	18.11	37.91	6-26°C

[(a)]Parameters obtained from a least square of lnP versus (1/T) with pressures determined from Antoine equation data, reference (8).
[(b)]Parameters obtained from vapor pressure data for solid and liquid compound, reference (9).

Table II contains the enthalpy and entropy changes for the determination of the vapor pressures used to determine the Henry's law constants K_H of equation (6). As defined in equation (6), the Henry's law constant equals the equilibrium vapor pressure of the solute in its hypothetical pure state. For the gases, the equilibrium vapor pressure of the solute in its pure hypothetical state was obtained from the solubility data as expressed in equation (14).

In the case of the gas solubilities, the mole fractions were established from reported solubilities in units of volume of gas at 0°C and 1 atmosphere dissolved in one volume of water at 1 atmosphere solute pressure at the measured temperature as:

$$10^4 X_2 = \left(\frac{M_{H_2O}\, \alpha}{2.2414\, \rho^\circ_t} \right) \tag{17}$$

where α is the reported solubility, M_{H_2O} is the molecular weight of water (18.016), ρ°_t is the density of water at t°C, and X_2 is the solute mole fraction.

For the solid and liquid solubilities, the mole fractions were established from reported molar saturation concentrations by using the relation:

$$10^n X_2 = 10^n \left[\frac{10^3 \rho^\circ_t}{M_{H_2O} m} - \frac{M}{M_{H_2O}} + 1 \right]^{-1} \tag{18}$$

where m is the reported concentrations, M_{H_2O} is the molecular weight of water (18.016), M is the molecular weight of the solute, ρ°_t is the density of water at t°C, and X_2 is the solute mole fraction.

Values of RTlnP versus T are shown in Fig. 2 for ethane. The solid line passing through the data points is parallel to the sublimation values for the solid in the triple point temperature range.

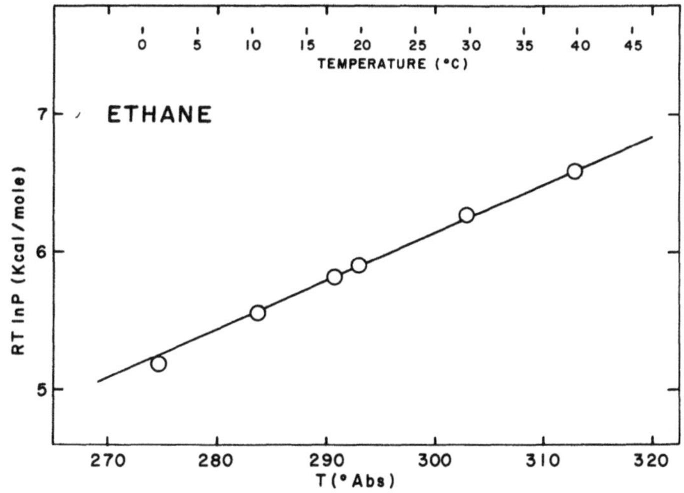

Fig. 2. RTlnP versus T for ethane. The solid line passing through the data points is parallel to sublimation pressure values for solid ethane in its triple point range. Data from reference (4).

The plots of RTlnP versus T from measurements for benzene, toluene, p-xylene, m-xylene, ethylbenzene, naphthalene, and biphenyl show a behavior similar to that shown in Fig. 2. Figures for the liquids have been reported by Herington[10]; however, no attempt was made to account for nonlinear behavior in his work.

Values of $\Delta\mu_e^o$ were calculated for methane, ethane, benzene, toluene, p-xylene, m-xylene, ethylbenzene, naphthalene, and biphenyl in the range of temperatures from 0° to 50°C. All the compounds gave a smooth function for $\Delta\mu_e^o$ which was found to follow the equation:

$$\Delta\mu_e^o = (a - c\,T + b\,T^2) + d\,T(T - e)^2. \qquad (19)$$

Thus, the values for ΔH_e^o and ΔS_e^o, as obtained from this equation, can be expressed as:

$$\Delta H_e^o = (a - b\,T^2) - 2\,d\,T^2(T - e) \qquad (20)$$

and

$$\Delta S_e^o = (c - 2\,b\,T) - d\,(3T - e)(T - e). \qquad (21)$$

Non-zero values for the constants of equations (19), (20), and (21) are given in Table III for all the compounds studied. These constants were used with the data in Table I and Table II to calculate saturation concentrations from equations (12)(for gases), (15)(for solids), and (16)(for liquids). The use of equations (17)(for gases)

TABLE III

CALCULATED PARAMETERS FOR EXCESS THERMODYNAMIC FUNCTIONS PARAMETER

	a	b	c
Gases			
Methane	1370	-	11.18
Ethane	533	-	2.341
Liquids			
Benzene	-	-	7.067
Toluene	-	-	9.027
p-Xylene	-	-	6.204
m-Xylene	-	-	8.238
Ethylbenzene	-	-	8.089
Solids			
Naphthalene	-	0.02527	17.393
Biphenyl	-	0.02097	16.128
	d	e	
All compounds:	0.000408	291.15	

and (18) (for solids and liquids) permitted a comparison of calculated and experimental saturation concentration values. This comparison is given in Table IV for all the compounds studied.

It is worthwhile to note that the temperature independent constant a in equation (19) is zero for the solids. Therefore, because the two constants d and e are independent of the solute, a value for the heat of fusion for a solid can be obtained from solubility versus temperature measurements by the use of equation (15) in the form:

$$f(T) = + RT\ln X_2 - d\ T(T - e)^2 = -\Delta H_f^o - c\ T + b\ T^2. \quad (22)$$

A fit of the calculated values of f(T) from measured values of X_2 over a range of temperatures to a quadratic function of T gives the three constants ΔH_f^o, c, and b. This procedure was followed for biphenyl to give a value of 5.31 (Kcal/mole) for ΔH_f^o as compared to a value of 5.18 (Kcal/mole) from data fround in Table II.

An examination of Table III shows that only two of the solute specific parameters are non-zero for gases and solids and only one solute specific parameter is non-zero for liquids. The remaining two parameters (d and e) do not depend upon the solute. Since changes in heat capacity differences with temperature and specific volume effects have been neglected in this procedure, a consideration of these factors might account for some of the differences in the parameters of Table III. Also, interactions of the solute molecules with the water can account for some observed differences. However, it is important here to emphasize the similarities in the behavior of the excess free energy term for the non-polar solutes as obtained from the parameters in Table III.

TABLE IV

SOLUBILITIES OF HYDROCARBONS IN WATER

Gases Temp. (°C)	Methane α[a] (obs.)	α (calc.)	Temp. (°C)	Ethane α[a] (obs.)	α (calc.)
1.6	0.0547	0.0546	1.5	0.0937	0.0933
2.0	0.0538	0.0540	10.5	0.0655	0.0663
10.5	0.0428	0.0428	17.5	0.0527	0.0526
19.8	0.0351	0.0348	19.8	0.0496	0.0491
30.4	0.0289	0.0291	29.8	0.0375	0.0377
39.6	0.0255	0.0261	39.7	0.0307	0.0307

Liquids Temp. (°C)	$m \times 10^2$[b] (obs.)	$m \times 10^2$ (calc.)	Temp. (°C)	$m \times 10^2$[b] (obs.)	$m \times 10^2$ (calc.)
	Benzene			Toluene	
0.4	2.22 (solid)	2.42 (super-cooled)	0.4	0.717	0.721
			3.6	0.704	0.706
			10.0	0.684	0.685
5.2	2.31 (solid)	2.35 (super-cooled)	11.2	0.680	0.683
			14.9	0.679	0.677
			15.9	0.676	0.676
10.0	2.30	2.30	25.6	0.681	0.682
14.9	2.26	2.27	30.0	0.697	0.694
21.0	2.28	2.27	30.2	0.699	0.694
25.6	2.28	2.29	35.2	0.716	0.714
30.2	2.35	2.33	42.8	0.763	0.761
34.9	2.41	2.39	45.3	0.781	0.780
42.8	2.55	2.55			
	p-Xylene			p-Xylene	
0.4	0.146 (solid)	0.197 (super-cooled)	21.0	0.184	0.185
			25.6	0.186	0.186
			30.2	0.188	0.190
10.0	0.176 (solid)	0.187 (super-cooled)	30.3	0.191	0.190
			34.9	0.194	0.195
			35.2	0.194	0.195
14.9	0.182	0.185	42.8	0.208	0.208
	m-Xylene			Ethylbenzene	
0.4	0.198	0.195	0.4	0.206	0.207
5.2	0.190	0.189	5.2	0.201	0.201
14.0	0.182	0.183	20.7	0.194	0.194
21.0	0.185	0.183	21.2	0.194	0.195

TABLE IV (continued)

Liquids Temp. (°C)	m x 10^2 (b) (obs.)	m x 10^2 (calc.)	Temp. (°C)	m x 10^2 (b) (obs.)	m x 10^2 (calc.)
	m-Xylene			Ethylbenzene	
25.6	0.185	0.184	25.6	0.196	0.196
30.3	0.188	0.188	30.2	0.199	0.200
34.9	0.193	0.192	34.9	0.207	0.205
39.6	0.206	0.199	42.8	0.217	0.219
Solids Temp. (°C)	m x 10^4 (b) (obs.)	m x 10^4 (calc.)	Temp. (°C)	m x 10^4 (b) (obs.)	m x 10^4 (calc.)
	Naphthalene			Biphenyl	
0.0	1.06	1.06	0.4	0.182	0.179
0.4	1.06	1.08	2.4	0.191	0.194
0.5	1.07	1.08	5.2	0.217	0.217
0.9	1.12	1.10	7.6	0.234	0.239
1.9	1.16	1.14	10.0	0.261	0.263
9.4	1.51	1.48	12.6	0.295	0.292
10.0	1.49	1.52	14.9	0.328	0.321
14.9	1.80	1.82	15.9	0.339	0.335
15.9	1.89	1.89	25.6	0.500	0.505
19.3	2.16	2.15	30.1	0.619	0.616
25.6	2.76	2.75	30.4	0.616	0.624
30.1	3.31	3.30	33.3	0.705	0.710
30.2	3.38	3.31	34.9	0.768	0.763
35.2	4.19	4.08	36.0	0.800	0.803
36.0	4.22	4.22	42.8	1.11	1.10
42.8	5.66	5.67			
	Benzene			p-Xylene	
0.4	222	223	0.4	14.6	14.0
5.2	231	233	10.0	17.6	17.2

(a) Observed values from reference (4).

(b) Observed values from reference (5).

Fig. 3 shows the behavior of the excess free energy term for all solutes as obtained from equation (19) using the constants in Table III. Fig. 4 shows the excess enthalpy term for the solutes from equation (20), and Fig. 5 shows the excess entropy term for the solutes from equation (21) as obtained also by using the constants in Table III. The behavior of the three thermodynamic parameters as a function of temperature leads to the general conclusions of this

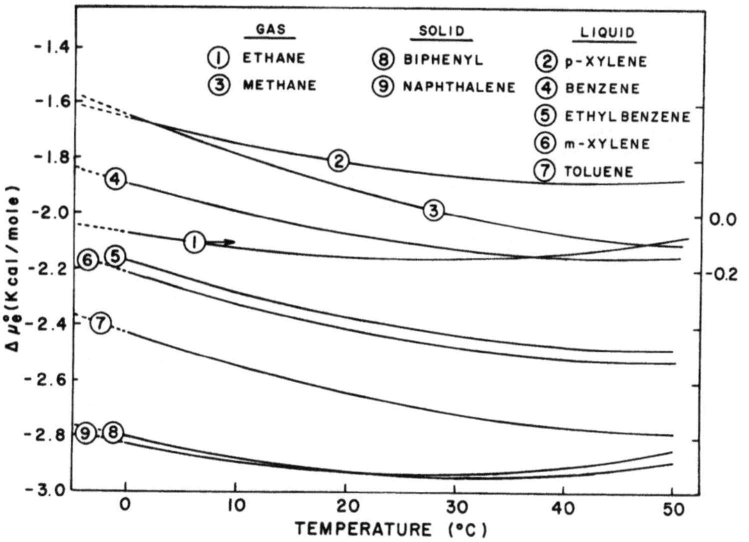

Fig. 3. Excess free energy of mixing as a function of temperature for non-polar solids, liquids, and gases in water. The free energy units for ethane are along the right vertical scale; the free energy units for the remaining compounds are along the left vertical scale.

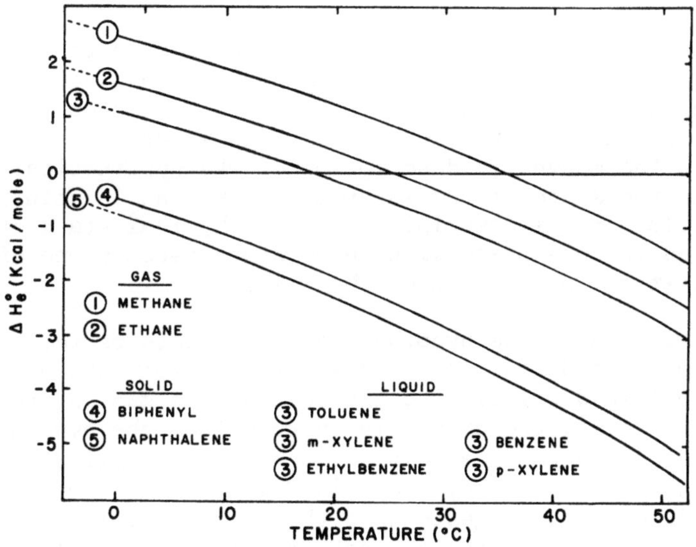

Fig. 4. Excess enthalpy of mixing as a function of temperature for non-polar solutes in water. Values obtained from equation (20) and the parameters in Table III.

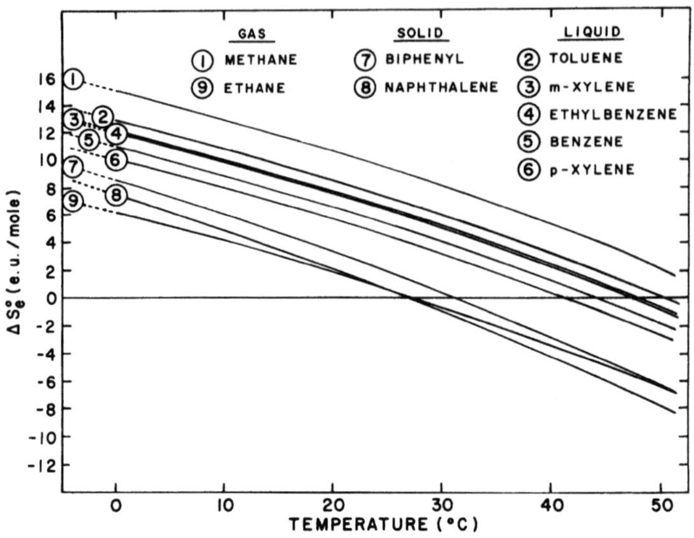

Fig. 5. The excess entropy of mixing as a function of temperature for non-polar solutes in water. The entropy units are (cal/°Abs).

work concerning the nature of the solutions of non-polar molecules in water and the use of their solubility behavior to obtain evidence of water structure.

DISCUSSION AND CONCLUSION

In view of the premise that aqueous solutions up to saturation values may be formed by a two step process which involves, first, a solute change in state to a hypothetical state and, second, a mixing of the solute in this hypothetical state with water following ideal mixing laws, a thermodynamic description of the hypothetical state should reflect the state of the water both pure and in solution. Thus, the problem of the description of the physical state of water and the changes in this state with temperature becomes one of describing the hypothetical pure state of the solute.

The description of the hypothetical pure state of the solute involves both energy and configuration and these terms can be most conveniently established relative to the solute as a gas in its standard state. In such terms, the change in entropy from the hypothetical pure state to the standard gas state is the entropy of sublimation of the solute plus the excess entropy shown in Fig. 5. The first term, the entropy of sublimation, suggests that the hypothetical state must be a highly ordered state and the second term, the excess entropy, reflects the decrease in this ordered structure with temperature. It should be evident from Fig. 5 that the decrease in structure with increasing temperature is that for water and not that for any real state of the solute. This is particularly evident for the liquid curves which are superimposed by a vertical displacement. Then too, there is no visible evidence of phase transition behavior for the two compounds which have melting points in the temperature range of the measurements (benzene and p-xylene). Thus, the conclusion here is that the solution behavior of non-polar molecules in water reveals not only the highly structured character of liquid water but also the decrease in this structure with increasing temperature.

For the change in state from the hypothetical pure solute to the standard gas state, the change in enthalpy is the enthalpy of vaporization of the pure liquid solute plus the excess enthalpy shown in Fig. 4. This is also the enthalpy change for the solute going from the real solution to the standard gas state because of the ideal mixing process. Thus, for the solution of non-polar molecules in water, the enthalpy change is that associated with the mixing of indifferent particles plus an additional enthalpy change which is related to the temperature dependent structural changes of the water as deduced from the similarities of enthalpy and entropy changes shown in Figs. 4 and 5. Here too, the excess change is due primarily to the characteristics of the pure water.

ACKNOWLEDGEMENTS

This work was supported in part by Grant ES 00177-03 of the U. S. Public Health Service.

REFERENCES

1. H. S. Frank and M. W. Evans, *J. Chem. Phys.*, **13**, 507 (1945).
2. K. Shinoda and M. Fujihira, *Bull. Chem. Soc. Jap.*, **41**, 2612 (1968).
3. J. H. Hildebrand and R. L. Scott "Regular Solutions," Prentice Hall, Inc., Englewood Cliffs, New Jersey (1962).
4. W. F. Claussen and M. F. Polglase, *J. Am. Chem. Soc.*, **74**, 4817 (1952).
5. R. L. Bohon and W. F. Claussen, *J. Am. Chem. Soc.*, **73**, 1571 (1951).
6. API Project 44, "Selected Values of Properties of Hydrocarbons and Related Compounds."
7. R. R. Dreisbach, "Physical Properties of Chemical Compounds," Advances in Chemistry Series No. 22, American Chemical Society (1959).
8. R. R. Dreisbach, *ibid.*, No. 15 (1955).
9. R. C. Weast, Ed., "Handbook of Chemistry and Physics," 47th Ed., Chemical Rubber Publishing Co., Cleveland, Ohio (1966-1967).
10. E. F. G. Herington, *J. Am. Chem. Soc.*, **73**, 5883 (1951).

PHOSVITIN, A PHOSPHOPROTEIN WITH POLYELECTROLYTE CHARACTERISTICS*

Gertrude E. Perlmann and Kärt Grizzuti

The Rockefeller University

New York, New York 10021

It is well established that phosphoproteins are widely distributed in all embryonic and developing tissue, and it has been suggested that several biological processes such as enzymic phosphoryl transfer or energy storage may involve the participation of the phosphorylated amino acid residues in these proteins (5, 13). Inasmuch as a knowledge of the chemistry and macromolecular conformation of phosphoproteins is an essential prerequisite for an understanding of how these materials enter metabolic processes, we have initiated a study on phosvitin, the phosphoprotein from hens' egg yolk first isolated in 1949 by Mecham and Olcott (10). Although in our work chemical and physico-chemical techniques are used, we shall present here only some of the physico-chemical studies currently carried out in our laboratory. We should like to show that the conformational characteristics of phosvitin are highly sensitive to pH and ionic strength and that this can best be explained by considering this protein as a polyelectrolyte.

Before discussing what we have been able to establish on the conformation of phosvitin, a few of the characteristics of this protein are reviewed. Phosvitin, isolated from fresh hens' egg yolk by the procedure of Joubert and Cook (9) has a molecular weight of 40,000.[1] It contains 10.6 per cent phosphorus[2] which is directly esterified to a hydroxyamino acid residue. The high content of phosphorus, i.e., 136 atoms of phosphorus per molecule, is balanced by a large amount of serine (1). Although the amounts of phosphorus and serine are nearly equimolar, Dr. Allerton has been able to demonstrate the presence of two residues of phosphothreonine per mole of protein in addition to non-phosphorylated serine (1). Furthermore, phosvitin is also a glycoprotein. The carbohydrate moiety corresponds to six per cent of the protein. Dr. Shainkin who currently is investigating to which amino acid residue the polysaccharide unit is attached has shown that phosvitin contains six moles of hexose of which there are three residues of galactose and three residues of mannose, in addition to four moles of glucosamine and two moles of sialic acid (14). Phosvitin has one NH_2-terminal amino acid, alanine, and leucine as the COOH-terminus.[3]

A closer examination of the amino acid distribution, given in Table I reveals that the content of phosphoamino acids corresponds to fifty four per cent, whereas that of the dicarboxylic acids is twelve per cent. In addition this protein has sixteen per cent basic constituents and is characterized by an unusually low content of non-polar and aromatic amino acids and proline (1). That this protein resembles in many respects a polyelectrolyte is easily understood if one considers the presence of 136 phosphomonoester groups. In alkaline solution, namely at a pH higher than 8.0, where two hydroxyls of each phosphate group are ionized the net electronic charge corresponds to about one negative charge per amino acid residue. Even in distinctly acid solution, namely at pH 3.0 to 5.0, where only one hydroxyl of the phosphate groups has lost its proton, the charge density is still relatively high. It corresponds to a negative charge of fifty per cent of all amino acid residues present in phosvitin. It is only below pH 2.0 where the phosphate groups are no longer ionized that phosvitin becomes electrically neutral - actually it carries a small positive net charge. In view of this unusual amino acid composition, phosvitin should have the characteristics of a polyelectrolyte. This supposition is fully supported by viscosity measurements in water and in sodium chloride solutions in concentrations

Table I

Amino Acid Composition of Hens' Egg Phosvitin*

Amino Acid	Number of residues per mole protein	Per Cent
Acidic (Asp, Glu)	31	12
Basic (Lys, His, Arg)	43	16
Non-polar (Gly, Val, Leu, Ile, Ala, Met)	25	10
Hydroxy (Ser, Thr)	11	4
Phosphoamino (PSer, PThr)	136	54
Proline	4	2
Aromatic	4	2

* Taken from Allerton and Perlmann (1).

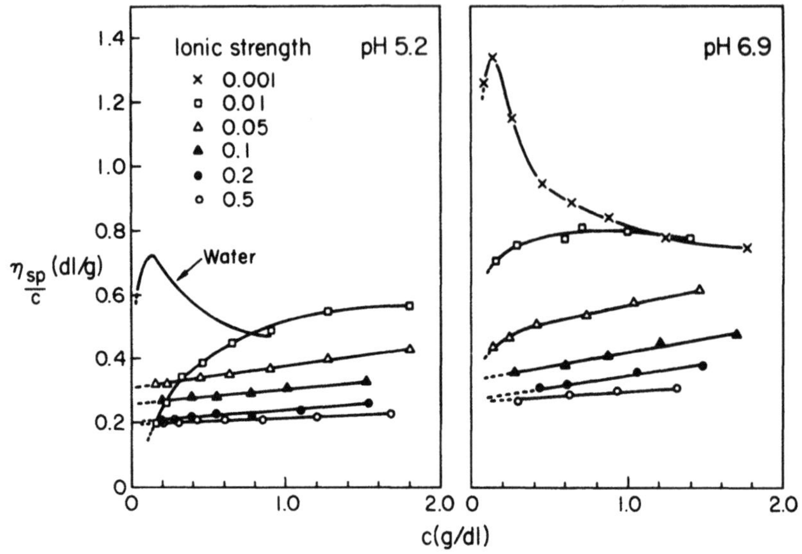

Fig. 1. Dependence of specific viscosity of phosvitin on protein concentration at various pH and ionic strength.

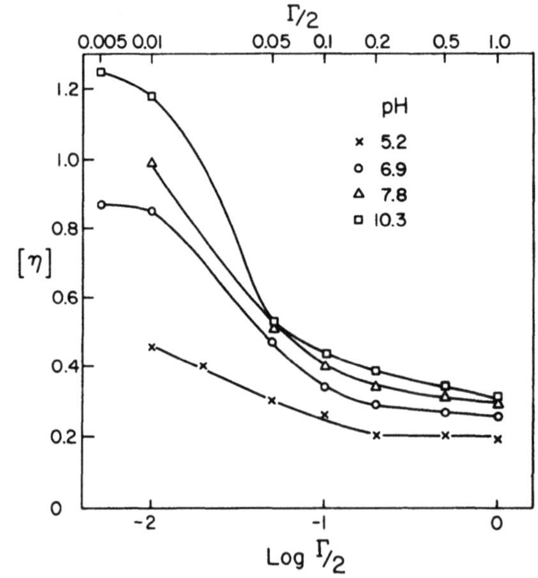

Fig. 2. Dependence of intrinsic viscosity of phosvitin on ionic strength and pH.

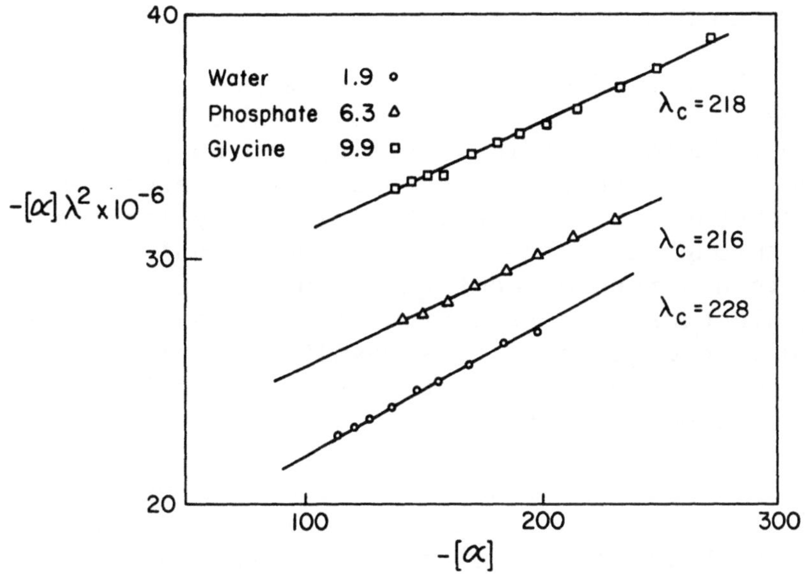

Fig. 3. Dependence of the optical rotatory dispersion constant, λ_c, of phosvitin on pH.

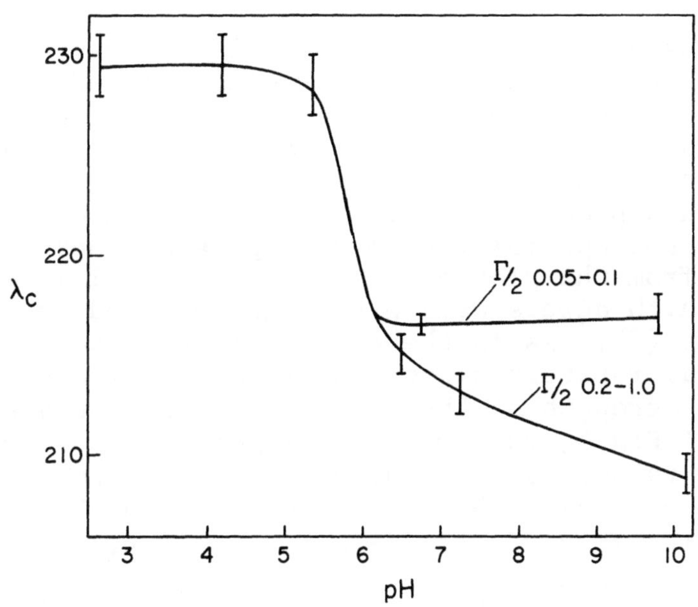

Fig. 4. Dependence of the optical rotatory dispersion constant, λ_c, of phosvitin on pH and ionic strength.

of 0.001 to 0.5 N.

As shown in Fig. 1, in the presence of sodium chloride of 0.1 to 0.5 N at pH 5.2 and 6.9, respectively, the reduced specific viscosity can be linearly extrapolated to a value of 0.2 to 0.3 dl/g, whereas at low ionic strength or in water, the viscosity increases and passes through a maximum.[4] The protein molecules seem to attain a stretched configuration as a result of strong electrostatic repulsion between the negatively charged acidic groups. This result is reminiscent of that reported previously for poly-L-methionine-S-methylsulfonium bromide (11).

If we now compare the results of measurements illustrated in Fig. 2, where the intrinsic viscosities at different pH values are plotted against the logarithm of the ionic strength, the intrinsic viscosity of phosvitin undergoes a sharp transition in the ionic strength interval of 0.05 to 0.1. Furthermore, the intrinsic viscosity of phosvitin is highly dependent on pH, further supporting the polyelectrolyte character of this protein.

Let us now consider the optical rotatory properties of phosvitin. In the spectral region removed from the absorption bands, namely in the wave length range of 350 to 600 mμ, the optical rotatory dispersion can be represented by the one-term Drude equation. As shown in Fig. 3, in a phosphate buffer of pH 6.3 and 0.1 ionic strength, the optical rotatory dispersion constant, λ_c, obtained from the plot of $\alpha.\lambda^2$ versus $[\alpha]$ (19) is 216 mμ, a value which agrees well with that reported by Jirgensons (7). As is the case with the viscosities, the optical rotatory dispersion is also affected by pH and ionic strength of the solvent. As illustrated with the aid of Fig.4, the rotatory dispersion constant, λ_c, at low pH values is independent of ionic strength. However, on raising the pH of the solution above pH 5.8, λ_c decreases from 228 to 218 mμ in the range of 0.05 to 0.1 ionic strength, but it is not affected by the pH of the solution. At higher ionic strengths, however, the optical rotatory dispersion constant decreases further to a value of 210 mμ.

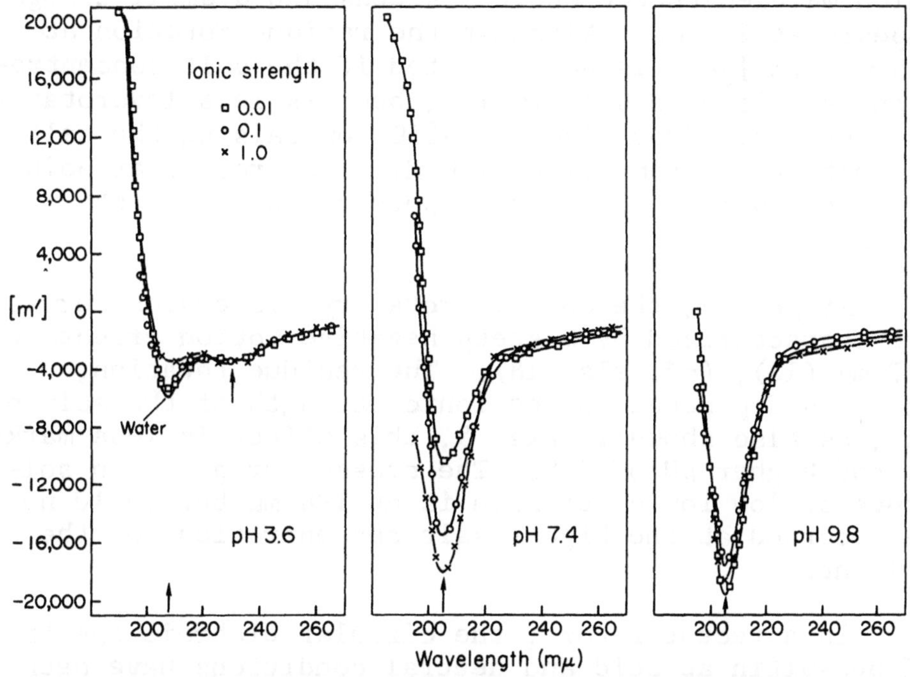

Fig. 5. Dependence of the optical rotatory dispersion of phosvitin on ionic strength and pH.

On comparing the effect of pH on the optical rotatory dispersion curves in the far-ultraviolet an unusual type of Cotton effect is obtained. In Fig. 5 are shown the optical rotatory dispersion curves at pH 3.6, 7.4 and 9.8 at an ionic strength of 0.01, 0.1 and 1.0, respectively. As we have shown elsewhere, at pH 3.6 the curves display a deep trough at 205 mμ (12). The rotation becomes less negative up to 225 mμ, above which wavelength the curve flattens and a small trough appears at 232 mμ. Although the residue rotation at 205 mμ, $[m']_{205}$, is not affected if the salt concentration is raised to 0.1, $[m']_{205}$ becomes less levorotatory and increases from -5400 to -3500 on raising the salt concentration from 0.1 to 1.0 M. The cross-over point of these optical rotatory dispersion curves is at 198 mμ.

At pH 7.4, the optical rotatory dispersion curves are characterized by a steep negative Cotton effect at 207 mμ (12), (cf. also 18). The residue rotation, [m'], is dependent on the ionic strength of the solvent but, as also shown in Fig. 5, this effect is less marked at the higher pH of 9.8. The cross-over point in solvents of low ionic strength is at 198 mμ but could not be recorded at the higher salt concentrations of the solvent.

In a recent report, the circular dichroic spectra of phosvitin at acid and neutral conditions have been described (17). These measurements have now been extended to a wider range of pH and ionic strength, respectively. The results in Fig. 6 reveal that at pH 3.8 a negative dichroic band is present at 198 mμ. The ellipticity, $[\theta']_{198}$, becomes less negative at higher salt concentrations. In addition, a weak dichroic band above 220 mμ becomes noticeable. At pH 7.6 and 9.8, the dichroic spectra are characterized by a strong negative band at 198 mμ (cf. 17).

Although not apparent from Fig. 6 where the dichroic bands are set upon a rather large and steeply changing background, is the occurrence of dichroic bands in the wavelength range of 210 to 250 mμ. As il-

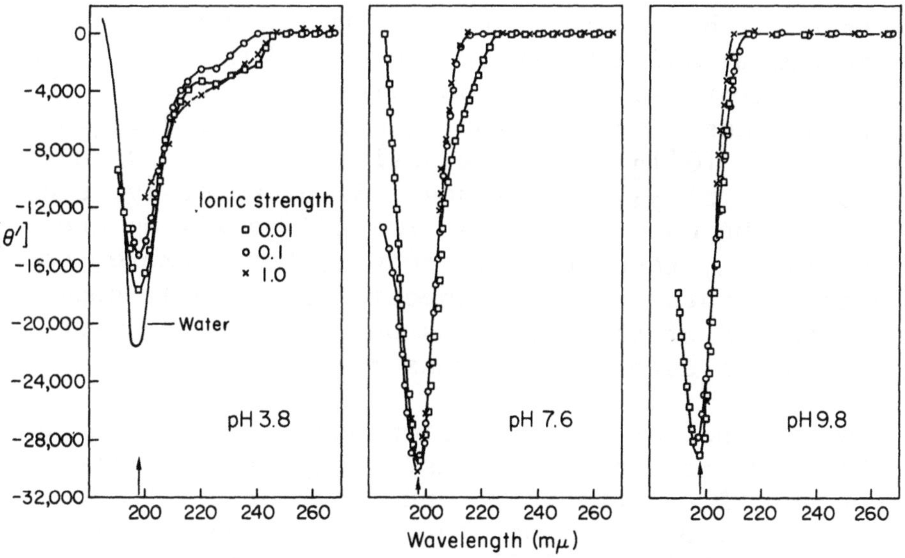

Fig. 6. Dependence of circular dichroism of phosvitin on ionic strength and pH.

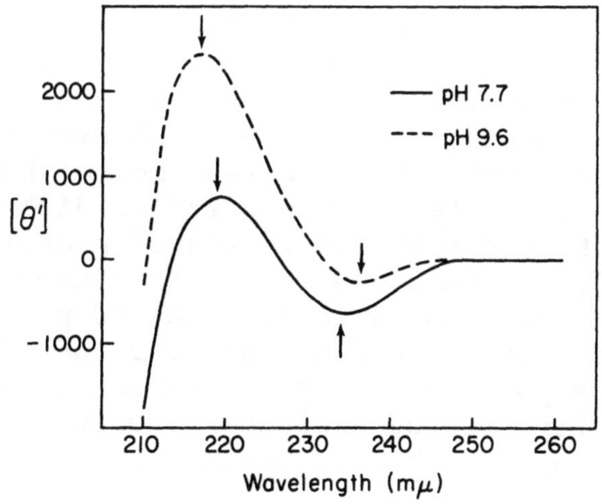

Fig. 7. Circular dichroism of phosvitin.

——— Na phosphate, pH 7.7 and Γ/2 0.1.

------- Na glycine, pH 9.6 and Γ/2 0.1.

lustrated with the aid of Fig. 7, two peaks, a positive one at 220 mμ and a negative one at 233 mμ, are present in the dichroic spectra of phosvitin. On raising the pH of the phosvitin solution from pH 7.7 to 9.6, the ellipticity, $[\theta']_{220}$, of the positive and of the negative bands, $[\theta']_{233}$, increases. Simultaneously, small shifts of the band positions from 220 to 217 mμ and 233 to 237 mμ, respectively, occur. Furthermore, the strength of these dichroic spectra is not only dependent on the pH of the solutions but also on the protein concentration used for the measurements.

We may now ask the question what information can be obtained from these results? Comparison of the ORD curves and CD spectra in the pH range of 6.0 to 10.0 with the data obtained on standard polypeptides (2, 3, 18) shows a striking similarity to curves observed with poly-L-glutamic acid and poly-L-lysine in the unordered conformation; these two polymers are also characterized by a trough at 205 mμ (6, 20).

Other structures which display similar optical rotatory dispersion characteristics include collagen with a trough at 207 mμ, a cross-over point at 197 mμ, and a flattening of the wave length dependence of the rotation above 200 mμ (4). Similarly, poly-L-proline I has a negative Cotton effect at 207 mμ (4). In view of the special structures of collagen and poly-L-proline I and the small amount of proline in phosvitin, namely 1.5 to 2.0 per cent, it would seem most likely that the conformation reflected by the circular dichroism and optical rotatory dispersion curves of phosvitin at pH 7.0 to 10.0 is an unordered conformation (cf. 12 and 17). Furthermore, the axially extended periodic side-chain charges may exert a rigidifying and regularizing effect on the polypeptide chain backbone, thus giving rise to locally ordered extended structures with characteristic circular dichroism spectra (16).

On the other hand, at pH 3.8 phosvitin in addition to the intrinsic Cotton effect at 205 mμ has also a second trough at 232 mμ and a corresponding dichroic band at 220 mμ (17). Since neither one corresponds to any known polypeptide conformation, it seems most likely that the changes in optical rotatory dispersion that occur as the pH is shifted from 6.6 to 3.4 are probably the result of a transition from an extended unordered chain to a more compact structure which may contain up to 20 per cent α-helix. This view is fully supported by the viscosity measurements (Figs. 1 and 2). Furthermore, it is readily understood since at the more acid pH, e.g. pH 3.6, not all carboxyls of the dicarboxylic amino acids are dissociated and only one hydroxyl of the phosphate group has lost its protons. Hence the electrostatic repulsion of the acidic groups must have weakened considerably. It is, therefore, feasible that at this pH a small fraction of the amino acid residues may be present as small helical segments, dispersed throughout the polypeptide chain of the protein. It should further be added that Taborsky has recently obtained evidence that below pH 2.0 phosvitin is present in β-structure (15).

Footnotes

* Work supported in part by the National Science Foundation Grants GB-2912 and GB-5582.

1. Using osmotic pressure measurements Allerton obtained a molecular weight of 35,000 (Allerton, S.E. personal communication).

2. The phosphorus content of our most recent phosvitin preparations varied between 11.0 and 11.6 per cent.

3. Rimon, S., unpublished.

4. cf. Jirgensons (8).

Literature Cited

1. Allerton, S.E., and Perlmann, G.E., J. Biol. Chem. 240, 3892 (1965).

2. Beychok, S. (1967), in Poly-α-Amino Acids, ed. G.D. Fasman, Marcel Dekker, Inc., New York, p. 293.

3. Blout, E.R., Schmier, I., and Simmons, N.S., J. Am. Chem. Soc. 84, 3193 (1962).

4. Blout, E.R., Carver, J.P., and Gross, J., J. Am. Chem. Soc. 85, 644 (1963).

5. Engström, L., Biochim. et Biophys. Acta 54, 179 (1961).

6. Iizuka, E., and Yang, J.T., Proc. Nat. Acad. Sci. U.S.A. 55, 1175 (1966).

7. Jirgensons, B., Arch. Biochem. Biophys. 79, 57 and 70 (1958).

8. Jirgensons, B., J. Biol. Chem. 241, 147 (1966).

9. Joubert, F.J., and Cook, W.H., Can. J. Biochem. and Physiol. 36, 399 (1958).

10. Mecham, D.K., and Olcott, H.S., J. Am. Chem. Soc. 71, 3670 (1940).

11. Perlmann, G.E., and Katchalski, E., J. Am. Chem. Soc. 84, 452 (1961).

12. Perlmann, G.E., and Allerton, S.E., Nature 211, 1089 (1966).

13. Rabinowitz, M., and Lipmann, F., J. Biol. Chem. 235, 1043 (1960).

14. Shainkin, R., to be published.

15. Taborsky, G., J. Biol. Chem. 243, 6014 (1968).

16. Tiffany, M.L., and Krimm, S., Biopolymers, in press.

17. Timasheff, S.N., Townend, R., and Perlmann, G.E., J. Biol. Chem. 242, 2290 (1967).

18. Timasheff, S.N., Susi.,H, Townend, R., Stevens, L., Gorbunoff, M.J., and Kumosinski, T.F., in Conformation of Biopolymers, ed. G.N. Ramachandran, Academic Press, London and New York, p. 173 (1967).

19. Yang, J.T., and Doty, P., J. Am. Chem. Soc. 79, 761 (1957).

20. Yang, J.T., Proc. Nat. Acad. Sci. U.S.A. 53, 438 (1965).

STRUCTURE AND PROPERTIES OF THE CELL SURFACE COMPLEX

Edmund Jack Ambrose

Chester Beatty Research Institute, London, England

J. S. Osborne and P. R. Stuart

National Physical Laboratory, Teddington, England

I. Introduction

Great advances have been made in recent years towards an understanding of the biochemistry and molecular biology of cellular synthesis. The role of deoxyribose nucleic acid as the carrier of genetic information, of ribose nucleic acids in the transcription and translation of this message and the function of the ribosomal complex in protein synthesis has become known in considerable detail. In spite of this progress the nature of life itself remains a mystery, because living cells, even the smallest cells of microplasms and bacteria, are able to maintain a highly integrated system of functioning units, which is nevertheless labile and capable of undergoing cyclic and other changes.

It has long been recognized that the types of molecular interaction studied by physical chemists in liquid crystals must play an important role in the maintenance of such organization within living cells. In recent years it has become clear that one of the major components maintaining such structures are the various cellular membranes. These include the plasma membrane, mitochondrial membranes, smooth and rough endoplasmic reticulum, nuclear membranes and chloroplast membranes. In this paper the account will be restricted to studies of plasma or outer membranes of mammalian cells, both normal tissue cells and malignant cells. "The cell surface complex" (Ambrose and Easty) is a useful term which is intended to include the outer coat, lipid layer and underlying protein components. It will be shown in this paper that the cell surface complex behaves as a functional component of the cell and that its behavior depends on an interaction between assemblies of molecules showing liquid crystalline properties.

Physico-chemical studies of liquid crystals mainly involves work with one component. Unfortunately this is not possible with biological systems. It is possible to isolate components from the cell surface complex and study their physico-chemical properties; such work is most valuable in attempts to explain biological behavior. But the properties of the materials are often changed by isolation procedures and the true biological behavior is only shown in the living cell. Various physical methods have therefore been devised to study the problem with living cells, and will be described in this paper.

II. Chemical constitution and structure of the cell surface complex in mammalian cells

The cell surface complex can be broadly divided into three parts: an outer coat, a lipid layer and a sub-surface structure as shown diagramatically in Fig. 1a.

(a) The outer coat is the part of the cell which makes immediate contact with the external environment. The most powerful tool for the study of this structure has proved to be cell electrophoresis. This method in which cells are suspended in a balanced saline medium and their velocity is measured in an applied electric field gives information at ionic strength of 0.1 about structures down to a depth of 10 A (The thickness of the double layer). At ionic strength of 0.01 it gives information down to a depth of about 30 A.

One method of investigating the chemical constitution of the outer coat is to carry out cell electrophoresis measurements before and after enzyme treatments which do not disrupt the cell or lead to disturbance of the permeability barrier provided by the lipid layer. An illustration of such an experiment is given in Fig. 2. It indicates the electrical mobility, measured as μ/volt/cm/sec. for a number of tumour cells. All the cells carry a net negative surface charge. In the second vertical line of each set, the effect of treatment with heat inactivated neuraminidase is shown; it will be seen that the mobility is unchanged, indicating that adsorption of the enzyme molecules does not affect the mobility. In the third line the effect of the enzyme is shown. The mobility of the cells is reduced in each case to a greater degree in some cells than others. This enzyme removes sialic acids N-acetyl neuraminic acid and N-glycolyneuraminic acid from the cell surface. The negative surface charge located at the outer surface is due almost entirely to the carboxyl groups of this sugar located as terminal groups in the case of human erythrocytes which suffer a 95% reduction in mobility after neuraminidase treatment. The various tumour cells as shown in Fig. 2 show variable proportions of this component at

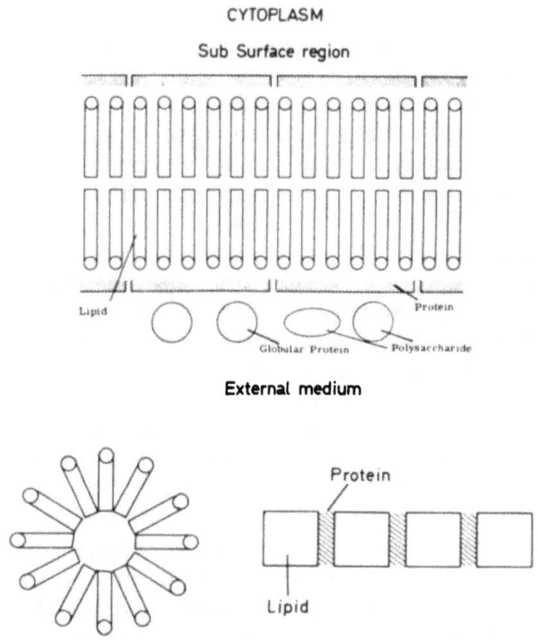

Fig. 1 The cell surface complex; a. lamellar, b. micellar.

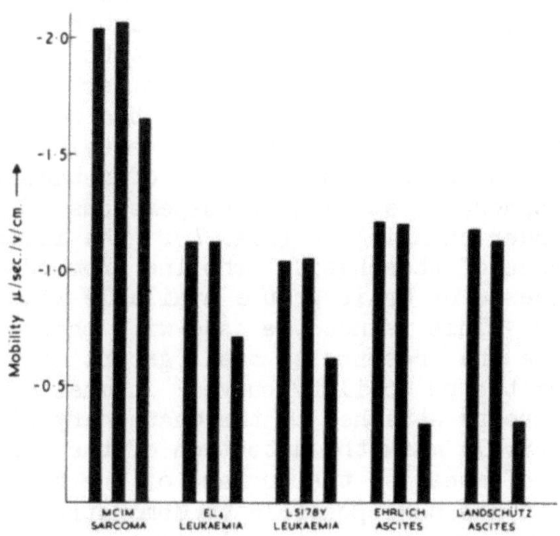

Fig. 2 Effect of neurominidase treatment on the electrophoretic mobilities of various types of murine tumor cells. First block-control, second block-treated with heat inactivated neurominidase, third block-treated with active enzyme.

the surface. In addition to sialic acid it is now known that N-acetyl glucosamine is also present as an available group to which certain specific plant agglutinins can be attached. The outer surface of the cell is a mosaic, containing enzyme molecules (ATPases and others), blood group substances and other molecules in small amounts. The bulk of the coat material appears to be a sialomuco-protein, which is responsible for the negative charge.

(b) The lipid layer.

The main components of the lipid layer in mammalian cells are phospholipids, phosphatidyl choline, phosphatidyl ethanolamine, sphingomyelin and cholesterol. Much work, which will not be described in detail here, has been done on the properties of lipid layers formed from these materials and from lipids isolated from plasma membranes. The original model proposed by Danielli (1) for the plasma membrane is shown in Fig. 1a. It indicates that the lipid molecules are arranged as a bilayer with the polar groups facing outwards. There is little doubt that such structures occur in suspensions of purified phospholipids in water, and in the myelin sheath of nerve fibres. There has been some controversy in recent years about whether a micellar form also exists in plasma membranes in living cells (2). At high dilutions, phospholipids form spherical micelles (Fig. 1b). These could be packed hexagonally or in other ways in the plasma membrane, being associated with the non-polar side chains of proteins. But recent work on comparisons between isolated cell membranes and phospholipids in relation to ion permeability, etc., indicate close similarities. The present favored view is that the lamella form is the stable form but that this can transform into the micellar form for short periods in certain types of living cells, and in certain local regions of the membrane. Protein molecules may fit into the lipid layer at certain points to provide pores for the regulation of entry of metabolites, etc.

Cell electrophoresis studies of suspensions of phospholipids have been made under changing pH (Fig. 4). The interesting plateau in the case of phosphatidyl choline from pH 12 to pH 4, indicates the absence of basic groups available within 10 A. of the outer surface. This is not the case with phosphatidyl ethanolamine where the presence of basic groups at the surface is indicated from the pH mobility curve. It must be concluded that the methyl groups attached to the quaternary nitrogen atom of choline effectively mask the detection of the basic group within the plane of shear at the surface of the bilayer. By cell electrophoresis it has been possible to demonstrate that the general arrangement of outer coat on lipid layers in the plasma membrane, as shown in Fig. 1a, is substantially correct.

It has been shown that sialic acids do not adsorb calcium ions in the case of cells in suspension. Adsorption of calcium ions is therefore an indicator of the presence of other available

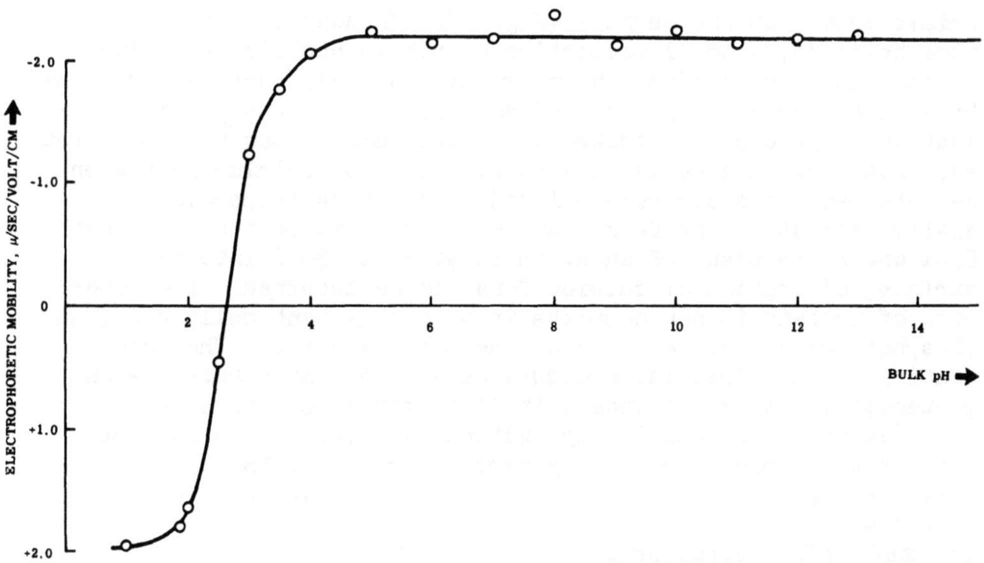

Fig. 3 Microelectrophoresis of suspensions of phosphatidyl choline under changing pH.

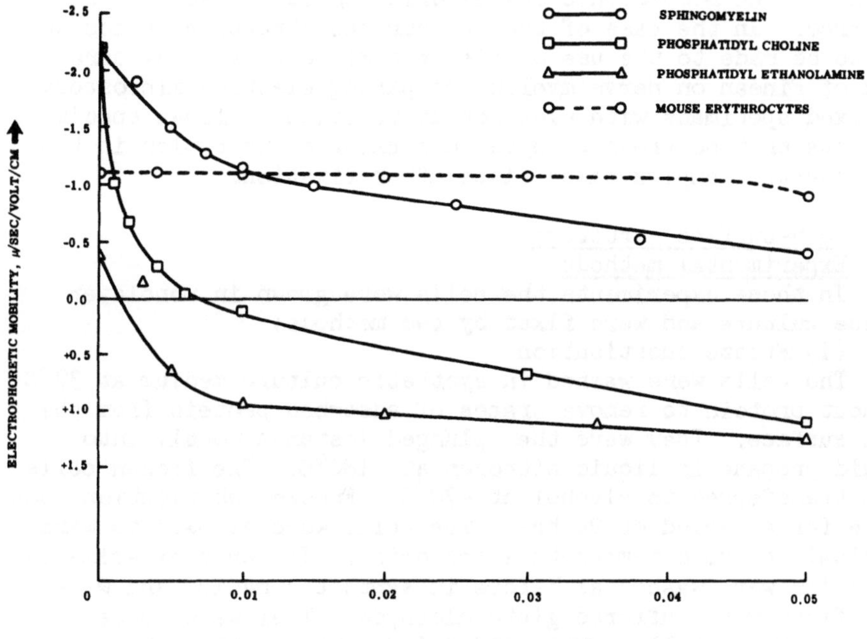

Fig. 4 Effect of varying calcium concentration on the electrophoretic mobilities of some phospholipids and mouse red blood cells.

acidic groups at the surface (Fig. 4). In addition to phospholipids, a small proportion of ribose nucleic acid (RNA) is now known to be firmly bound within the cell surface. But the bulk of the charged groups available are due to phospholipid. That these groups are located below the surface can be demonstrated. EL4 leukaemia cells show no affinity for calcium ions when measured at ionic strength 0.1 and have a high proportion of sialic acid at the surface. But when measured at ionic strength 0.01 where the plane of shear penetrates to 30 A into the surface, adsorption of calcium ions can be detected. The outer coat of protein is not complete in some malignant cells and some phosphate groups are exposed at the outer surface. The outer coating of the lipid layer occurs as a mosaic structure. With proteolytic enzymes (pronase) it is possible to remove the outer layer almost completely, without killing the cells. As long as the lipid permeability barrier remains intact, the cell can repair the outer coating by renewed synthesis or secretion.

(c) Sub surface structures

Studies of the subsurface structures are a recent development, and will be described in detail.

The work described above has been carried out on living cells in which it is possible to be certain that no major structural change in the mesomorphic states existing at the surface has occurred. In the case of the sub surface structure, recourse has to be made to the use of electron microscopy. The careful work of Finean on nerve myelin, comparing electron microscopy of fixed specimens with electron diffraction of fresh specimens suggests that no gross changes in packing of molecules in the cell surface complex takes place after fixation.

III. Sub-surface structure

(a) Experimental methods

In these experiments the cells were grown in monolayer tissue culture and were fixed by two methods:

(i) Freeze substitution

The cells were washed in synthetic culture medium at 37°C without protein to remove traces of adsorbed protein from the cell surface. They were then plunged instantaneously into liquid propane in liquid nitrogen at -180°C. The frozen cells were transferred to alcohol at -70°C. Freeze substitution took place for a period of 24 hrs. The cells were allowed to warm up slowly to room temperature and dried. In other experiments, the cells were washed as before in synthetic medium and were then fixed with buffered glutaraldehyde. They were subsequently washed with distilled water and dried. The results obtained according to the two methods of fixation were compared. The specimens were placed in an apparatus for the generation of a gaseous discharge in hydrogen (3). The hydrogen ions so produced bombarded and etched the cell surface. By etching the surface for progressively increasing periods of time, it is possible to study the structure of the cell surface in depth by this method. In Fig. 5a and b are shown such a series of electron micrographs taken with a scanning electron microscope (Cambridge Instrument Company). After etching, the surface was coated with gold-palladium and pictures were taken up to a magnification of 15,000x.

(e) Results of the ion etching investigation

It can be seen that, before ion etching, the cell surface of the fibroblasts is comparatively smooth, although pits and other irregularities are sometimes seen at high magnification. In this case the outer coat of the cell is seen in the microscope. After a short period of etching, the sub surface structure can be seen with layers of plasma membrane still attached in places. Once the whole of the plasma membrane has been revealed, the sub surface structure can be seen. It is immediately apparent that an organized structure is present, consisting of fibres or of aggregates of fibres. These are extremely well orientated in the case of the fibroblast cell shown in Fig. 5b. A systematic study of a number of cell types has revealed this fibrous sub surface structure, although the degree of orientation varies very much with the cell type. It is much reduced in highly malignant cells. The presence of fibres within the cytoplasm of cells has been demonstrated by transmission electron microscopy but it has not previously been possible to demonstrate a high concentration of fibrous material in an orientated state near the inner surface of the plasma membrane.

IV. Model for the structure of the cell surface complex

The conclusions to be drawn from present evidence are summarized in Fig. 6.

The outer coat is complete in some cells but not in others. Beneath this lies the lipid phase, the bimolecular leaflet or micelles, and below this lies an assembly of fibrous material, which is probably protein in nature.

It now seems likely that the biological properties of the plasma membrane involve the interaction between two systems that can be looked upon as being in a liquid crystalline state. These are the lipid phase and the sub surface layer. In the lipid phase the molecules lie predominantly perpendicular to the surface whereas in the region below the surface the fibrous aggregates are tangentially arranged.

V. Dynamics of the cell surface complex

In most work on liquid crystals attempts are made to achieve equilibrium conditions but in biological systems an equilibrium of this kind is not achieved in living, but only in dead cells. In living cells there is a constant transport of

Fig. 5a Surface of C_{13} strain of cultured hamster fibroblasts as seen under stereoscan microscopy (glutaraldehyde fixed).

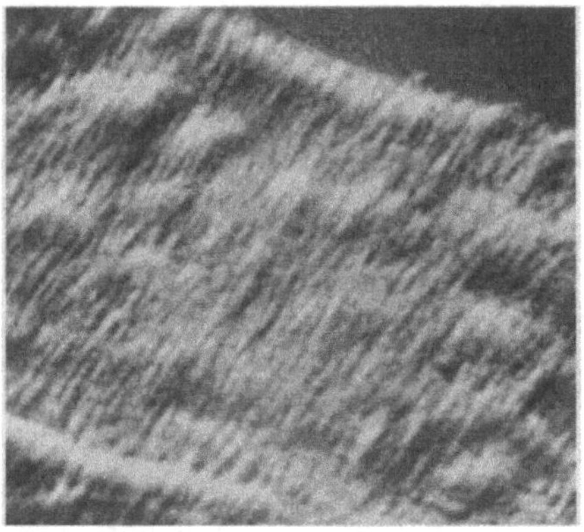

Fig. 5b Similar cells after ion etching (glutaraldehyde fixed).

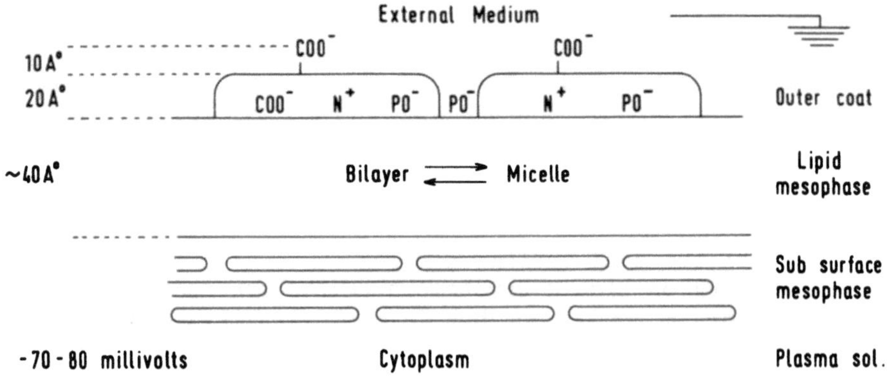

Fig. 6 Structure of the cell surface complex.

ions against a concentration gradient which maintains an electrical potential across the surface and the membrane itself undergoes constant shape changes. Some of these shape changes have been instantaneously fixed and are shown in the stereoscan pictures of Figs. 7 and 8.

In Fig. 7a and b is shown the spread out region at the end of a moving cell from a human brain tumor. This region of the membrane is showing an extremely active wave like movement. Such waves or muffled membrane movements are also occurring on the lower surface of the cell and enable the cell to migrate on the solid surface. The plasma membrane forms these waves on its lower surface also, which is in contact with the glass or other substrate. These enable it to make intermittent contacts and move forwards like an earthworm. Another type of transient change seen at the cell surface is the generation of microvillae; these are between 500 and 1000 A in diameter. They look somewhat like myelinic forms and are shown clearly in the stereoscan picture of Fig. 8. They are generated rapidly; once they have formed they become stiff and are then withdrawn. They are thought to play a role in enabling the cell to make initial contacts with substrate and with other cells, because the repulsive forces between surfaces are reduced in regions of high curvature. Cells also produce much longer pseudopodia which may be stable for long periods. Good examples of these are the extremely long and narrow pseudopodia produced by cells of the nervous system. This may be many microns in length. A group of such pseudopodia are shown in the stereoscan picture of Fig. 9a. These are not ion etched specimens and the plasma membrane is still intact. Nevertheless, the presence of fibrous elements can be clearly seen in the region where one pseudopodium is overlapping another in Fig. 9b. This is because the fibres are pressed against the flexible membrane like tendons beneath the skin.

VI. Relationship of structure to function in the cell surface complex

Extensive studies using cell electrophoresis, chemical isolation, transmission and surface electron microscopy have been used to attempt to elucidate the structure of the cell surface complex in mammalian cells. Only the ion etching work has been described in this paper in detail. The general conclusions of these studies are summarized in Fig. 6. The functioning of living cells and the transient phenomena seen at the surface and described above relate to this supermolecular structure of the cell surface complex. Let us first consider the question of shape changes. It is well known, in the case of muscular contraction, that the bivalent cations Ca^{++} and Mg^{++} play an important role in association with ATP in the initiation of contraction. The Mg^{++} dependent ATPase activity is triggered by small changes in Ca^{++} concentration. Ca^{++} is released into the

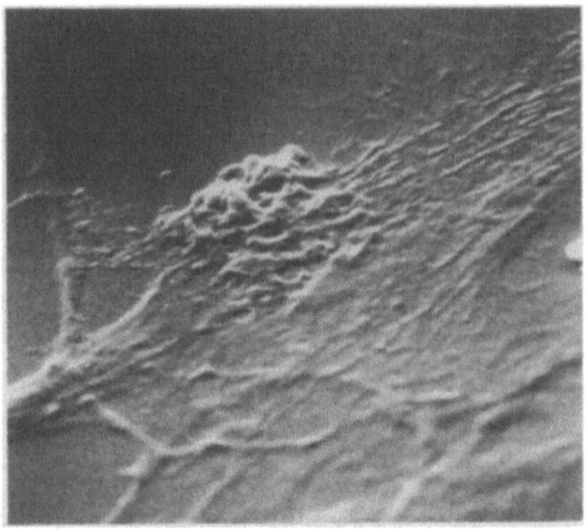

Fig. 7a Stereoscan picture of spread out pseudopodium of human astrocyte. Low power.

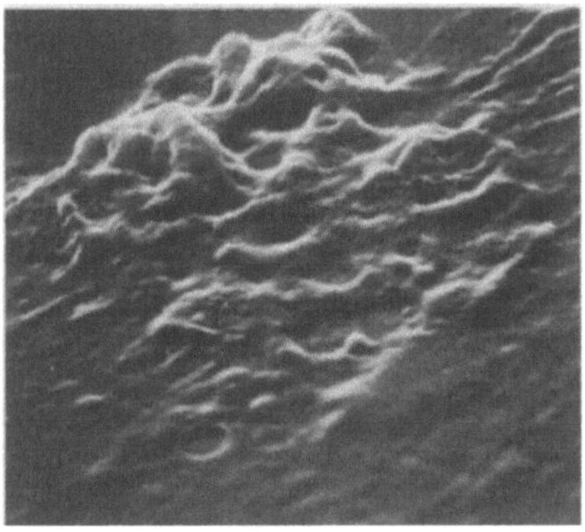

Fig. 7b High power of the same.

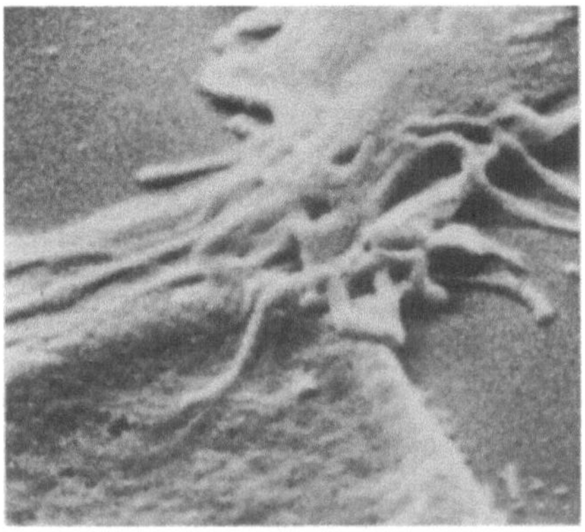

Fig. 8 Stereoscan picture of microvillae in contact between two C_{13} cells.

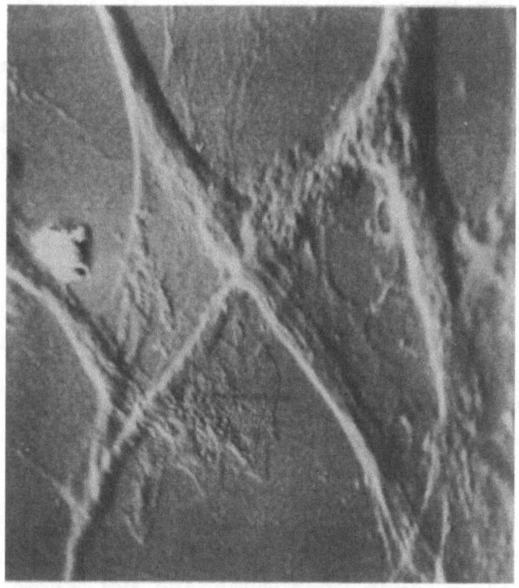

Fig. 9a Stereoscan picture of overlapping pseudopodia of astrocytes. Low power.

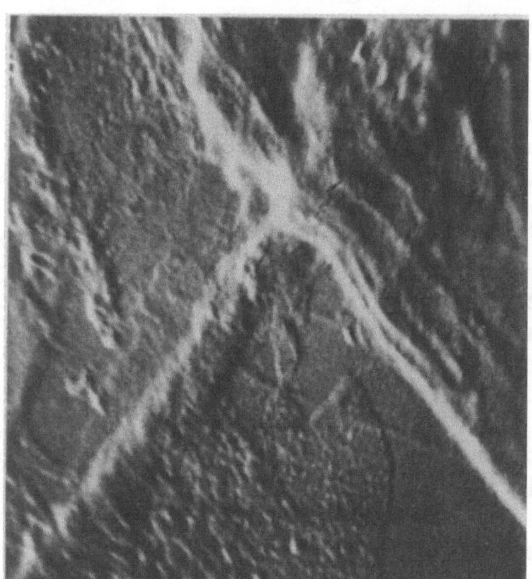

Fig. 9b High power of the same.

cytoplasm from the sarcolemma, by rapid changes in ion permeability. Studies of the effect of polyvalent cations on the cell surface complex have also shown that it is possible to produce contraction by local application, but a general change in ion concentration round the whole cell also produces osmotic changes which complicate the phenomena. The cell nuclear membrane can also behave in somewhat the same way; within the living cytoplasm it can rotate and cause cytoplasmic movements in its neighborhood, which are probably connected with the presence of surface fibrillar complexes. Isolated nuclei do not show the same disturbance due to osmotic changes and provide a convenient model for study. In Fig. 10 is shown the effect of changing the concentration of polylysine round an isolated but intact nucleus of Amoeba proteus. These reversible contractions and expansions of the membrane complex can be brought about many times. Addition of polyanions such as polyglutamic acid or heparin reverses the effect of the poly-base and causes expansion. ATP alone has no effect, Ca^{++} or Mg^{++} alone do not have an effect. But Ca^{++} or Mg^{++} with ATP causes contraction. Similarly with other polyanions. Here we see the important role of the bivalent Ca^{++} probably forming bridges between two poly-anionic surfaces as shown in Fig. 11. These phenomena are not shown by the nucleoprotein of the cell nucleus, showing that we are dealing with a membrane surface phenomenon. The region of the nuclear membrane being affected is the outer surface which is in contact with the cytoplasm. This suggests that changes in Ca^{++}, Mg^{++} and ATP could also be expected to affect contractile changes within the mesophase of long filaments in the sub surface region of the plasma membrane. It is not possible at the present time to suggest a detailed mechanism for the generation of the transient surface changes observed in living cells, but it is interesting to point out that all the components capable of generating local contractile movements and relaxations are present.

(1) High Ca^{++} ion concentration outside, low inside.
(2) Rapid transient changes in ion permeability, as in nerve action potentials are known to occur.
(3) The presence of ATPases.
(4) An assembly of fibrous elements in a mesomorphic phase adjacent to the lipid permeability barrier.

Another most important aspect of the properties of the cell surface complex is the ability to remain stable after a transient disturbance. When this property is lost, the cell will die. The mesomorphic phases of lipid and sub surface structure are not in equilibrium, but in a steady state; failure to supply energy for the working of the active transport mechanisms responsible for maintaining the high electrical potential leads to rapid changes in the membrane. In absence of oxygen or glucose, irregular pseudopodia are formed, which soon break up into drop-

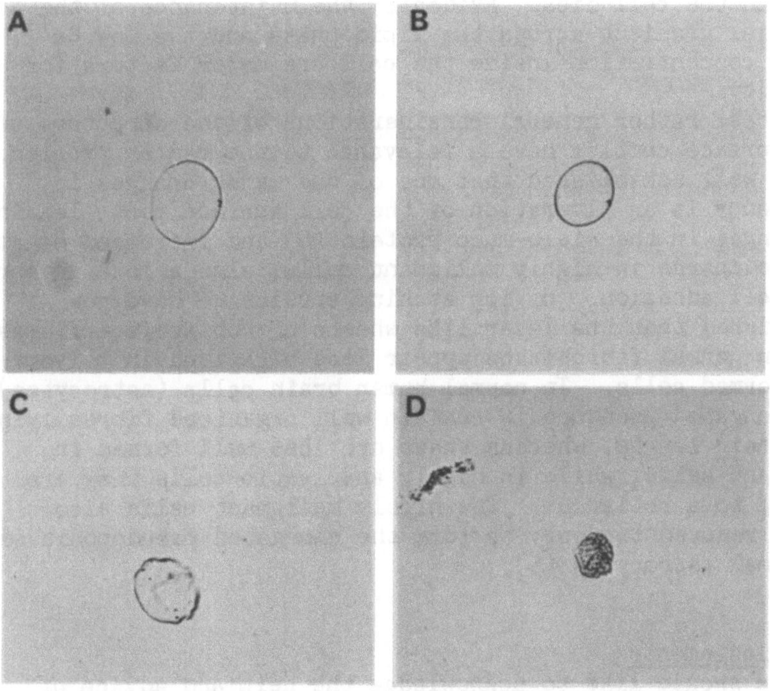

Fig. 10 Effect of increasing polylysine concentration on isolated but intact nucleus.
A. 0.14 M. sodium chloride central.
B. +0.005% poly-L-lysine.
C. +0.05% polylysine.
D. +0.15% polylysine.

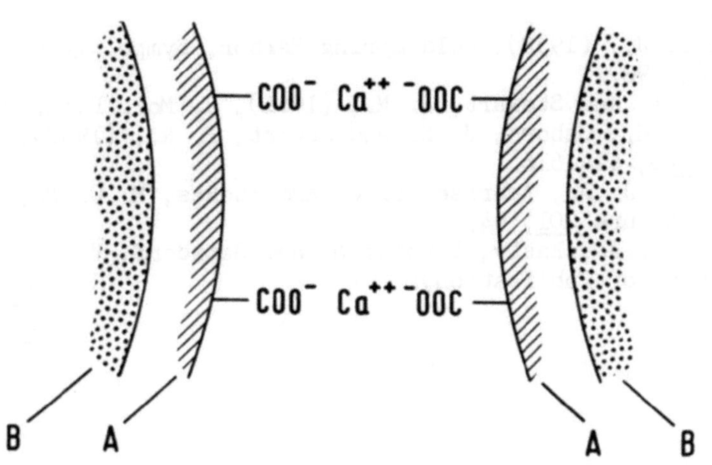

Fig. 11 Calcium bridges.

lets and the cell dies. Evidently the maintenance of the potential gradient across the lipid phase and the low Ca^{++}, high K^+ ion concentration inside the cell are major factors for stability.

These rather general considerations of the structure of the cell surface complex have a relevance to the cancer problem. It is now well established that one of the major changes in malignancy is an alteration of the cell surface coat, leading to changes in the sialo-muco protein (4) and increased negative surface charge in highly malignant cells, also a loss in surface cell-cell adhesion. By ion etching studies we have now established that the layer like sheets of sub surface fibres seen in normal fibroblasts appear less organized in polyoma transformed cells. In normal human brain cells (astrocytes) the elongated pseudopodia contain well organized fibres lying along their length, whereas these are less well formed in malignant cells, while in highly anaplastic cells they are reduced to a reticulum. The highly malignant cells also show a reduced tendency to form the elongated pseudopodia seen in normal astrocytes (5).

Acknowledgements

We should like to acknowledge the help and advice of Dr. Dorothy M. Easty, Professor V. Batzdorf, University of California (L.A.), Dr. J. A. Forrester, Dr. P. Wand and Dr. W. Korohoda (University of Krackow).

This work has been supported by grants to the Chester Beatty Research Institute, Institute of Cancer Research, Royal Marsden Hospital from the Medical Research Council and the British Empire Cancer Campaign.

References

(1) Danielli, J., (1938), Cold Spring Harbor, Sympo. Quant. Biol. 6, 190.
(2) Lucy, J. A. and Stewart, J. M., (1964), J. Mol. Biol. 8, 727.
(3) Lewis, S. M., Osborn, J. S. and Stuart, P. R., (1968), Nature 220, 614-616.
(4) Forrester, J. A., Ambrose, E. J. and Stokes, M. G. P., (1964), Nature 201, 945.
(5) Ambrose, E. J., Easty, Dorothy M. and Batzdorf, V., (In Course of Publication).

FLOW INDUCED ORDERED STATE OF HELICAL POLY-BENZYL-L-GLUTAMATE

SHIRO TAKASHIMA

ELECTROMEDICAL DIVISION, MOORE SCHOOL OF ELECTRICAL

ENGINEERING, UNIVERSITY OF PENNSYLVANIA, PHILADELPHIA

PA.19104

SUMMARY

Dielectric constant of α-helical poly-benzyl-L-glutamate (PBLG) is measured in the presence of a velocity gradient using rotating cylindrical electrodes. The electric field is applied across the annular gap between electrodes and the vectors of the mechanical and the electrical forces are perpendicular to each other. The dielectric constant decreases with the increasing velocity gradient, as long as the frequency is sufficiently low. When the frequency is increased, the dielectric constant increases slightly and moreover, the dielectric constant is no longer affected by the velocity gradient if the frequency is further increased. In the presence of a sufficiently high velocity gradient, PBLG solutions exhibit a considerably different dielectric relaxation from that without the velocity gradient. A semi-empirical explanation is presented to account for the behavior of polar molecules in a mechanical field and an electrical field.

Introduction

Although the dipole orientation of polar molecules in liquid states is restricted by the viscosity of the medium, the molecules are still free to rotate along the major axis as well as along the minor axis. In solid states, the molecular orientations severely restricted by the lattice energy and the dielectric constant usually decreases substantially on freezing liquids. In ordered fluids or liquid crystals, the dipole orientation is partially hindered and it is quite reasonable to expect that the dielectric constant of a liquid or a solution would decrease on the formation of ordered structures or liquid crystals.

The behavior of polar molecules in gaseous states and in dilute solutions can be described by the Debye theory(1) to the first order approximation. In this theory, the mean moment is essentially defined by the interaction energy $\mu . E$, that is, the interaction energy between the dipole moment μ and the electric field E and by the thermal energy kT. Therefore, the mean moment can be written as follows:

$$<\mu> = \mu \int \cos\theta \, e^{\mu E \cos\theta / kT} d\Omega \, / \int e^{\mu E \cos\theta / kT} d\Omega \qquad (1)$$

where $d\Omega$ is the segment of the solid angle. This equation is valid only for ideal systems where there is no potential energy between the molecules or there is no other external force. In the case of ordered fluids, the Debye theory is no longer applicable in its original form because of the strong inter molecular forces or because of the presence of a shearing stress to create an ordered structure.

The ordered structure of the helical poly-benzyl-L-glutamate (PBLG) is studied by Hermans(2) and it is known that PBLG forms an ordered fluid when a moderate shearing stress is applied. We can, therefore, investigate the dielectric properties of ordered states of PBLG by applying a shearing stress in the dielectric cell. This can be done most effectively by rotating the cylindrical dielectric electrode and create a velocity gradient in the annular gap. The rotating dielectric cell was devised by Jacobsen(3) some years ago. In the present experiment, the rotating cell is incorporated into a flow birefringence measuring apparutus and therefore we can measure the orientation of the molecules as well as the dielectric constant as function of the mechanical shear. By so doing, we can analyse the change of the dielectric constant in terms of the ordering of the molecule.

EXPERIMENT

Construction of a rotating dielectric cell: The electrodes consist of two concentric stainless steel cylinders whose diameters are 39 mm and 40 mm respectively. The gap between cylinders is therefore 0.5 mm. The outer cylinder is rotated using a d.c.motor in order to produce a velocity gradient. The velocity gradients used in this measurement is from 0 to 4400 sec^{-1}. The inner cylinder is static and is connected to the high potential side of the impedance bridge. The rotating outer cylinder is connected to the ground. The connection between the rotating cylinder and the ground is made by a brush with a silver plating. A conducting grease is applied at the contact in order to reduce the electrical noise due to the friction. The cell assembly was fitted into the Rao flow birefringence measuring apparatus B-22 and is placed between a polarizer and an analyzer. They are carefully aligned so that the extinction angle and the birefingence of the solution can be measured with the maximum resolution of the apparatus.

The solution of PBLG is placed in the annular gap using a syringe and the gap is completely filled with the solution to the top without noticeable gas bubbles which interfere the electrical as well as the optical measurements. An a.c. electric field is applied across the annular gap and the gap is illuminated from the bottom with a tungsten lamp. The dielectric constant and the conductivity are measured with and without the velocity gradient. At the same time, the extinction angle is measured using the optical system.

The impedance bridge used for the dielectric measurement is designed by Schwan(4) and is of a very high precision. The frequency range is between 50 Hz and 200 KHz. The resolution of the capacity reading is 0.1 $\mu\mu$F and that of conductance reading is 0.001 μmho. These resolutions decrease slightly by rotating the cylinder because of the frictional noise. The temperature is kept at 20°C but the temperature in the annular gap is not measured. A slight temperature rise in the gap is, however, indicated by the slow increase of the conductivity of the solution during the prolonged rotation. The capacity of the solution is unaffected by the small temperature rise.

Poly-benzyl-L-glutamate was purchased from the Pilot Cehmical Comp. Two samples with different molecular weights were actually used in this work. The weight average molecular weights of these samples are 335,000 and 200,000. PBLG was dissolved in helix forming organic solvents but it was found that dioxane is by far the best for the purpose of this investigation because of the small dielectric constant and the low conductivity. The results obtained in other solvents will be reported elsewhere.

RESULTS

The frequency profile of the dielectric constant of PBLG in the absence of the velocity gradient is shown in Fig.1. From this curve, we obtain the dielectric increment 0.285/gm/liter which gives a dipole moment 2670 D.U. Also we obtain a relaxation frequency 250 Hz and this gives a relaxation time 6.3×10^{-4} sec. These values are obtained for the molecular weight 335,000 and are in good agreement with the values obtained by Wada(5). The dispersion curve deviates considerably from the Debye theory as shown in the same figure by the broken line. The curve is calculated by the following equation:

$$\epsilon = \epsilon_\infty + \frac{\Delta\epsilon}{1 + (\omega\tau)^2} \qquad (2)$$

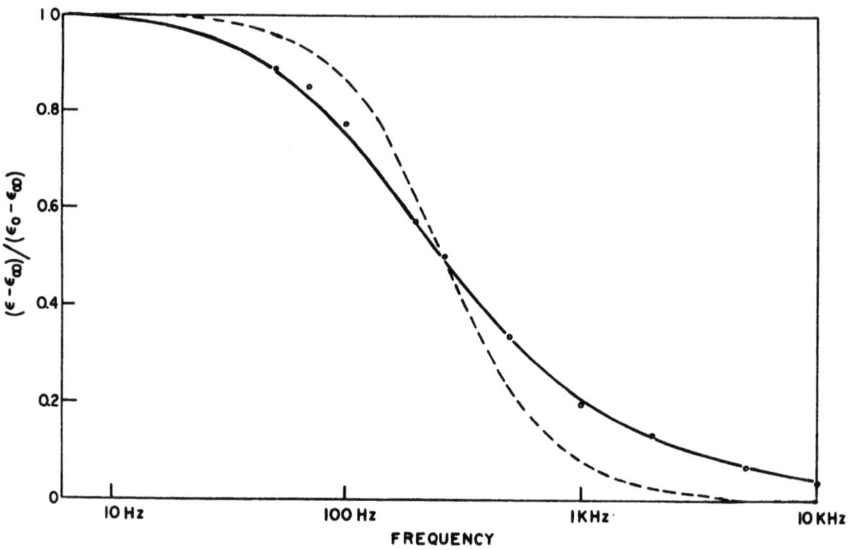

Fig. 1. Normalized dielectric constant of PBLG in dioxane as function of the frequency. The dotted line is calculated using the Debye theory.

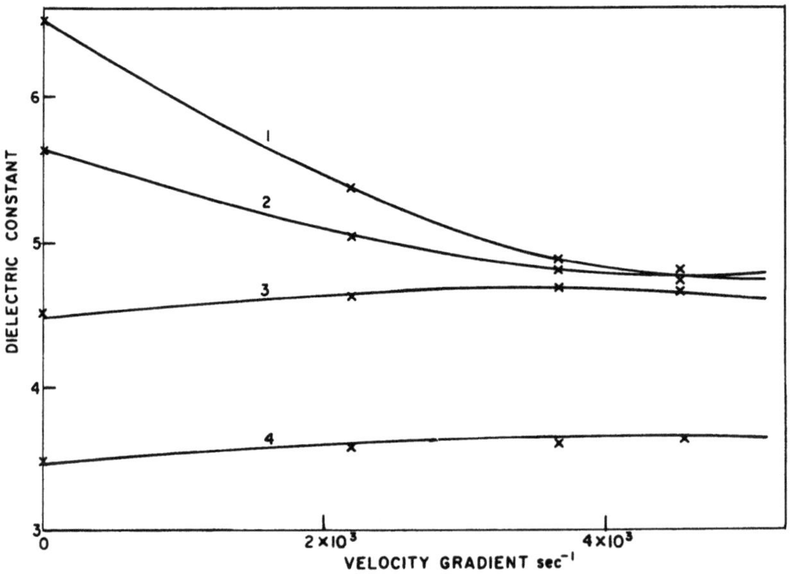

Fig. 2. Dependence of the dielectric constant on the velocity gradient. Curve 1,50 Hz, curve 2 100 Hz, curve 3, 200 Hz and curve 4, 500 Hz.

where ϵ_∞ is the high frequency dielectric constant, ω is the angular frequency $2\pi f$, τ is the relaxation time and $\Delta\epsilon$ is the dielectric increment and is given by $\epsilon_0 - \epsilon_\infty$ (ϵ_0 is the low frequency dielectric constant).

The change of the dielectric constant with the velocity gradients is shown in Fig.2. Curve 1 in this figure shows the behavior of the dielectric constant at 50 Hz in the velocity gradients. For a given velocity gradient, there is a definite limiting value and no further change is observed once this value is reached. Each point in these curves is obtained after the transition period and the limiting value is established. The extinction angle which is simultaneously measured is shown in Fig.3. The purpose of the simultaneous measurement of the extinction angle may better be understood if we convert it into a parameter α using the following equation.

$$6/\alpha = \tan 2\chi \tag{3}$$

where χ is the extinction angle. The parameter α is related to the angular distribution function $\rho(\phi)$ which shows the angular density of the molecule with respect to the direction of the flow. The relationship between the distribution function and α is obtained by solving the following partial differential equation.

$$\frac{\partial \rho}{\partial \phi} + \alpha\rho \sin^2 \phi = C \quad C = \text{constant} \tag{4}$$

This equation was solved by Boeder(6) and the result of this calculation indicates that as the value of α increases, the distribution function has a sharper maximum and that the angle at which the value of the distribution function is maximum approaches zero asymptotically. The calculation also reveals that the angle of the maximum distribution function is equal to the extinction angle. Therefore, the measured extinction angle indicates the average angle of inclination of molecules with the stream line. As the velocity gradient decreases, the distribution function becomes broad and the peak of the distribution function converges to 45°. In other words, the extinction angle cannot be greater than 45° and, by definition, this represents the state of a random distribution without any external force. The perfect alignment in the direction of the flow cannot be achieved because of the presence of the thermal energy and there is a certain equilibrium value for a given velocity gradient for a molecule. The typical values of χ for the PBLG used in the present work is approximately 20 - 30 degree with velocity gradients 2000 - 4000 sec^{-1}. This means that under these conditions, the PBLG molecules have an inclination of 20 - 30 degrees with the stream line. The angle of the molecular orientation will be used later for the analysis of the change of the dielectric constant in the velocity gradient.

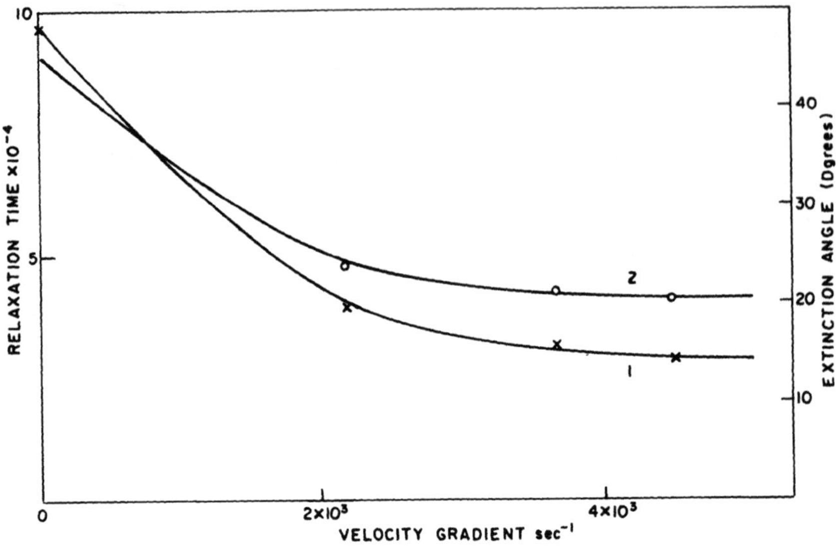

Fig. 3. Dependence of the extinction angle (curve 2) and the relazation time (curve 1) upon the velocity gradient.

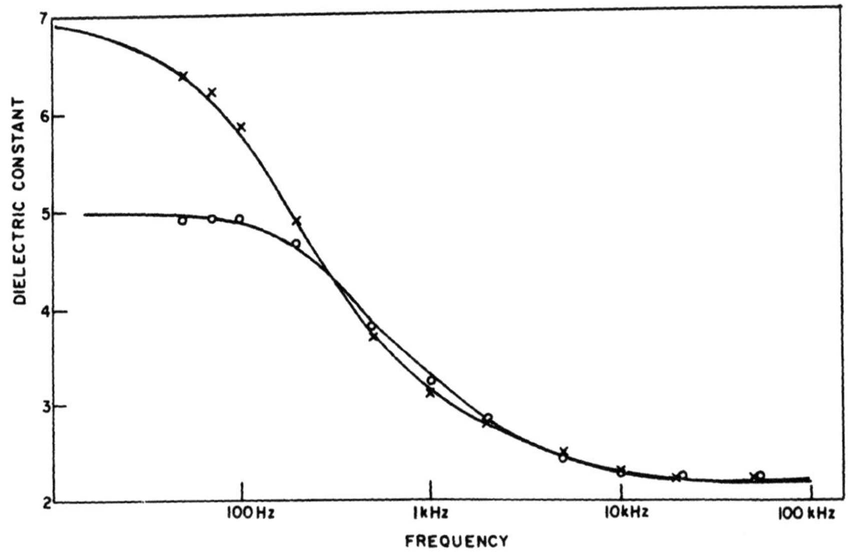

Fig. 4. Dielectric dispersion curves of PBLG. Curve 1, without the velocity gradient and curve 2, with a velocity gradient 4400 sec^{-1}.

The measurements of dielectric constant were carried out as function of the frequency as well as the velocity gradients. Fig. 2 shows the results of these measurements. It must be noted that the dependence of the dielectric constant on the velocity gradient varies considerably with the frequency of the a.c.field. For example, curve 1 was obtained with 50 Hz. At this frequency, the dielectric constant of PBLG is very close to the static dielectric constant(see Fig.1). The dielectric constant decreases monotonically and finally approaches a lower limiting value. However, if a higher frequency is used (for example, curve 2 at 500 Hz), the behavior of the dielectric constant is more complicated. One observes, in this case, a slight increase at lower velocity gradients and then decreases with the further increase in the velocity gradient. If the frequency is further increased, one observes only a slight increase. Obviously, the change in the dielectric constant in the mechanical and electrical fields consists of decreasing and increasing parts and the proportion of them varies with the frequency.

Fig.4 shows the dielectric dispersion curves with and without the velocity gradient. Curve 2 which shows the dispersion curve with a velocity gradient was constructed from the limiting dielectric constant at the velocity gradient of 4400 sec^{-1} at each frequency. The difference between these two curves is evident. First of all, the static dielectric constant ϵ_0 is considerably lower when a mechanical shear is present. Secondly, the relaxation frequency shifts from 250 Hz to 700 Hz indicating that the effective entity of the rotation is changing.

The relaxation time can be obtained by plotting the dielectric dispersion curve at various velocity gradients and the results are shown in Fig.3. Similar to the change in the dielectric constant, the relaxation time decreases with a velocity gradient and approaches a limiting value asymptotically.

DISCUSSION

Fig.5 shows the orientation behavior of the molecules in the annular gap. The arrow along the gap indicates the force vector due to the mechanical flow and the arrow across the gap indicates the vector of the electrical field. In the ordinary dielectric theory, the dipolar orientation is investigated only in the presence of the electric field and the thermal energy which tends to randomize the molecule. In the present treatment, however, we have to take the mechanical energy into account which restrict the dipolar orientation.

The calculation of the mean moment in the presence of an external force in addition to the electric field has not really been done and this turns out to be a very difficult calculation.

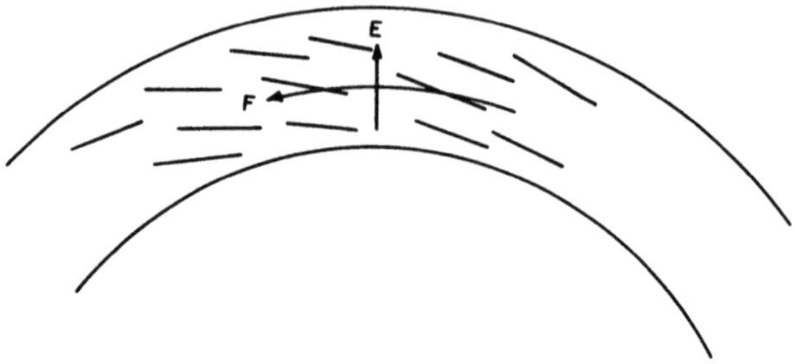

Fig. 5. Orientation of rod-like molecules in the presence of a mechanical shear (f) and an electrical field (E). The curved surfaces represent the surface of outer and innter electrodes.

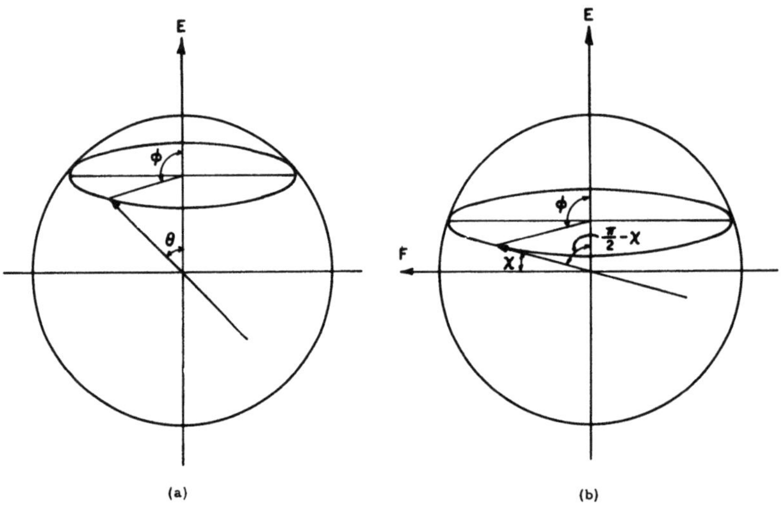

Fig. 6. Dipolar orientation of a molecule. (a) Without the velocity gradient and (b) with a velocity gradient. The mechanical force vector is perpendicular to the electric field.

The treatment described below is more or less semiquantitative and is based on the modification of the original Debye theory

According to the Debye theory, the probability of finding a molecule between and an angle θ and $\theta + d\theta$ and between $\varphi + d\varphi$ is given by:

$$\frac{1}{2} \sin \theta \, d \theta \, d \varphi \qquad (5)$$

if no electric field is applied. If an electric field is applied, the probability is modified by adding the Boltzman factor (see Fig. 6a).

$$\frac{1}{2} e^{\mu E \cos \theta / kT} \sin \theta \, d\theta \qquad (6)$$

where μ is the dipole moment, E is the intensity of the electric field. The mean dipole moment is obtained by integrating the function with respect to θ and φ. The limit of the integrations is 0 to 2π for φ and from 0 to π for θ. If a mechanical shear is applied, the dipole orientation is restricted and the limit of the integrations must be altered. Fig. 6b shows the orientation of molecules in the presence of the velocity gradient as well as the electric field. The arrow E indicates the direction of the electric field and F indicates the direction of the flow. The angle χ is the angle between the major axis of the molecule and the direction of the flow. At this point, we will divide the problem in two parts. 1) The behavior of the longitudinal component of the dipole moment and 2) The behavior of the transverse component. Longitudinal component ... As shown previously, the mean dipole moment is obtained by integrating the equation (6) with respect to the angles θ and φ from 0 to π for θ and from 0 to 2π for φ if no mechanical force is present. If a mechanical shear is applied, the molecule is oriented by the force and the angle between the molecule and the direction of the flow will be either χ or $-\chi$. If we apply an electric field, and the field is normal to the direction of the flow, the molecules with a dipole moment rotate in the direction of the field but, because of the presence of the mechanical shearing stress, the range of the rotation is restricted to a narrow region. This is true only when the electric field intensity is much smaller than the mechanical force and the application of the field does not change the angular distribution of the molecule greatly. Since we find the maxima of the distribution function at χ and $-\chi$, we set the limit of the integration from $\frac{\pi}{2} -\chi$ to $\frac{\pi}{2} +\chi$, that is between the maxima of the distribution function. The angle χ converges to 45° and can never be greater than this limiting value. By definition, this limiting value of 45° means the state without any mechanical shear. Therefore, the integration from 45° to 135° must represent the case of the Debye theory. It is easy to show, however, that the integration from 45° to 135° is approximately the half of the mean moment value of Debye's original calculation as long as the field intensity is small compared to the thermal energy. The limit of the angle φ is still from 0 to 2π and unaltered.

With these facts into account, we can rewrite the Debye expression for the mean moment (more exactly, a half of it) as follows:

$$\langle \mu \rangle_{11} = \frac{\mu_{11} \cdot \int_{\frac{\pi}{2}-\chi}^{\frac{\pi}{2}+\chi} \cos\theta \, e^{\mu E \cos\theta/RT} \sin\theta \, d\theta}{\int_{\frac{\pi}{2}-\chi}^{\frac{\pi}{2}+\chi} e^{\mu E \cos\theta/RT} \sin\theta \, d\theta} \qquad (7)$$

Carrying out the integration of these integrals, we obtain the following solution:

$$\langle \mu \rangle_{\parallel} = \frac{\mu\left[e^{a\cos(\pi/2-\chi)}(a\cos(\pi/2-\chi)-1) - e^{a\cos(\pi/2+\chi)}(a\cos(\pi/2+\chi)-1)\right]}{a\left[e^{a\cos(\pi/2-\chi)} - e^{a\cos(\pi/2+\chi)}\right]} \qquad (8)$$

wher $a = \mu E/kT$. Numerical calculations show that the mean dipole moment indeed decreases as the angle χ decreases. It means that the more the molecules are aligned in the direction of the flow, the less will be the contribution of the longitudinal component. The numerical values of $\langle \mu \rangle_{\parallel}$ are shown in Fig.7. As shown, the value of $\langle \mu \rangle_{\parallel}$ decreases from the Debye's value to zero as χ decreases from 45° to 0°. This curve is calculated by assuming the electric field intensity to be 0.5 volt cm^{-1}. The value of the mean moment at the angle 45° indeed turns out to be nearly the half of the original Debye's value. Therefore, we can safely conclude that the integration between two peaks of the angular distribution function assuming this value to be one half of the mean moment is justified at the field intensity of 0.5 v/cm. Transverse moment .... Contrary to the longitudinal moment, the behavior of the transverse component is quite opposite. The mean moment of the transverse component in the absence of the velocity gradient will be very small because of the partial concellation. This can be understood easily if we consider the fact that the range of the integration should be from χ to $\pi - \chi$. If we apply the velocity gradient and restrict the orientation of the molecules the range of the integrals will be wider as the angle χ decreases. The value reaches a maximum value when the molecules are completely oriented parallel to the direction of the flow and the dipole orientation is represented only by the rotation of the molecule around the long axis. The mean moment of the transverse component is therefore give by the following equation instead of eq.(7).

$$<\mu>_{\perp} = \frac{\mu \int_{\chi}^{\pi-\chi} \cos\theta\, e^{a\cos\theta} \sin\theta\, d\theta}{\int_{\chi}^{\pi-\chi} e^{a\cos\theta} \sin\theta\, d\theta} \qquad (9)$$

The calculated value of $<\mu>_{\perp}$ assuming $\mu_{\perp}$ to be 1.8 D.U. is shown in Fig.7.

The experimentally observed decrease in the dielectric constant(Fig.3) as function of the velocity gradient is actually a composite of the decreasing $<\mu>_{\parallel}$ and the increasing $<\mu>_{\perp}$. Therefore, the curve can be reconstructed by calculating $<\mu>_{\parallel}$ and $<\mu>_{\perp}$ separately and add them vectorially. We can obtain the theoretical curve for various ratios of $\mu_{\parallel} : \mu_{\perp}$. First of all, the longitudinal component $\mu_{\parallel}$ is calculated as the combination of the dipole moment of the peptide group H - N - C = O and the moment of C = O group in the side chains. According to the Pauling and Corey model (7), the main chain peptide group is most likely to have a longitudinal moment 3.6 D.U. The moment of the side chain C=O group is so directed as to oppose the main chain moment according to Wada's analysis(5). Therefore, if the peptide C = O group is directed to the negative Z direction, the side chain C = O must be directed to the positive z direction with, however, a certain inclination with the z axis. Again using the analysis by Wada, the net longitudinal moment of a repeating unit turns out to be 2.7 D.U. This value is used in eq. (8) to calculate $<\mu>_{\parallel}$. For comparison, the value of 3.6 D.U. is also used for $\mu_{\parallel}$ and the result is also shown in the same figure. The transverse component of a repeating unit is so assumed as to give the best fit with the experimental observation and the value turns out to be 1.8 D.U. These results are shown in Fig. 7. Obviously, the experimental points fit the curve 3 where the ratio $\mu_{\parallel} : \mu_{\perp}$ is assumed to be 2.7:1.8 and deviated considerably from curve 4 where the ratio is 3.6:1.8.

The origin of the longitudinal moment is, as mentioned, well accounted for but the origin of the transverse component is not well known. The possible transverse component of the main chain peptide groups and those of the side chains compensate each other because of the spiral symmetry of the helix. It perhaps arises, although this is still a conjecture, from the high pitch winding of the helix. This tentative conclusion is reached from the analysis of the relaxation time. As shown in Fig. 3 the relaxation time decreases as the velocity gradient increases. This behavior can best be accounted for by expanding the Debye equation into a

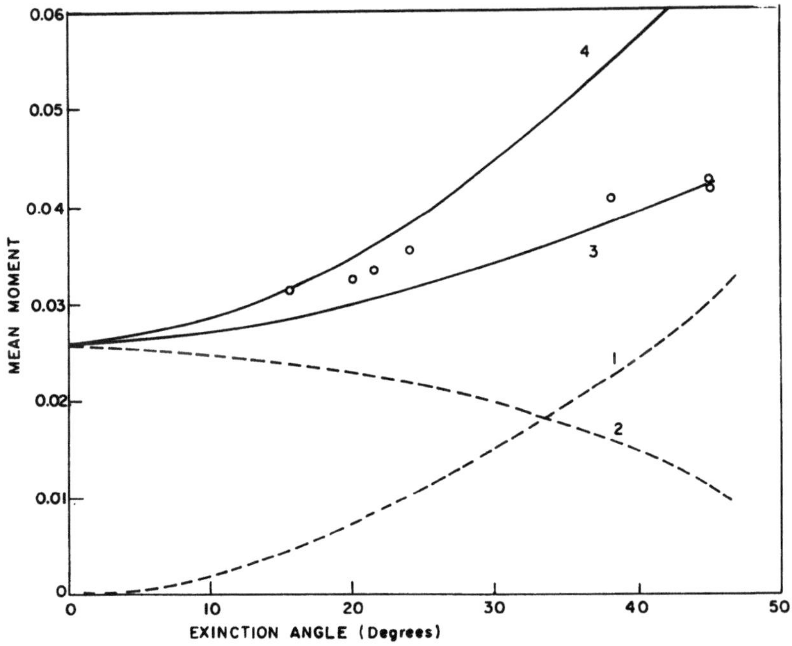

Fig. 7. The change of the longitudinal moment (curve 1) and the transverse moment (curve 2) as function of the extinction angle. The longitudinal moment of a repeating unit is assumed to be 2.7 D.U. and the transverse moment to be 1.8 D.U. Curve 3 is the vector sum of curves 1 and 2. Curve 4 is similarly constructed assuming the longitudinal moment to be 3.6 D.U. and the transverse moment 1.8 D.U.

linear combination of two dispersion terms (8):

$$\epsilon - \epsilon_{\infty} = \frac{\Delta \epsilon 1}{1+(\omega/\omega_1)^2} + \frac{\Delta \epsilon_2}{1+(\omega/\omega_2)^2} \quad (10)$$

where the first term represents the longitudinal component and the second term, the transverse component. $\Delta \epsilon_1$ is the longitudinal dielectric increment and ω_1 is the relaxation frequency ($\omega_1 = 1/2\pi\tau_1$) of the longitudinal dipolar rotation. Likewise, the ΔE_2 is the transverse dielectric increment and ω_2 ($\omega_2 = 1/2\pi\tau_2$)is the relaxation time of the transverse dipole orientation. The relaxation time shown in Fig. 3 is actually representing the combination of τ_1 and τ_2 but as the velocity gradient increases, the contribution of the second term in eq. (10) becomes more and more pre-dominant. The limiting value which is obtained from the curve in Fig. 3 must be close to the relaxation time of the transverse orientation. According to the theory of a viscous rotation by Debye the relaxation time of a sphere with a radius r is given by:

$$\tau = \frac{4\pi r^3 \eta}{kT} \quad (11)$$

where η is the viscosity of the medium in poise. Using this equation and assuming the radius of the helix including side chains to be 12.10^{-8} cm, we obtain a relaxation time for the rotation around the long axis of the helix to be $0.3.10^{-9}$ sec. This is far from that observed experimentally. Obviously, we have to assume a much larger effective radius in order to obtain a better agreement with the experimental result. This perhaps indicates either the bending or winding of the a-helix. The bended a-helical model may be more practical than a straight rod model although no other decisive experimental evidence has been provided for the compounded helix or a α-helix with a higher pitch winding.

The author is indebted to Dr. H. P. Schwan for his interest and advice. This work is supported by NSF GB 8475, NIH-HE-01253, and ONR-Nonr-551(52). The numerical calculations were done at the University of Pennsylvania Computer Center (supported by NIH FR 15).

Reference

1) P.Debye, Polar Molecules, Dover Publisher, New York, 1929.
2) J.Hermans, Jr., Advances of Chemistry Series, 63, 282, 1968.
3) B.Jacobsen and M.Wenner, Biochim.Biophys.Acta., 13, 3604,1956.

4) H.P.Schwan, Phys.Techniques in Biological Research,6,323,1963.
5) A.Wada, Polyamino Acids, Polypeptides and Proteins, Editor: M.A.Stahman, University of Wisconsin Press, Madison, 1962.
6) M.Boeder, Z.Physik, 75, 273, 1932.
7) L.Pauling and R.B.Corey, Proc.Natl.Acad.Sci.(U.S.), 39, 253, 1953.
8) J.L.Oncley, Proteins, Amino Acids and Peptides, Edited by E.J.Cohn and J.T.Edsall, Reinhold Pub.Co., New York, 1945.

CHOLESTERIC AND NEMATIC STRUCTURES OF POLY-γ-BENZYL-L-GLUTAMATE

E.T. Samulski and A.V. Tobolsky

Department of Chemistry, Princeton University, and

Textile Research Institute, Princeton, New Jersey

Fifteen years ago it was observed by Doty et al.[1] that synthetic polypeptides, $(\text{—NH—CHR—CO—})_n$, in solution can exist in a rigid rodlike α-helical conformation, in contrast with the random coil shape assumed by most other synthetic polymers in solution. This observation has stimulated a large body of investigation of the dilute solution properties of this class of polymers. In more concentrated polypeptide solutions (in the range of ten to fifty per cent polymer), Elliott and Ambrose[2] found that poly-γ-benzyl-L-glutamate (PBLG; $R = CH_2CH_2CO\text{—}O\text{—}CH_2C_6H_5$), a readily available synthetic polypeptide, forms a lyotropic liquid crystal. Robinson[3] extensively characterized the molecular arrangement in the liquid crystalline phase of PBLG and found it to be similar to the helicoidal structure of the liquid crystalline phases of many esters of cholesterol. We shall refer to the structure found by Robinson as 'cholesteric'. The structure is easily recognized with a polarizing microscope. The birefringent PBLG solutions present an image very reminiscent of a finger print. [As shown in Figure 1 (a)]. The spacing between the alternating bright and dark retardation lines, S, is equal to one-half of the pitch of the 'cholesteric' structure. The solutions that Robinson studied were quite fluid as are the liquid crystalline phases of smaller molecules whether thermotropic or lyotropic. The unusual arrangement of the molecules in these concentrated polypeptide solutions gives rise to properties heretofore not obtainable with polymeric systems.

Solid PBLG Films With 'Cholesteric' Structure

In the 1930's it was shown by Vorlander[4] that some liquid crystals could be quick-frozen to a metastable, brittle glassy state. More recently, Chistyakov and Kosterin[5] have studied vitri-

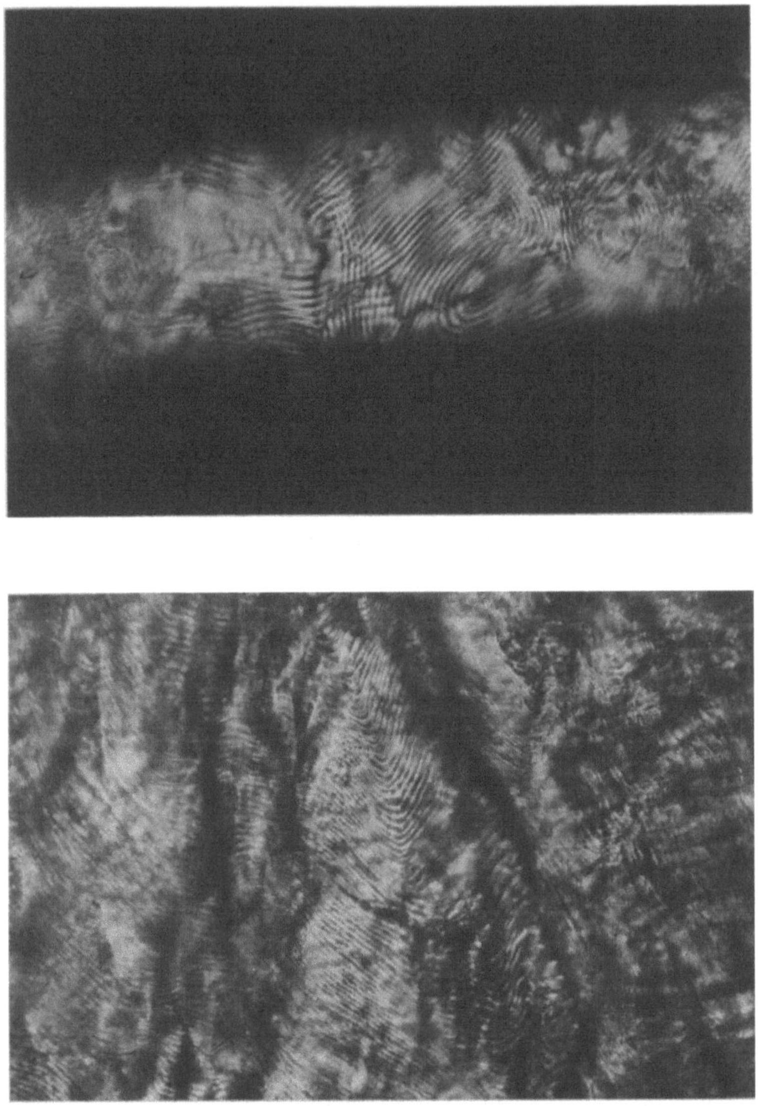

(a)

(b)

Figure 1. Retardation lines characteristic of a helicoidal supramolecular structure are observed in the photomicrographs of both the liquid and solid states of Poly-γ-benzyl-L-glutamate (PBLG).

(a) Birefringent fluid liquid crystalline solution of PBLG in dioxane, S = 50 microns, 10% PBLG (vol.).

(b) Birefrigent solid film of PBLG plasticized by 3,3' -dimethyl biphenyl, S = 2 microns, 30% PBLG (vol.).

fied liquid crystals with smectic, nematic and cholesteric structures, prepared by supercooling thermotropic liquid crystals. Specimens prepared by this technique are usually stable for several days at room temperature before they gradually crystallize. By working with a polymeric molecule we have been able to obtain thermodynamically stable solid films with a liquid crystalline local structure. The solid films are either pure PBLG or contain predetermined amounts of non-volatile liquids which act as plasticizers for the film. They are prepared by the evaporation of solutions of high molecular weight PBLG (Pilot Chemical Co., MW = 10^5) in various volatile solvents and mixed solvents (volatile solvent + plasticizer). The films are cast on a clean mercury surface and can be obtained in conditions describable as rubbery, leathery, or glassy, as is common for polymer films.

When solid films of pure PBLG are cast from solvents such as $CHCl_3$ or CH_2Cl_2, x-ray evidence[6] and anisotropic swelling characteristics[7] clearly indicate that the rodlike PBLG molecules lie in the plane of the film (parallel to the casting surface), but the helix axes are randomly oriented in this plane. These observations led us to intuit that these cast films retained the 'cholesteric' structure found in the fluid liquid crystalline solutions of PBLG[8]. A compressed 'cholesteric' molecular arrangement with the optical axis preferentially alligned normal to the film plane satisfactorily describes the uniplanar orientation of the helices and is consistent with the observed x-ray diffraction patterns and swelling studies.

Carefully prepared plasticized films do in fact exhibit the optical retardation lines characteristic of the 'cholesteric' structure in the liquid crystalline phase of PBLG. Consideration of the molecular structures of solvents which promote the rapid formation of the 'cholesteric' structure in PBLG solutions prompted the selection of the non-volatile liquid 3,3'-dimethylbiphenyl (DMBP) as a plasticizer for the PBLG films. Using this plasticizer it is possible to obtain solid films even at relatively low concentrations of PBLG (less than twenty per cent polymer). Controlled evaporation of chloroform solutions of PBLG + DMBP showing the 'cholesteric' structure resulted in solid films which retained the optical characteristics of the fluid liquid crystalline phase. [See Figure 1 (b)].

The lateral spacing between PBLG helices in the plasticized films has also been examined. X-ray studies show that this spacing increases continuously with increasing plasticizer concentration until phase separation occurs in the films (fifty per cent PBLG)[9]. Figure 2 shows that the concentration dependence of this spacing is the same as that observed by Robinson et al.[10] in the fluid liquid crystalline phase of PBLG. The spacing between the optical retardation lines is also dependent on PBLG concentration

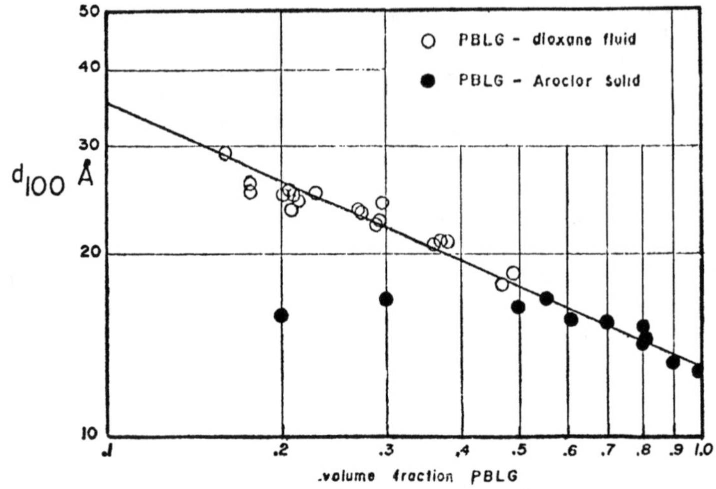

Figure 2. d_{100} spacing vs. PBLG concentration.

o solution data of Robinson et al. Discuss. Faraday Soc. 25, 29 (1958)

● d_{100} spacing in solid plasticized PBLG-Aroclor films (Aroclor = chlorinated polyphenyls).

in the plasticized films. Analogous behavior has been reported for liquid crystalline solutions of PBLG. This evidence is quite convincing that the unusual supramolecular arrangement of the fluid liquid crystalline phase of PBLG does exist in the solid state of mixtures of PBLG plus plasticizer. The fact that the PBLG films are solid with regard to mechanical properties and possess a stable liquid crystalline molecular arrangement is different from the fluidity previously associated with liquid crystals and probably results from the high molecular weight of the PBLG molecules.

Fluid PBLG Liquid Crystal With Nematic Structure

It has been known for some time that magnetic fields of the order of several hundred orsteds cause spontaneous ordering of nematic liquid crystals.[11] The local order present in a nematic liquid crystal allows the magnetic field to interact with the diamagnetic anisotropy of a large number of molecules (domain of molecules) in a cooperative manner, causing the direction along which the diamagnetism is the smallest to allign parallel to the applied magnetic field. Recently, magnetic fields have been used to orient thermotropic liquid crystals with a cholesteric structure.[12,13] In these cases, a sufficiently strong magnetic field will untwist the cholesteric structure to form an oriented nematic structure with the optical axis aligned parallel to the field. We have observed the influence of a magnetic field on the 'cholesteric' structure of PBLG solutions indirectly using nuclear magnetic resonance (NMR).

Three sets of workers (Samulski and Tobolsky[7]; Sobajima[14]; and Panar and Phillips[15]) independently found that the NMR spectrum of CH_2Cl_2 is split into a doublet in the liquid crystalline phase, PBLG + CH_2Cl_2. No spectrum is observed for PBLG in these concentrated solutions because nuclear dipolar interactions broaden the signal of the polymer beyond the limits of detection in high resolution NMR. On the other hand, the CH_2Cl_2 molecules, though partially oriented by interactions with the polymer, retain a high degree of mobility and exhibit a spectrum with linewidths similar to those observed in ordinary liquids. The partial orientation or anisotropic tumbling of the CH_2Cl_2 molecule in the liquid crystal matrix produces a non-zero average of the direct dipole-dipole interactions between the pair of protons on a CH_2Cl_2 molecule and, hence, a doublet in the CH_2Cl_2 spectrum.

The optical axis of the PBLG 'cholesteric' structure is randomly oriented throughout the macroscopic sample when the liquid crystal is first placed in the NMR spectrometer. This means that on a microscopic scale there are randomly oriented domains of nearly parallel PBLG helices. As a result of this microscopic arrangement in the liquid crystal, the initial NMR spectrum of CH_2Cl_2 is reminiscent of the "powder diagram" spectra observed for partially oriented water in powdered crystalline hydrates.[16] The "powder diagram" changes into a discrete doublet after the PBLG liquid crystal has been in the NMR spectrometer for a suitable length of time. The time dependent spectra of CH_2Cl_2 in the PBLG liquid crystal is shown in Figure 3. The time required for the appearance of the doublet is dependent on the viscosity of the liquid crystal (e.g. the concentration of PBLG in the solution) and is of the order of several hours for solutions containing ten to twenty per cent polymer.

The optical retardation lines in 'cholesteric' PBLG solutions are no longer visible in solutions subjected to a strong magnetic field. This observation, the x-ray studies of magnetically oriented films of PBLG (described below) and the NMR results indicate that a sufficiently strong magnetic field untwists the 'cholesteric' structure and forms oriented nematic structure with the rodlike PBLG molecules parallel to the applied field. Splitting of the NMR absorption of CH_2Cl_2 due to intramolecular direct dipole-dipole coupling would be expected in this kind of oriented nematic environment. (See reference 17 for a review of this phenomenon.) The high viscosity of the PBLG liquid crystal makes it possible to observe the angular dependence of the dipolar interactions by turning the oriented nematic axis to a given angle with the magnetic field and re-

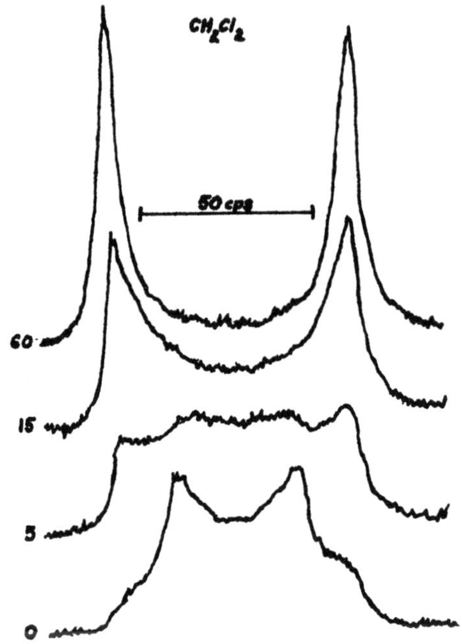

Figure 3. Time dependence of CH_2Cl_2 spectrum in PBLG liquid crystal (PBLG + CH_2Cl_2, 10%PBLG vol; time is given to the left of the spectra in minutes).

cording the spectrum before the liquid crystal reorients. We have observed similar behavior in the nematic lyotropic liquid crystal: an equi-molar mixture of the D and L isomers of polybenzyl glutamate.

The magnitude of the splittings observed in the NMR spectra depend on the geometry of the solute molecule (small mobile molecule) and the extent of orientation which the solute experiences in the liquid crystal. The splittings are dependent on the concentration of polymer in the PBLG solutions but in general are one to two orders of magnitude smaller than those observed in thermotropic liquid crystals. The smaller splittings observed in the lyotropic polypeptide systems have been attributed to chemical exchange between isotropic and oriented solute molecules, the latter molecules being more closely associated with the oriented polypeptide.[18] This interpretation is plausible since in the concentration range used for the NMR experiments there are approximately 5 to 10 solute molecules per polypeptide monomer, and obviously all of the solute molecules can not be directly associated with the oriented polypeptide matrix simultaneously. On the other hand, in NMR experiments using thermotropic liquid crystals, the number of solute molecules is only a very small fraction of the total number of molecules, hence each solute molecule is oriented to the same extent in the liquid crystal matrix.

NMR experiments using the same solute molecule but different polypeptides point out the importance of specific polymer-solute interactions on the extent of solute orientation. Figure 4 shows the NMR spectrum of dimethylformamide (DMF) in the liquid crystalline phase of poly-L-glutamic acid (PGA; R = $-CH_2CH_2COOH$). The

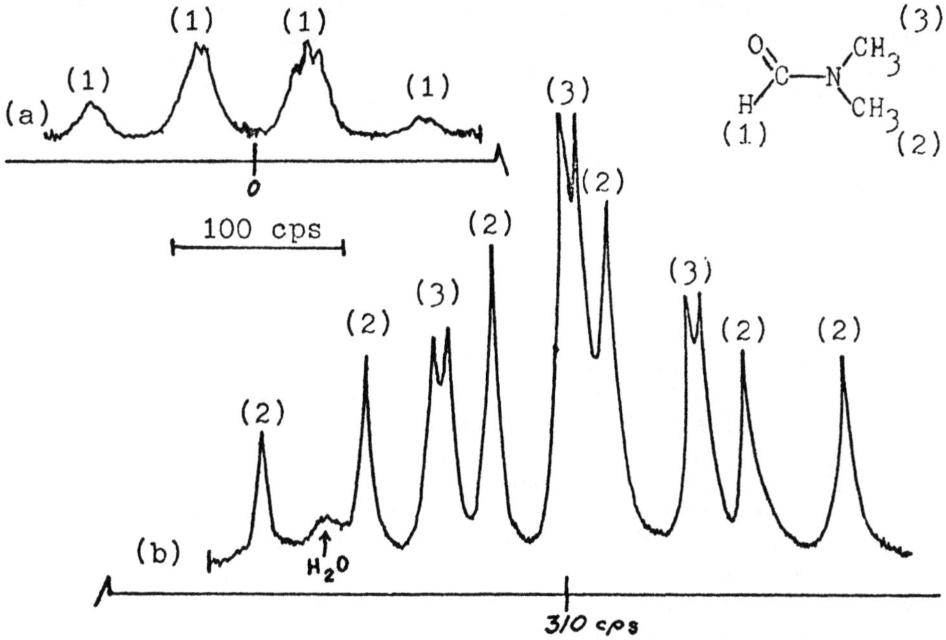

Figure 4. NMR spectrum of DMF in oriented PGA liquid crystal. DMF:D_2O:PGA monomer, 5.5:2:0.6 moles x 10^{-3}.
(a) spectrum of proton (1)
(b) spectrum of methyl groups (2) and (3).

splittings observed in the PGA liquid crystal are quite different from those reported for PBLG + DMF.[15] On the other hand, a DMF spectrum similar to that reported in PBLG was observed using the polypeptide poly-ε-carbobenzoxy-L-lysine (PCLL; R=$(CH_2)_4NHCOOCH_2C_6H_5$). The groups terminating the sidechains are the same for PBLG and PCLL. The polymer backbone chain is identical both in composition and spatial conformation for all three polypeptides. The only differences occur in the sidechains R. The observed differences in the direct coupling constants for DMF in different polypeptide liquid crystals therefore suggest that specific interactions between the functional groups on the polypeptide sidechain and the DMF molecule make significant contributions to the orientation of DMF in these liquid crystals.

Solid PBLG Films With Nematic Structure

Samulski and Tobolsky[7] and Sobajima[14] discovered that when solutions of PBLG in a volatile solvent are evaporated in the presence of a strong magnetic field, highly oriented films are obtained. The orientation occurs while the solutions pass through the concentration range in which they are liquid crystalline and becomes permanently locked in when the mixture of solvent plus PBLG becomes solid. The nature of the orientation in the films indicates that the PBLG liquid crystal in a strong magnetic field has an ordered nematic structure. In this structure the PBLG helices are oriented parallel to the direction of the applied magnetic field. We found that by suitable positioning of the magnetic field during the evaporation procedure, the rodlike molecules could be oriented in any given direction in the plane of the cast film or even perpendicular to the film plane.

The uniaxial orientation in these films produces x-ray diffraction patterns very similar to the fiber patterns obtained from mechanically oriented samples (fibers drawn from solution). Figure 5 (a) shows the type of diffraction pattern obtained from magnetically oriented PBLG films cast from CH_2Cl_2. The layer line pattern consists of streaks with Bragg reflections occurring on the equator only, implying that the lateral spacing between helices is fairly regular but there are probably random displacements parallel to the helix axis. The layer lines can be fitted to the theory of Cochran, Crick and Vand[19] giving helix parameters, 18 residues in 5 turns or 3.6 residues per turn. This is the α-helix with exactly the same helical parameters as those proposed by Pauling, Corey and Branson[20]. The same helical conformation has been reported for PBLG fibers drawn from dioxane solution.[21,22]

Figure 5 (b) shows the diffraction pattern of a magnetically oriented PBLG film cast from $CHCl_3$. The layer line pattern is markedly different from that observed for the normal α-helix. The strong layer line appearing at 10.4 Å and the absence of other layer lines between the equator and the "turn" layer line clearly demonstrates that the helical parameters are 7 residues in 2 turns (3.5 residues per turn). Prolonged heating of films containing the 3.5-helix at 140°C does cause layer lines characteristic of the normal α-helix to appear in the diffraction pattern.

X-ray diffraction patterns of drawn fibers containing a 50:50 mixture of the D and L isomers of polybenzyl glutamate are indicative of distorted helical conformations.[22,23,24] There is not exact agreement among these workers on the amount of distortion (values range from 3.5 to 3.58 residues per turn), but it is generally agreed that such distortions from the normal α-helical parameters facilitate sidechain-sidechain interactions between neigh-

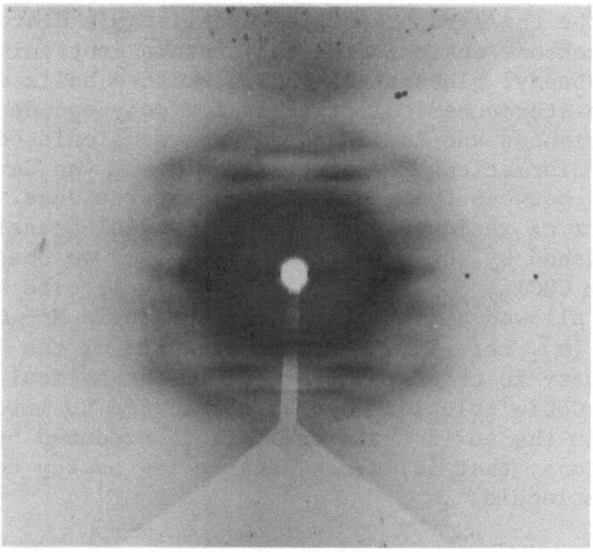

(a)

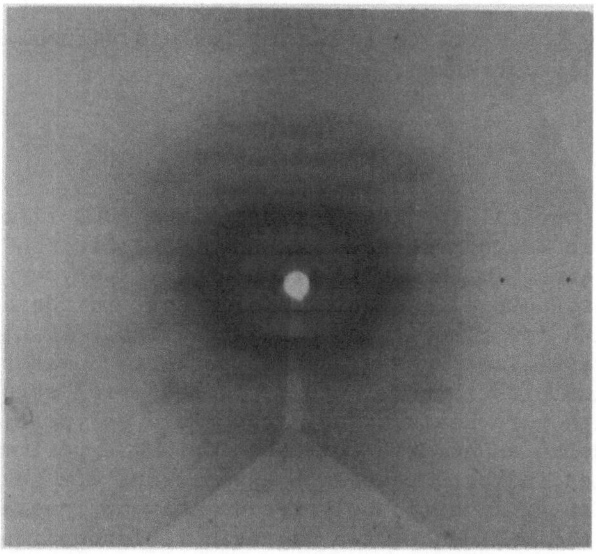

(b)

Figure 5. X-ray diffraction patterns of magnetically oriented PBLG films. The photographs were recorded with the fiber axis perpendicular to the cylindrical camera axis and the incident x-ray beam using a 200 micron collimator.

(a) PBLG cast from CH_2Cl_2.
(b) PBLG cast from $CHCl_3$.

boring D and L helices in the crystal lattice. However, our results seem to be the first report of considerable distortion of the polypeptide backbone chain in the solid state containing only the L isomer of polybenzyl glutamate. This distorted helix does appear to satisfy the stereochemical criteria for polypeptide conformations. Ramakrishnan and Ramachandran[25] have calculated allowed polypeptide conformations for specified minimum van der Waals contact distances between the atoms in adjacent residues. In a diagram of 'number of residues per turn' vs. 'axial translation per residue' published by these workers, the values we observed for PBLG cast from $CHCl_3$, 3.5 and 1.5 A respectively, lie very close to the region of allowed conformations for the angle $N{-}_{\alpha}C{-}C'\ \tau = 110^{\circ}$ (see Figure 4 (b), ref. 25). Further analysis of the diffraction data is necessary in order to determine if the helical conformation obtained from $CHCl_3$ solutions can be attributed to intermolecular interactions in the solid state or if it is produced by intramolecular interactions, that is, a new, stable low energy conformation for the PBLG molecule.

Acknowledgement

We wish to thank Dr. Yukio Mitsui for his valuable discussions of the x-ray diffraction patterns.

Summary

X-ray and optical studies of plasticized PBLG films show that the 'cholesteric' structure found in the lyotropic PBLG liquid crystal can be obtained in the solid state of this polypeptide. The 'cholesteric' structure is stable indefinitely in the homogeneous solid films with low plasticizer concentrations. At higher concentrations of plasticizer, there appears to be two phases in the solid films, PBLG plus plasticizer (with a 'cholesteric' structure) in equilibrium with pure plasticizer. The fluid polypeptide liquid crystals can serve as nematic solvents for NMR experiments. Preliminary studies indicate that such experiments are potentially useful for elucidating specific polypeptide-solvent interactions. Utilization of the influence of a magnetic field on liquid crystalline polypeptide solutions enables one to produce highly oriented solid films. In these films the degree of orientation of the polypeptide is sufficient to examine the polypeptide conformation by x-ray diffraction. Our observations suggest that small differences in polypeptide conformations occurring in different solvents might be detected by this new technique for orienting polypeptides. It is quite possible that similar studies could be extended to other rodlike biological macromolecules exhibiting a liquid crystalline phase such as DNA.

References

1. P. Doty, A.M. Holtzer, V.H. Bradbury and E.R. Blout, JACS 76, 4493 (1954).
2. A. Elliott and E.J. Ambrose, Faraday Soc. Discuss. 9, 246 (1950).
3. C. Robinson, Molecular Crystals, 1, 467 (1966) and references cited therein.
4. D. Vorlander, Trans. Faraday Soc., 29, 907 (1933).
5. I.G. Chistyakov and Ye. A. Kosterin, Rost Kristallov, 4, 68 (1964).
6. A.J. McKinnon and A.V. Tobolsky, J. Phys. Chem. 70, 1453 (1966).
7. E.T. Samulski and A.V. Tobolsky, Macromolecules 1, 555 (1968).
8. E.T. Samulski and A.V. Tobolsky, Nature, 216, 997 (1967).
9. E.T. Samulski and A.V. Tobolsky, Molecular Crystals, in press.
10. C. Robinson, J.C. Ward and R.B. Beevers, Faraday Soc. Discuss. 25, 29 (1958).
11. G.W. Gray, "Molecular Structure and the Properties of Liquid Crystals", Academic Press Inc. (London)Ltd. 1962, Chap. 4.
12. E.S. Sackman, S. Mieboom and L.C. Snyder, JACS 89, 5981 (1967).
13. G. Durand, L. Leger, F. Rondelez and M. Veyssie, Phys. Rev. Letters, 22, 227 (1969).
14. S. Sobajima, J. Phys. Soc. Japan, 23, 1070 (1967).
15. M. Panar and W.D. Phillips, JACS 90, 3880 (1968).
16. G.E. Pake, J. Chem. Phys. 16, 327 (1948).
17. G.R. Luckhurst, Quarterly Reviews 22, 179 (1968).
18. B.M. Fung, M.J. Gerace and L.S. Gerace, JACS, in press.
19. W. Cochran, F.H.C. Crick and V. Vand, Acta Cryst., 5, 581(1952).
20. L. Pauling, R.B. Corey and H.R. Branson, Pro. Nat'l. Acad. Sci., U.S., 37, 205 (1951).
21. Y. Mitsui, Acta Cryst. 20, 694 (1966).
22. A. Elliott, R.D.B. Fraser and T.P. MacRae, J. Mol. Biol., 11, 821 (1965).
23. Y. Mitsui, Y. Iitaka and M. Tsuboi, J. Mol. Biol., 24, 15(1967).
24. M. Tsuboi, A. Wada and N. Nagashima, J. Mol. Biol, 3, 705(1961).
25. C. Ramakrishnan and G.N. Ramachandran, Biophysical J., 5, 909 (1965).

References

THE PROTON MAGNETIC RESONANCE SPECTRA OF ACETYLENE AND ITS ^{13}C-ISOMERS IN NEMATIC LIQUID CRYSTALLINE SOLUTIONS

H. Spiesecke

EURATOM CCR, Ispra, Italy

ABSTRACT

The nmr spectra of the two carbon-13 containing isomers of acetylene are analysed. The perpendicular allignment to the magnetic field is confirmed. The geometrical structure is determined and compared with the results obtained by other methods.

INTRODUCTION

In his paper on the proton magnetic resonance spectra of partially orientated acetylenic compounds (1) Englert deduced a perpendicular orientation of acetylene in the magnetic field from a small (-0.077 ppm) negative shift in the nematic solvent. Since the degree of order (S = -0.014) was very small too, we tried to verify these results on an independent way, relating the orientation of the molecule to the positive scalar coupling between a proton and the adjacent carbon-13 atom in 1-^{13}C-acetylene and 1,2-^{13}C-acetylene. Also the degree of order was increased by lowering the temperature and concentration.

EXPERIMENTAL RESULTS

The spectra were recorded with a Varian HA 100 spectrometer operating at 34°C with 4 kHz sidebands.

The spectra were calibrated either by andiomodulation of the transmitter frequency or by substitution of the sample for one containing chloroform and tetramethylsilane. The two signals are separated by 726 Hz at 100 MHz. Only upfield traces were used for the determination of the experimental line positions which were taken as averages over three to five different runs. The statistical measuring error amounted to ±1.5 Hz. The average full line width at half height was about 10 Hz.

The carbon-13 containing species of acetylene were synthesized starting from 58 % carbon-13 enriched $BaCO_3$ which was converted to the carbide using metallic barium (2). The liberated acetylene was dried over phosphorous pentoxide and purified by several bulb to bulb destillations at controlled temperature and pressure. All samples were prepared under vacuum using a mixture of 60 mol% Butyl p-(p-Ethoxyphenoxycarbonyl)-phenyl carbonate and 40 mol% p-capronyloxy-p'-ethoxyazo- or azoxybenzene (3).

The experimental spectrum is compared with the theoretical one in Figure 1. The latter is the weighted sum of the three different isomers. The B-values given in table 1 were derived in the following way: For the ^{12}C-species it was taken directly from the experimental spectrum since $\Delta\nu = |3\ B_{HH}|$.

Assuming that the orientation for all isomers of acetylene would be the same the spectrum of 1-^{13}C-acetylene was iteratively approximated using a dipolar version of LAOCN 3 and keeping B_{HH}, taken from normal acetylene, constant (4). This led to a very good agreement with the experimental spectrum. The root mean square error amounts to 0.06 Hz. Using these values to calculate the spectrum of 1,2-^{13}C-acetylene-iterating on B_{CC} only - led to a poor agreement with the experimental frequencies with a rms error of 3.5 Hz. Iteration on all dipolar coupling constants slightly improved the results, but the rms error still remained high (2.8 Hz). It was with the latter results that the geometrical data in table 2 were calculated.

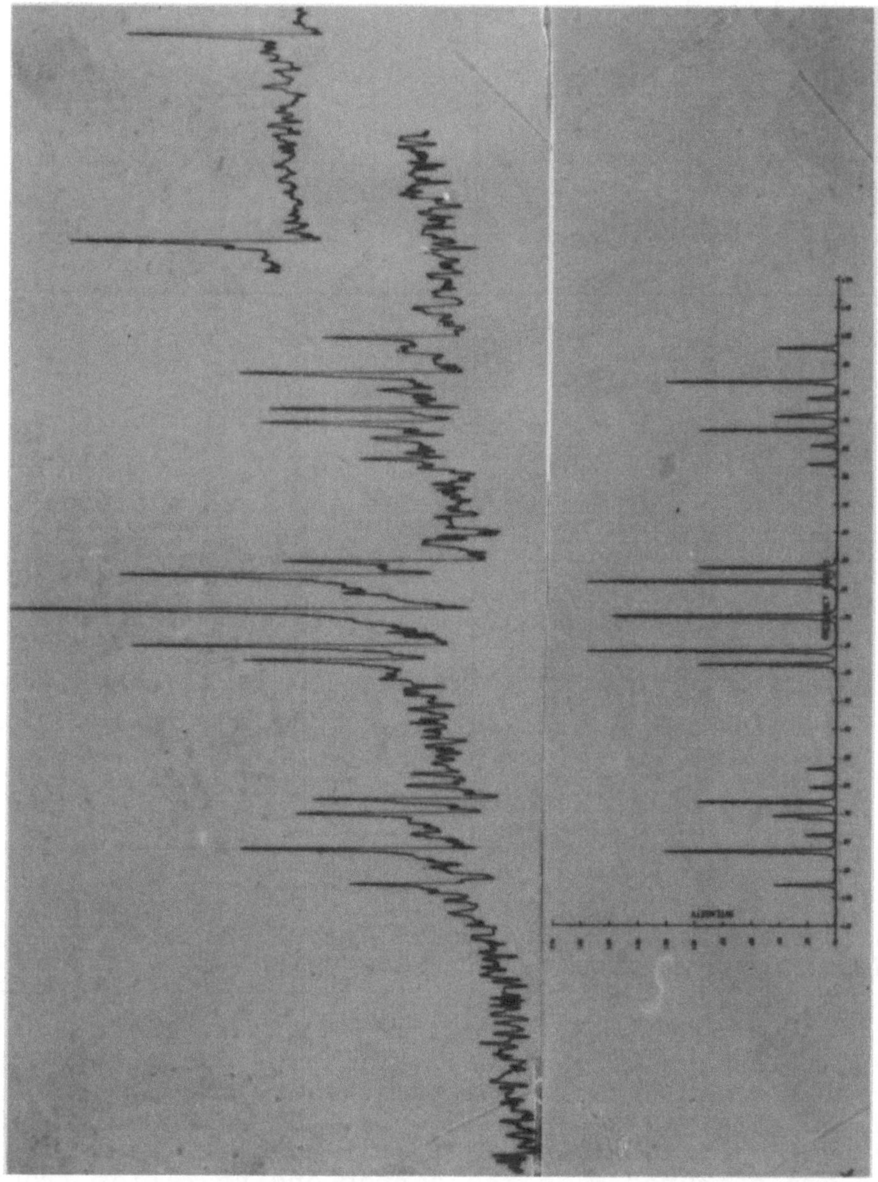

Figure 1 Experimental and calculated spectra of carbon-13 isomers of acetylene

	H-C≡C-H (1, 2)	H-^{13}C≡C-H	H-^{13}C≡^{13}C - H
B_{12}	81.87	81.87	81.87 (80.85)
B_{13}	-	616.37	616.37 (617.94)
B_{23}	-	62.46	62.46 (62.50)
B_{34}	-	-	103.57 (103.71)

Table 1. Dipolar coupling constants of acetylenic isomers

In the ^{13}C containing species all line positions depend on the scalar coupling constants. A sign variation showed that the spectra could satisfactorily be described only when the sign for all indirect coupling constants was positive. Since $J_{^{13}CH}$ is generally accepted to be positive this gives a negative "S" for the orientation of the long molecular axis

$$B_{HH} = - \frac{\gamma_H^2 h}{4\pi^2 r_{HH}^3} \cdot S_{HH}$$

which confirms the results found by Englert, Saupe, and Weber from chemical shift measurements.

In molecules belonging to the point group $D_{\infty h}$ the ratios of internuclear distances are independent on the orientation of the molecule. The values derived from the experimental dipolar coupling constants are compared with the corresponding ratios from r_o and r_e bond lengths.

	r_o(5)	r_e(5)	r_e-δ(6)	r_{nmr}	$r_{nmr}^{calc.}$
r_{13}/r_{23}	0.4664	0.4682	0.4720	0.4662	0.4640
r_{13}/r_{12}	0.3181	0.3189	0.3229	0.3216	0.3161
r_{13}/r_{34}	0.8741	0.8808	-	0.8771	0.873
r_{23}/r_{12}	0.6819	0.6811	0.6840	0.6844	0.6818
r_{23}/r_{34}	1.875	1.880	1.865	1.876	1.880
r_{12}/r_{34}	2.749	2.761	2.728	2.727	2.759

Table 2. Bond length ratios for 1- and 1,2-^{13}C-acetylene

As can be seen from the expression given above, relating B, S and r, the bond length ratio derived from a nuclear magnetic resonance experiment is not equal to the one using the equilibrium distance r_e, but it is still modified by the effect of thermal vibrations. Under the assumption of small vibrations of a harmonic oscillator one can derive the following equation (7), (8),

$$r_{nmr} = r_e(1 - 2\frac{a^2}{r_e^2})$$

relating the nmr determined bond length with the equilibrium bond length "r_e" and the mean square amplitudes of vibration "a" of this distance. For acetylene the following values result: r_{12} = 3.318 Å, r_{13} = 1.048 Å, r_{23} = 2.260 Å, r_{34} = 1.202 Å (6). They have been used to calculate the last column of table 2. It is obvious that in the case of acetylene the correspondance with the experimentally determined ratios has not improved. Very good results, however, are obtained if one takes into account the Bastiansen-Morino shrinkage effect (9) (10). The corresponding values are given in the third column of table 2. From spectroscopic data it can be shown with the aid of symmetry and valence coordinates(11) that in linear structures the effect of anharmonicity cancels in the computation of the shrinkage effect and it seems that this is the reason for the better correspondence as compared with the results obtained from r_e correcting for the mean amplitudes of vibration.

CONCLUSIONS

From the study of carbon-13 substituted acetylenes in a nematic liquid crystal it can be concluded that the average orientation of the guest molecules is perpendicular to the long axis of the host substance. This must be due to some specific type of interaction as has been found with acetylene in isotropic spectra too (12).

In spite of this interaction the ratios of internuclear distances derived from the direct dipolar coupling constants are in good agreement with the values derived from infrared spectroscopic measurements, especially, if one takes into account the linear shrinkage effect.

LITERATURE

1) G. Englert, A. Saupe, J.-P. Weber, Z. Naturf. 23a, 152 (1968)
2) M. Calvin, Isotopic Carbon, John Wiley + Sons, 1949, page 205
3) H. Spiesecke, J. Bellion-Jourdan, Angew. Chem. 79, 475 (1967)
4) P.J. Black, The Procter and Gamble Company, Cincinnati, Ohio, Private Communication
5) W.J. Lafferty, E.K. Plyler, E.D. Tidwell, J. Chem. Phys. 37, 1981 (1962)
6) E. Meisingseth, S.J. Cyvin, Acta Chem. Scand. 15, 2021 (1961)
7) L.A. Bartell, J. Chem. Phys. 23, 1219 (1955)
8) L.C. Snyder, S. Meiboom, J. Chem. Phys. 47, 1480 (1967)
9) O. Bastiansen, M. Traetteberg, Acta Cryst. 13, 1108 (1960)
10) Y. Morino, Acta Cryst. 13, 1107 (1960)
11) Y. Morino, J. Nakamura, P.W. Moore, J. Chem. Phys. 36, 1050 (1962)

12) J.A. Pople, W.G. Schneider, H.J. Bernstein, High Resolution Nuclear Magnetic Resonance, McGraw Hill, New York 1959

STUDIES OF THE HELIX-COIL TRANSITION AND AGGREGATION IN POLYPEPTIDES BY FLUORESCENCE TECHNIQUES

Thomas J. Gill III, Charles T. Ladoulis,
Martin F. King and Heinz W. Kunz
Department of Pathology
Harvard Medical School
Boston, Massachusetts

The technique of polarization of fluorescence using dye-macromolecule conjugates is a very sensitive hydrodynamic method for studying intramolecular transitions and intermolecular interactions. Several considerations influence the choice of the dye in such conjugates: extinction coefficient, lifetime, effect of pH, and interaction with the macromolecule to which it is coupled. Fluorescein has a higher molar extinction coefficient (3.4×10^4) than DNS (1-dimethylaminonaphthalene-5-sulfonyl chloride) (4.3×10^3); therefore, it is useful when working with small amounts of material. The lifetime of DNS is constant from approximately pH 2 to 14 (2), whereas that of fluorescein is strongly dependent upon pH. The latter varies in an approximately linear fashion from pH 2 (<1 ns) to pH 8 (6 ns); thereafter, it gradually increases (7 ns at pH 11.8). In addition, the lifetime of the dye can also be influenced by the molecule to which it is coupled. The degree of polarization of DNS and its conjugates is not significantly influenced by pH over the range 2.5 to 14; hence, DNS is useful for studying the effect of pH on intramolecular transitions. On the other hand, fluorescein and its conjugates are quite sensitive to the effects of acid and base, so fluorescein is not useful for such studies (3). Finally, the interaction between the dye and the molecule to which it is attached is different for fluorescein and DNS conjugates. This factor influences measurement of the rotational relaxation time more than that of the transition temperature, i.e., the temperature at which internal rotation or intermolecular aggregation occurs. This effect may be due to the capabilities of the various dyes to interact with different portions of the macromolecule; thus, they would reflect the behavior of different rotational units (3).

Many interesting phenomena involve the effect of urea on various conformations of macromolecules, and polarization of fluorescence is especially useful for such studies. However, urea solutions cause strong Rayleigh scattering, which increases as the instrumental sensitivity used for the measurements increases. In addition, the amount of scattering increases as the temperature increases. Hence, the effect of urea scattering on both the horizontally and vertically polarized light must be measured and used to correct the experimental measurements. The error in the calculation of the rotational relaxation time introduced by not correcting for urea scattering can be as large as 40%. For example, the uncorrected rotational relaxation time of DNS-lysozyme in 8 M urea at 25° is 32 ns, which is the same as that in 0.2 M NaCl + 0.1 M buffer. The corrected value in 8 M urea is 23 ns, which is a significant decrease and indicates that the structure of molecule is, indeed, disrupted by urea.

The fluorescence measurements reported here were made with a modified Brice-Phoenix light scattering photometer and with polypeptides coupled to either fluorescein isothiocyanate or DNS (3). Fluorescein was excited by unpolarized light at 436 mμ and the intensity of the fluorescent radiation was measured at 90° from the direction of the exciting light after passing through a saturated sodium nitrite solution (1 mm thick) and a 520 mμ cut-off filter. The saturated sodium nitrite solution was used as an additional precaution to insure that no scattered ultraviolet light from the excitation source reached the cut-off filter and caused intrinsic fluorescence in the glass. In the case of DNS, the excitation filter was 365 mμ and the cut-off filter, 416 mμ; the sodium nitrite filter was also used.

The helix-coil transition in polylysine (molecular weight = 51,000) measured by polarization of fluorescence is shown in Figure 1, and the pH values at 50% and 90% completion of the transition are summarized in Table 1. The studies carried out in NaCl were done at salt concentrations where the polyelectrolyte effect was almost completely suppressed (Figure 2). In water, the helix reached maximal rigidity (pH[50%] = 9.64) before all of the residues were in the helical conformation (pH[50%] = 10.0), whereas in 0.2 M NaCl + 0.1 M buffer, the development of rigidity in the molecule (pH[50%] = 10.3) followed the same pattern as the increase in helical content (pH[50%] = 10.2). The titration of the molecule followed the changes measured by polarization in all solvents studied. The helix was destroyed by 8 M urea within 3 hours, as measured by polarization of fluorescence.

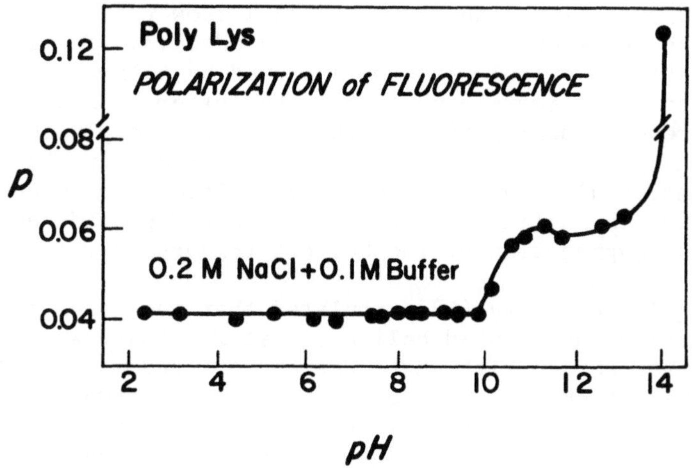

Figure 1. Helix-coil transition in DNS-poly Lys

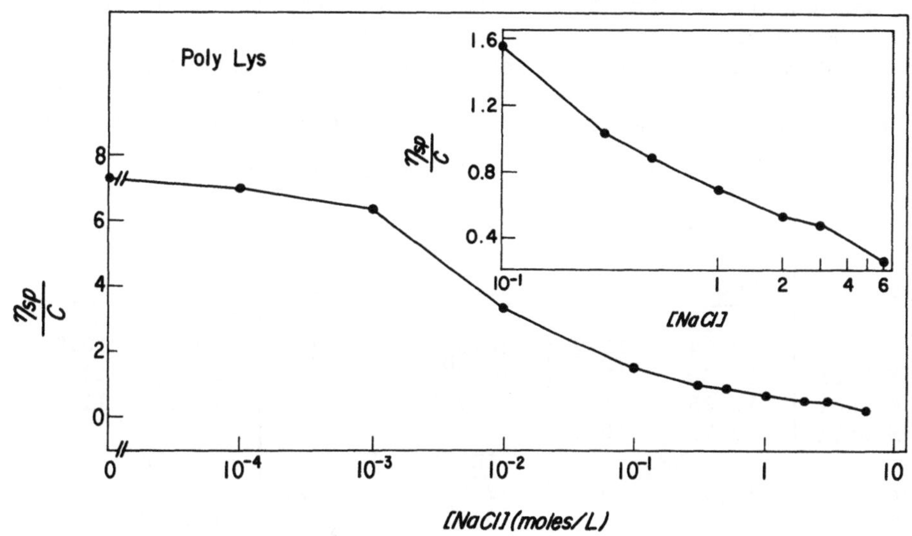

Figure 2. Effect of NcCl concentration on the specific viscosity of poly Lys (1% solution, pH 7, M = 100,000)

The polarization changes in Figure 1 show that the polylysine molecule is completely helical by pH 11.3. Then, there is a slight decrease in polarization followed by a large increase beginning at approximately pH 13.5. The latter change most likely represents aggregation of the uncharged polylysine helices; the possible cause of the prior decrease in the degree of polarization will be discussed later. The other techniques used to study the helix-coil transition do not provide any evidence for aggregation of the helices (Figure 3).

Applequist and Doty (5) postulated that the polylysine molecule formed interrupted helices, that is, there were helical segments interspersed with short regions of random coil, during the transition from the coil to the helical conformation. Their arguments were based on studies of the equilibrium between the helical and coil forms, including the breadth of the transition; on the loss of rigidity as measured by flow birefringence when the molecule was 90% helical; and on the presence of a minimum in the intrinsic viscosity and a maximum in the sedimentation constant in the transition region. The data from polarization measurements support this concept, and they also provide three lines of evidence that very short regions of the coil persist after the completion of the coil-helix transition. (a.) In water, the molecule reached maximal rigidity after all of the residues were in the helical conformation, suggesting that short segments of coil must be completely titrated before the molecule can become rigid (Table 1). (b.) The rotational relaxation time

Table 1. Helix-coil transition in poly Lys[a]

	Polarization		Optical Rotation		Titration	
Solvent	50%	90%	50%	90%	50%	90%
water	9.64	10.2	10.0	10.6	9.5	10.5
0.2 M NaCl	10.30	10.7	10.2[b]	10.6	10.2[c]	10.9
1.0 M NaCl	10.35	11.0	9.7[d,e]	10.8	10.3[d]	11.2
3.0 M NaCl	10.72	10.9				
	NMR		IR (L/Lo)		[η]	
water	10.2	10.3	9.0	9.5		
0.2 M NaBr			10.5[b,f]	11.1	11.0[d,g]	11.7

[a]References (4-9); [b]0.2 M NaBr; [c]0.5 M NaCl; [d]1.0 M NaBr + 0.2 M $NaHCO_3$; [e]10.4 in 1 M KBr in H_2O and 11.0 in 1 M KBr in D_2O by $[m']_{233}$ (16); [f]9.8 and 10.5 in 1.0 M NaBr + 0.2 M $NaHCO_3$; [g]10.9 and 11.6 from sedimentation constant measurements.

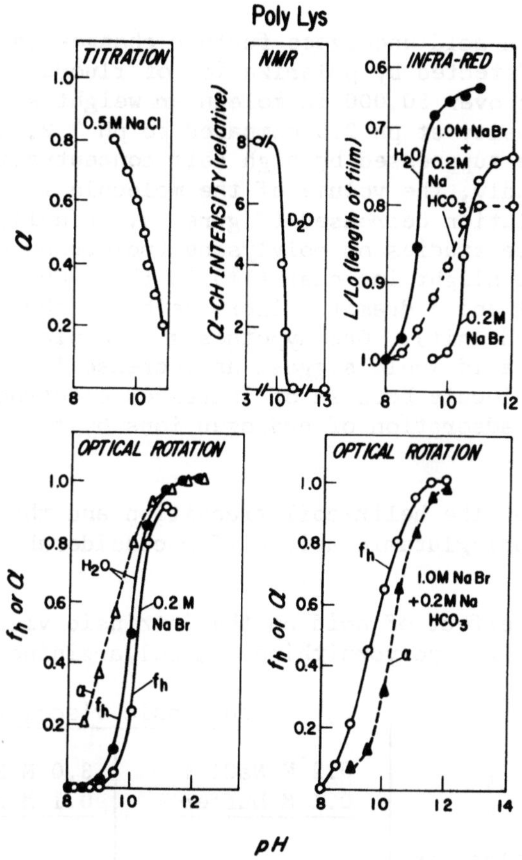

Figure 3. Helix-coil transition in poly Lys (4-9).

was only 36% of the theoretical value at the pH where the molecule attained the maximal helical conformation (Table 5). (c.) The molecule had to be titrated far past the point where the coil-helix transition was complete before aggregation of the helices occurred (Figure 1), suggesting that the molecule remained somewhat flexible until extremely high pH values were attained.

Polylysine in acid undergoes further changes in structure which cannot be detected by polarization of fluorescence measurements. Molecules over 50,000 in molecular weight showed an increase in viscosity at pH 2.5 compared to pH 7.2, and this increase could be suppressed by high salt concentrations (Table 2). Secondly, the volume of the molecule increased as the pH of the solution decreased (Figure 4). Finally, nuclear magnetic resonance studies on polylysine (molecular weight = 100,000) showed a slight increase (4%) in the α-CH intensity from pH 7 to pH 1 and a dramatic increase below that (80% in approximately 1 pH unit). One hypothesis to explain the data on polylysine in acid would suggest an increase in the volume of the polylysine molecule because of increased electrostatic repulsion due to adsorption of hydrogen ions by the peptide backbone.

The nature of the helix-coil transition and the structure of the helix in polyglutamic acid differ considerably from those

Table 2. The effect of acid on the intrinsic viscosity of polylysine, polyornithine and polyarginine.

		Intrinsic viscosity			
		0.2 M NaCl + 0.1 M buffer		3.0 M NaCl + 0.1 M buffer	
Polypeptide	Molecular Weight	pH 7.2	pH 2.5	pH 7.2	pH 2.5
poly Lys	40,000	0.6	0.6	0.3	0.3
	51,000	0.7	1.6		
	100,000	1.4	2.5[a]	1.2	1.4
poly Orn	45,000	0.6	0.6	0.4	0.4
poly Arg	28,000	0.2	0.2		

[a]Treatment with acid, base, or high salt concentrations prior to study did not alter the increased intrinsic viscosity at pH 2.5.

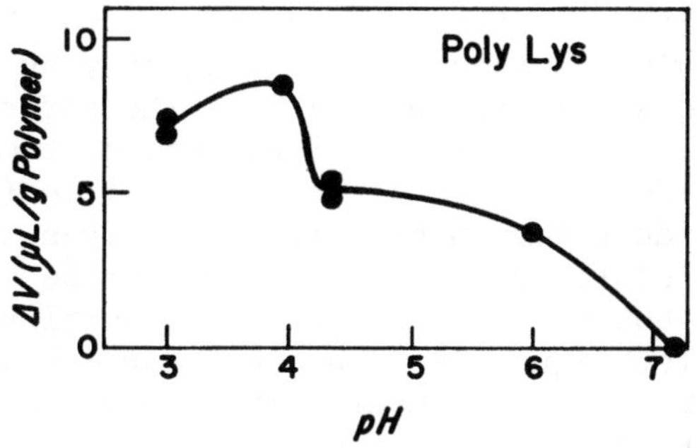

Figure 4. Volume increase in poly Lys (molecular weight = 100,000) in water as a function of pH

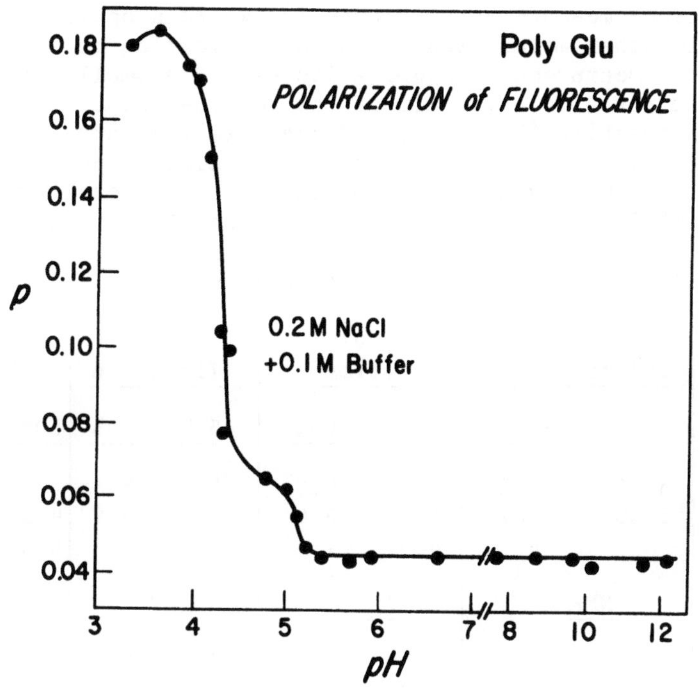

Figure 5. Helix-coil transition in DNS-poly Glu.

of polylysine (poly $Glu^{97}Lys^{3}$, molecular weight = 25,000, was used as a model for polyglutamic acid in order to provide a few lysine residues for conjugation with the fluorescent dyes). Just after the coil-helix transition was complete, the helices aggregated. This behavior is shown quite distinctly by polarization of fluorescence measurements (Figure 5). The entire molecule became helical, since its rotational relaxation time was the same as that predicted for a completely rigid particle (Table 5). Thus, glutamic acid residues form more complete helices than lysine residues, but the glutamic acid helix can also be destroyed by 8 M urea within 3 hours. The helix-coil transition measured by a variety of other techniques is shown in Figures 6 and 7 and the pertinent data for all of the measurements are summarized in Table 3. In water, the helix reached maximal rigidity (pH[50%] = 5.80) at approximately the same time that all of the residues assumed the helical conformation (pH[50%] = 6.0), and the same was true in 0.2 M NaCl + 0.1 M buffer (pH[50%] = 5.05 and 5.2, respectively). In both solvents, the titration changes followed the changes in polarization and optical rotation. The aggregation of the helices was shown dramatically by polarization of fluorescence measurements (Figure 5), whereas optical rotation was the only other method that gave any indication of aggregation. In the latter measurement, aggregation caused a small decrease in the specific rotation after all of the residues went into the helical conformation (Figure 6). These data demonstrate the molecular phenomena which each method emphasizes: polarization is basically a hydrodynamic measurement which is sensitive to changes in rigidity and aggregation, whereas optical rotation

Table 3. Helix-coil transition in poly Glu[a]

	Polarization[b]		Optical Rotation		Titration		[η]	
Solvent	50%	90%	50%	90%	50%	90%	50%	90%
water	5.80	5.62	6.0	5.8	5.7[c]	5.5		
0.2 M NaCl	5.05	4.82	5.2[d]	5.0	5.1	4.9	4.1	<4.1
3.0 M NaCl	e	e	4.6	4.4			4.3	4.2
	NMR		ΔH		ΔV		$\overline{\Delta n}$	
water	4.8	4.3			5.7[h]	5.0	5.7[h]	4.9
0.2 M NaCl	5.2[f]	5.0	5.1[g]	4.7	5.2	4.6	5.2	4.6

[a]References (8, 9, 10-15); [b]poly $Glu^{97}Lys^{3}$; [c]0.005 M NaCl; [d]4.8 in 0.2 M NaCl in H_2O and 5.4 in 0.2 M NaCl in D_2O by $[m']_{233}$ (16); [e]none detected; [f]1.0 M NaCl; [g]0.1 M KCl; [h]0.01 M NaCl.

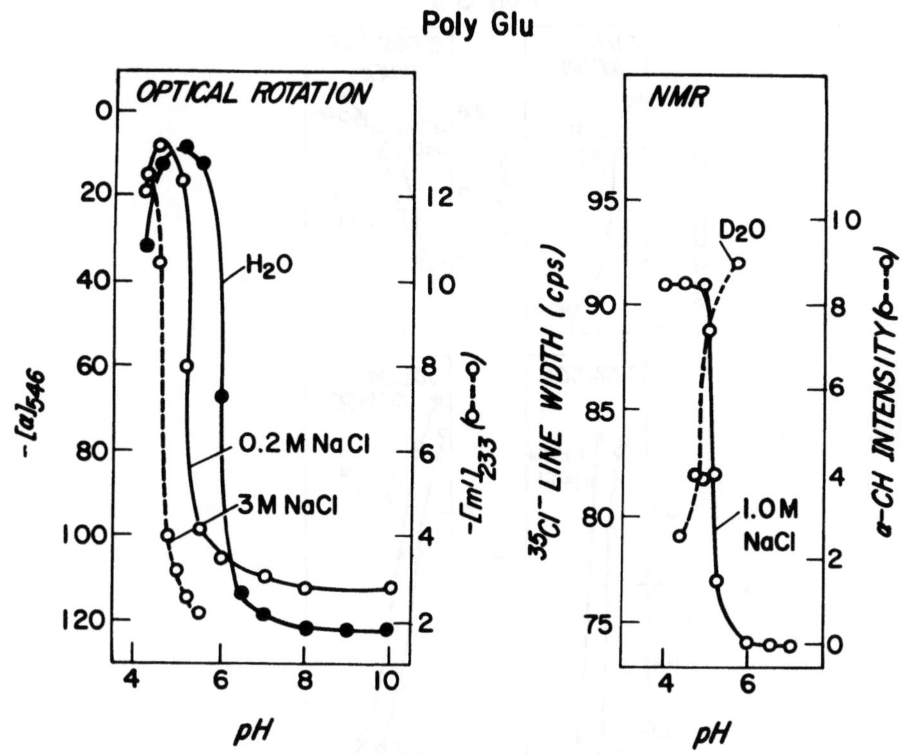

Figure 6. Helix-coil transition in poly Glu (9, 12-14).

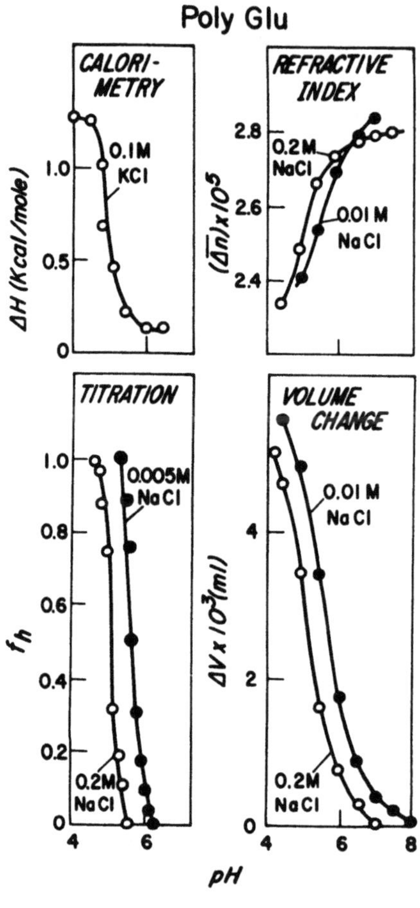

Figure 7. Helix-coil transition in poly Glu (8, 10, 11, 15).

emphasizes conformational changes and is not very sensitive to aggregation.

The properties of the copolymer of glutamic acid and lysine (molecular weight = 48,000) are dominated by the behavior of the glutamic acid residues (Figures 8 and 9). The hydrodynamic behavior of the molecule as a function of pH showed a clearly defined helix-coil transition for the glutamic acid residues, but only a vestigial transition for the lysine residues (Figure 8). In water, the maximal rigidity (pH[50%] = 6.12) occurred before all of the residues went into the helical conformation (pH[50%] = 5.2), but in 0.2 M NaCl + 0.1 buffer, the changes in rigidity (pH[50%] = 5.00) and in conformation (pH[50%] = 5.2) paralleled each other. The data for the transitions are summarized in Table 4. Since the titration behavior of each type of residue in the copolymer is influenced by a variety of complex environmental factors, it is not possible to correlate titration behavior with that measured by polarization of fluorescence or by optical rotation.

The copolymer molecule can form only short helical segments of predominantly one kind of residue, since the growing helix soon encounters a sequence of the opposite type of residue which will prevent further helix formation: thus, the molecule will be composed of alternating segments of helix and coil. This conclusion is strengthened by the finding that the rotational relaxation time of the molecule was only 86% of that of a comparable rigid particle (Table 5). Therefore, the copolymer

Table 4. Helix-coil transition in poly $Glu^{60}Lys^{40}$ [a]

	Polarization		Optical Rotation	
Solvent	50%	90%	50%	90%
water:Glu	6.12	5.7	5.2	4.4
Lys	13.3[b]			
0.2 M NaCl:Glu	5.00	4.6	5.2[c]	4.4
Lys	13.7[b]		11.0[c]	12.0
3.0 M NaCl:Glu	4.80	4.2		
Lys	13.6[b]			
	Titration[d]		[η][b]	
0.2 M NaCl:Glu	4.18	3.4	2.3[e]	1.2[e]
Lys	10.72	11.8	11	12.5

[a]References (7, 17, 18); [b]approximate values; [c]poly $Glu^{50}Lys^{50}$;
[d]0.15 M KCl; [e]in 3 M NaCl: 2, 10.5 (50%) and 1, 11 (90%).

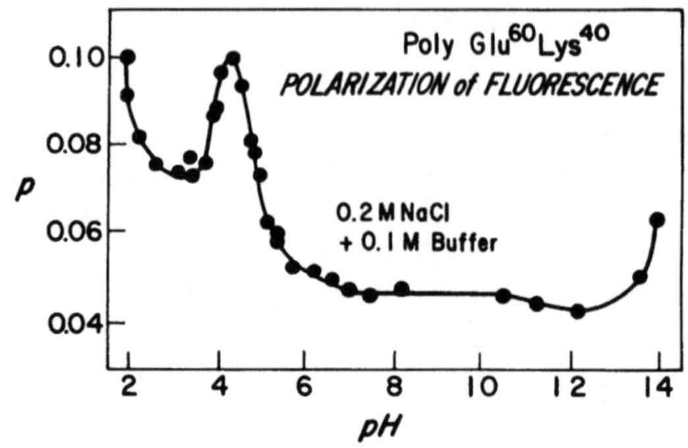

Figure 8. Helix-coil transition in DNS-poly $Glu^{60}Lys^{40}$

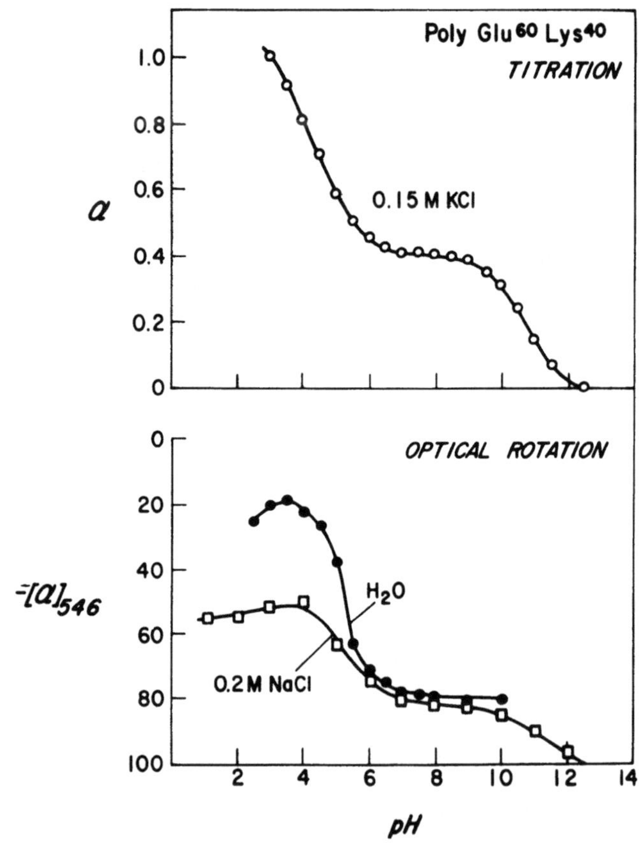

Figure 9. Helix-coil transition in poly $Glu^{60}Lys^{40}$ (17, 18).

has the same type of structure (interrupted helices) that polylysine has, and, indeed, the behavior of both macromolecules is similar in many respects. The helical segments were less stable than in polylysine, however, since they could be destroyed by 6 M urea within 3 hours.

The presence of interrupted helices in a macromolecule causes several characteristic types of behavior. Both in the copolymer and in polylysine, the maximal rigidity of the molecule was attained when the majority of the residues that were capable of forming the helix had done so: a few more helical segments did not significantly alter the hydrodynamic behavior of the molecule, but they did change its optical rotatory behavior. Secondly, the addition of salt delayed the onset of maximal rigidity in poly $Glu^{60}Lys^{40}$ and in poly Lys so that it coincided with the entrance of the residues into the helical conformation. The explanation for this finding probably lies in the influence of the salt on the non-helical portions of the molecule: a localized polyelectrolyte effect. In water, the rigidity of the molecule is due both to the helical segments and to repulsion between like charges in the coil portions of the molecule. In polylysine, the coil portions would be lysine residues that were not in the helical segments, and in the copolymer they would be the lysine residues that lay between the predominantly glutamic acid helices. In salt solutions, the electrostatic repulsions in the coil regions would be neutralized to a large extent (Figure 3), and this would cause a decrease in the overall rigidity of the molecule. This mechanism would leave the helical segments as the sole source of the rigidity; hence, the development of rigidity would parallel the amount of helix in the molecule. Finally, the onset of aggregation occurred at pH's considerably after the completion of the coil-helix transition: approximately pH 13.5 for poly Lys and approximately pH 2 for poly $Glu^{60}Lys^{40}$. In addition, there is a decrease in polarization between the completion of helix formation and the increase due to aggregation, and this effect is most marked in the copolymer. The magnitude of this decrease in polarization may correlate with the amount and/or length of the coil segments, which are longer in the copolymer than in polylysine; such a decrease in polarization is not seen in polyglutamic acid which is completely helical. The decrease in polarization could be due to contraction in the size of the polymer molecule caused by intramolecular interactions between the non-polar helical segments prior to aggregation.

Some of the characteristics of the helix and the coil obtained by polarization of fluorescence measurements are summarized in Table 5. The ratio of the measured to the calculated relaxation time was unity when the whole molecule was helical (poly $Glu^{97}Lys^{3}$) and less than unity for the polypeptide.

with interrupted helices. In all cases, the rotational relaxation time was longer for the helix than for the coil: this indicates that the former is a more rigid structure. The transition temperature, which represents the onset of internal rotation, was higher for the helical form, since it takes a larger amount of thermal energy to disrupt the highly organized helical structure than to induce rotation in the coil form. The low transition temperatures for both the helical and coil forms of polyglutamic acid may be due, at least in part, to the relatively small size of the molecule.

Table 5. Structural parameters of the helix and the coil[a]

Polymer	Molecular Weight	Helix pH	Helix ρ_h^5 (ns)	Helix ρ_h^5/ρ_o	Helix T_T(°C)
poly Lys	51,000	11.3	28	0.36	32
poly $Glu^{97}Lys^{3}$[b]	25,000	4.6	31	1.00	18
poly $Glu^{60}Lys^{40}$	48,000	4.2	53	0.86	42

Polymer	Coil pH	Coil ρ_h^5(ns)	Coil ρ_h^5/ρ_o	Coil T_T(°C)	ρ_h^5 (helix) / ρ_h^5 (coil)
poly Lys	7.3	8	0.11	13	3.5
poly $Glu^{97}Lys^{3}$	7.0	<1	<0.03	<0	>30
poly $Glu^{60}Lys^{40}$	7.2	12	0.19	10	4.4

[a]Solvent is 0.2 M NaCl + 0.1 M buffer (citrate, phosphate, carbonate); [b]used as a model for poly Glu to provide lysine residues for conjugation with a fluorescent dye.

Summary

Polarization of fluorescence measurements reflect the hydrodynamic properties of macromolecules; thus, they are particularly useful for investigating the helix-coil transition, the structure of helices, and aggregation. Polarization offers sensitivity comparable to a variety of optical methods for studying helix formation, and it is more useful for studying aggregation. The polarization of fluorescence and the optical methods best complement one another in investigating the structure of the helix. The interaction between the dye and the macromolecule to which it is conjugated and the influence of the solvent, especially the pH, on the lifetime of the dye are major experimental considerations in using the polarization of fluorescence method.

In the helical form, polylysine and a copolymer of glutamic acid and lysine contain multiple discrete helical segments (interrupted helices), whereas the entire polyglutamic acid molecule is in the helical conformation. The glutamic acid residues, which are strong helix formers, dominate the behavior of the copolymer in solution. The molecules containing interrupted helices do not aggregate until the pH is considerably past the point of maximal helix formation. In contrast, the polyglutamic acid helices aggregate just after all of the residues go into the helical conformation. The behavior of the molecules with interrupted helices can be explained by the influence of the coil segments on the overall properties of the polypeptide chain. In addition to the coil-helix transition in alkali, the polylysine molecule also undergoes a structural change in acid; this change may be due to adsorption of hydrogen ions by the polylysine molecule.

References

1. Gill, T.J. III, in Protides of the Biological Fluids, 16th Colloquium (1968), ed. H. Peeters, Pergamon, 1969, p. 21.

2. Weber, G., Biochem. J. 51, 155 (1952).

3. Gill, T.J. III, McLaughlin, E.M., and Omenn, G.S., Biopolymers 5, 297 (1967).

4. Katchalsky, A., Shavit, N., and Eisenberg, H., J. Polymer Sci. 13, 69 (1954).

5. Applequist, J. and Doty, P., in Polyamino Acids, Polypeptides, and Proteins, ed. M. A. Stahmann, Univ. of Wisconsin Press, Madison, 1962, p. 161.

6. Noguchi, H., Biopolymers 4, 1105 (1966).

7. Doty, P., Imahori, K. and Klemperer, E., Proc. Nat. Acad. Sci. U.S. 44, 424 (1958).

8. Ciferri, A., Puett, D., Rajagh, L. and Hermans, J., Jr., Biopolymers 6, 1019 (1968).

9. Bradbury, E.M., Crane-Robinson, C., Goldman, H. and Rattle, H.W.E., Biopolymers 6, 851 (1968).

10. Nagasawa, M. and Holtzer, A., J. Am. Chem. Soc. 86, 538 (1964).

11. Rialdi, G. and Hermans, J., Jr., J. Am. Chem. Soc. 88, 5719 (1966).

12. Blout, E.R. and Idelson, M., J. Am. Chem. Soc. 80, 4631 (1958).

13. Yang, J.T. and Iizuka, E., Biochemistry 4, 1249 (1965). (see also Yang, J.T. and McCabe, W.J., Biopolymers 3, 209 (1965)).

14. Bryant, R.G., J. Am. Chem. Soc. 89, 2496 (1967).

15. Noguchi, H. and Yang, J.T., Biopolymers 1, 359 (1963).

16. Appel, P. and Yang, J.T., Biochemistry 4, 1244 (1965).

17. Gould, H.J., Gill, T.J. III and Kunz, H.W., J. Biol. Chem. 239, 3083 (1964).

18. Blout, E.R. and Idelson, M., J. Am. Chem. Soc. 80, 4909 (1958).

This paper is number 6 in the series "Studies of Polypeptide Structure by Fluorescence Techniques"; the preceding paper is reference 1. The work was supported by grants from the National Science Foundation (GB-5866 and GB-8379) and from the National Institutes of Health (HE-1771 and 5-T01-5274). Thomas J. Gill III is the recipient of a Research Career Development Award (K3-AM-5242).

STRUCTURAL STUDIES OF THE CHOLESTERIC MESOPHASE

Furn F. Knapp* and Harold J. Nicholas

Institute of Medical Education and Research, and Department of Biochemistry, St. Louis University School of Medicine, St. Louis, Missouri 63104

INTRODUCTION

Mesomorphic substances were discovered in the latter part of the last century and Reinitzer is generally accredited with the early work on cholesteric liquid crystals (34). The cholesteric mesophase is unique in that it can scatter white light to form brilliant colors. Nearly all cholesteric substances are esters of cholesterol. These compounds have a negative optical activity with an unusually large rotation. This large rotatory power is due to the unusual arrangement of molecules and not to the asymmetry of the individual steroid molecules. The colors formed by the cholesteric mesophase are due to the scattering of white light (circular dichroism). While one circularly polarized component of the incident beam is reflected, the other is scattered, resulting in the formation of the classical iridescent colors generally associate with this mesophase (13).

The arrangement of molecules in the cholesteric mesophase has naturally been the subject of numerous studies. The molecules are arranged in layers with their long axes parallel to the molecular plane. The orientation of the molecular axes is displaced regularly in successive layers such that the overall change in orientation describes a helix (18).

*NASA Predoctoral fellow. This work represents part of the research partially fulfilling the requirements for the Degree of Doctor of Philosophy at St. Louis University.

Oseen originally suggested the helical structure to explain the unique optical properties of cholesteric substances (30). Such a model system has more recently been used by de Vries (10). Assuming a stacking of parallel plates of molecules containing random Bragg scattering sites Fergason developed an equation with which to calculate the pitch of the helicoidal structure (14). The effect of chain length of the fatty acyl moiety of various cholesteryl esters on the helical pitch has been studied by Adams and co-workers (1,2). The observed dependence of pitch on acyl chain length was attributed to geometric factors.

Although cholesteric compounds were originally believed to represent a special class of nematic liquid crystals (17), Gray proposed that they more closely resembled the smectic mesophase (18). Recent calorimetric data, however, have shown the cholesteric mesophase to indeed be more closely related to the nematic mesophase (33). Several calorimetric techniques (DSC, differential scanning calorimetry, or DTA, differential thermal analysis) and light intensity methods have been used to study the properties of cholesteric substances (4,5,6).

Thus, while there have been many studies of the physical properties of the cholesteric mesophase using a limited number of compounds, there has been no systemmatic investigation to determine the critical structural parameters required for mesophase formation. The transition temperatures of the homologous series of fatty acid esters of cholesterol have been reported by Gray (18). These have been recently determined using calorimetric methods and good agreements in the two techniques have been found (5). It is known that mesophase formation is dependent upon critical structural requirements and that small structural changes can disrupt mesophase formation (14,19). The esters are broad, flat molecules and seem to be associated by weak attractive forces. The alkyl side chain of the steroid molecule projects from the molecular plane and is probably responsible for the twisted structure (16).

Studies of steric effects with nematic compounds have been reported (9) and the effect of terminal substitution on the nematic properties of some benzyleneanilines have been studied (7). Wiegand investigated the effect of nuclear unsaturation on the cholesteric properties of the benzoic acid esters of various cholestanes, cholestenes and cholestadienes (37). Recent studies have demonstrated the cholesteric properties of the palmitic acid esters of 24ξ-methyl cycloartanol, 4α,14α,24ξ-trimethyl-9,19-cyclocholestan-3β-ol and cycloartenol (21,23). These studies have been expanded by the synthesis of the homologous series of fatty acid esters of these three triterpenes and their corresponding unsaturated forms. The palmitic acid esters of a number of representative triterpenes and sterols have also been prepared. The results of these studies suggest very critical roles played by the nucleus, fatty acyl residue and side chain in mesophase formation.

EXPERIMENTAL SECTION

General

All solvents and reagents were analytical grade, purchased from Fisher Scientific Company. The acyl chlorides were obtained from the Eastman-Kodak Company and were of the highest purity available. Thin-layer chromatography (TLC) was performed on silica gel G spread 250 microns thick on 20 cm by 20 cm glass plates prepared in the usual manner (22). Triterpene and steryl esters were chromatographed on silica gel G plates impregnated with 12 per cent silver nitrate. The solvent systems used for the development of the spotted plates were as follows: S-1, for triterpene and steryl esters, hexane-ether (93:7, v/v); S-2, for free triterpene and sterol alcohols, 2,2,4-trimethyl pentane-ethyl acetate-acetic acid (40:20:0.4, v/v/v). Both homologous triterpene esters and saturated-unsaturated analogs could be separated using S-1 and the silver nitrate plates (cycloartenol acetate, R_f 0.27; cycloartenol palmitate, R_f 0.54; cycloartanol palmitate, R_f 0.90). Using S-2 triterpenes and sterol alcohols were separated relative to the number of C-4 methyl groups (4-des methyl sterols such as β-sitosterol, R_f 0.48; 4α-methyl triterpenes like cycloeucalenol, R_f 0.54; 4,4-dimethyl triterpenes such as cycloartenol, R_f 0.62). The spot colors were detected by heating the plates after they were sprayed with anisaldehyde reagent. The gas-liquid chromatographic analyses (GLC) were performed with a Barber-Colman Model 5000 instrument utilizing a hydrogen flame detector. The phase and support material (Gas Chrom Q, 100/120 mesh) were purchased from Allied Science Laboratories. Triterpene and steryl esters were separated on a 0.50 meter column of 1% SE-30 prepared as described by Kuksis et al (29). The column was operated isothermally at 290° with a carrier gas (N_2) flow rate of 100 cc/min (26). Triterpene and sterol alcohols were chromatographed on a 2 meter column of 1% SE-30 or 3% XE-60 at 238° and 248°, respectively. In this case the carrier gas was Argon at 46 cc/min.

Table I. Esters of 24-Methylene Cycloartanol and 24ξ-Methyl Cycloartanol

Ester	Melting Point, °C	Phase Transition Temperature, °C	Associated Color
Acetate	111-112	...	...
Reduced*	110-111.5	...	...
Butyrate	72-73	sm→iso 110.5	...
Reduced	77.5-78	sm→iso 114.5	...
Hexanoate	106-107	sm→iso 112	...
Reduced	107-108	sm→iso 122	...
Octanoate	74-75	sm→iso 101.5	...
Reduced	63-64	sm→iso 83.5	faint blue
Decanoate	65-67	sm→iso 55	...
Reduced	82.5-84	sm→iso 83.5	...
Laurate	54-57	...	...
Reduced	58-60	ch→iso 65.5	green or violet
Myristate	61-62	...	...
Reduced	65-66	ch→iso 67.5	green or violet
Palmitate	57-59	...	...
Reduced	64-65	ch→iso 64	violet
Stearate	55-63	...	...
Reduced	65-68	...	...

*Reduced refers to the C-24 dihydro form, 24ξ-methyl cycloartanol.

Table II. Esters of Cycloeucalenol and 4α,14α,24ξ-Trimethyl-9,19-cyclocholestan-3β-ol

Easter	Melting Point, °C	Phase Transition Temperature, °C	Associated Color
Acetate	106-107	...	...
Reduced*	109-110	...	...
Butyrate	...	...	Red
Reduced	...	...	...
Hexanoate	73-74	ch→iso 59	Red
Reduced	89-90	ch→iso 77	Red
Octanoate	77.5-78	ch→iso 61	Green
Reduced	61-63	ch→iso 77	Red or Blue
Decanoate	75-77	ch→iso 56	Green
Reduced	76-77	ch→iso 69	Blue-Green
Laurate	63-65	sm→iso 55	...
Reduced	58-59.5	ch→iso 68	Dull Blue
Myristate	58-59	sm→iso 49	...
Reduced	63-67	ch→iso 61	Dull Blue
Palmitate	60-62	...	...
Reduced	62-64	ch→iso 52	Green or Violet
Stearate	76-77	sm→iso 56	...
Reduced	75-76	sm→iso 69	...

*Reduced refers to the C-24 dihydro form, 4α,14α,24ξ-Trimethyl, 9,19-cyclocholestan-3β-ol.

Melting Points and Phase Transitions

Melting points and phase transition temperatures were determined with a Nalge-Axelrod hot-stage polarizing microscope. In some cases the melting points were also determined with a Thomas-Hoover uni-melt apparatus. The phase transition temperature is definied as that temperature at which the birefringence disappeared from the melt, generally coincident with the temperature at which birefringence appeared upon cooling the isotropic liquid. Routinely, phase transition temperatures were determined on cooling. When a viscous, birefringent liquid is observed through the polarizing microscope it may represent either a smectic or cholesteric mesophase. A color is often not seen with these cholesteric mesophases when viewed through the polarizing microscope when the color is easily detected with the naked eye. It is known that cholesteric mesophases can adopt a homeotropic texture which is generally either a dull blue-gray or purple to the naked eye but optically extinct when viewed through crossed polaroids (18). In addition, cholesteric mesophases may even display no color at all, as is the case with cholesteryl stearate (19). The mesomorphic compounds described in this paper have only been examined through the polarizing microscope. For this reason we have not differentiated between a colorless cholesteric mesophase and a smectic mesophase. Thus, mesophases that are colored to the naked eye have been termed cholesteric and those which are colorless have been designated as smectic. If the optical rotatory power of the mesophases in question had been determined, the colorless cholesteric mesophases could have been easily detected. Future studies will be directed toward resolving these cases. As indicated in Tables I through III, the colors of the cholesteric mesophases often varied. The colors were dependent upon both the light source and the background material. Most of the cholesteric mesophases are monotropic with respect to the crystalline solid (e.g. colors are observed only upon cooling from the isotropic liquid). The formation of colors by many of these cholesteric mesophases is extremely sensitive to impurities. As an example 24-dihydro cycloeucalenol palmitate must be highly purified before the colors associated with the Gradjean plane are observed. On the other hand, the color associated with the mesophase of cycloartenyl palmitate is observed when this ester is present in a mixture of non-mesomorphic esters to the extent of only 10 per cent (25). The following abbreviations are used throughout this paper: iso = isotropic liquid, sm = smectic mesophase, ch = cholesteric mesophase; ch→iso means a transition from the cholesteric mesophase to the isotropic liquid. Similarly, sm→ iso means a transition from the

Table III. Esters of Cycloartenol

Ester	Melting Point, °C	Phase Transition Temperature, °C	Associated Color
Acetate	119-120	...	...
Butyrate	85-87	...	...
Hexanoate	...	...	...
Octanoate	56-58	ch→iso 59	Blue or Green
Decanoate	69-71	ch→iso 59	Blue or Green
Laurate	52-53.5	ch→iso 58	Deep Blue or Green
Myristate	49-50	ch→iso 56	Deep Violet
Palmitate	62-64	ch→iso 51	Green or Violet
Stearate	63-64	sm→iso 56	...

smectic mesophase to the isotropic liquid.

Triterpenes and Sterols

Plant material was obtained from the following sources: banana peels (Musa sapientum), obtained from bananas purchased locally; Strychnos nux-vomica seeds, purchased from S. B. Penick and Company; Fucus vesiculosus, obtained from the Marine Biological Laboratory, Woods Hole, Massachusetts; Salvia sclarea was grown locally. Cycloartenol was isolated from Strychnos nux-vomica seeds (25); 24-methylene cycloartanol and cycloeucalenol from banana peel as previously described (22); fucosterol from Fucus vesiculosus, and β-sitosterol from Salvia sclarea. Ergosterol, cholesterol and stigmasterol were purchased from Sigma Chemical Company. Cholestanol, stigmastanol and 24α-methyl cholestanol were prepared by catalytic reduction of cholesterol, stigmasterol and ergosterol, respectively. Lanosterol was purchased from Sigma Chemical Company and found by GLC to contain approximately 50 per cent lanosterol, 49 per cent 24-dihydro lanosterol and 1 per cent more polar sterols. These components were conveniently separated by reaction of the mixture with palmitoyl chloride followed by chromatography of the mixed palmitates on a column containing Celite-silica gel-silver nitrate (10:10:8) as described elsewhere (24). The 24-dihydro lanosteryl palmitate was eluted from the column with 4 per cent benzene in petroleum ether and the lanosteryl palmitate with 25 per cent benzene in petroleum ether. The more polar esters were retained on the column and were not further investigated.

Procedures

Triterpene and steryl esters were prepared as previously described (21,23). The alcohol (30-60 mg) was dissolved in anhydrous benzene. A small amount of pyridine was added and the mixture refluxed with a 1.5 molar excess of the acyl chloride under anhydrous conditions for one hour. After addition of 100 ml of ether the mixture was washed successively with 5 per cent hydrochloric acid and water and dried over anhydrous sodium sulfate. Evaporation of the solvent left a residue which was added in benzene to a short alumina column. The ester was eluted from the column with benzene and crystallized from either acetone or methanol-ether. In general long chain esters crystallized best from the latter solvent. Esters were catalytically reduced in ethyl acetate solution by shaking under a pressure of 40 psi of hydrogen for one hour.

The absolute configuration of the C-24 methyl group in the C-24 dihydro cycloeucalenol and 24ξ-methyl cycloartanol esters is not known.

RESULTS AND DISCUSSION

The present study was initiated by our original observation that while 24-methylene cycloartanyl palmitate (Figure I) is not a liquid crystal, the C-24 dihydro form (24ξ-methyl cycloartanol) forms a brilliant blue mesophase (21). One objective of this work was to further study the disruptive effect of the C-24 (31) double bond. The homologous series of fatty acid esters of 24-methylene cycloartanol, cycloartenol and cycloeucalenol have been prepared to study the effect of side chain alterations, nuclear modifications and acyl chain length on mesophase formation. These extensive studies have been limited to these three 9,19-cyclopropane triterpenes. In an effort to investigate the potential that other triterpene and steryl esters may form a cholesteric mesophase, the palmitic acid esters of representative members of these two structural groups were prepared. It should be stressed, however, that the lack of mesomorphic behavior of an ester of palmitic acid does not necessarily imply that either higher or lower homologs or some other esters could not form a liquid crystalline transition state. Synthesis of esters of palmitic acid merely represented a convenient means of studying the possible cholesteric properties of esters of many compounds. The palmitic acid esters were choosen since for the triterpenes previously studied the longer chain esters form a well-defined cholesteric mesophase.

Tetracyclic Triterpenes

Phase Transitions. Although the details of all of the phase transitions are too numerous to list, the most interesting and unique cases are summarized below. These will acquaint the interested reader with the types of mesophases formed by the esters discussed in this paper. All of the esters were of high purity as determined by TLC and GLC. Another criterion of purity was the near coincidence of transition temperatures on both heating and cooling (18,19). All of the compounds studied were reasonably pure by this criterion; these temperatures in all but a few cases differed by no more than two degrees. Saturated esters generally melted at a higher temperature than the corresponding unsaturated forms. There is a seemingly erratic change in melting

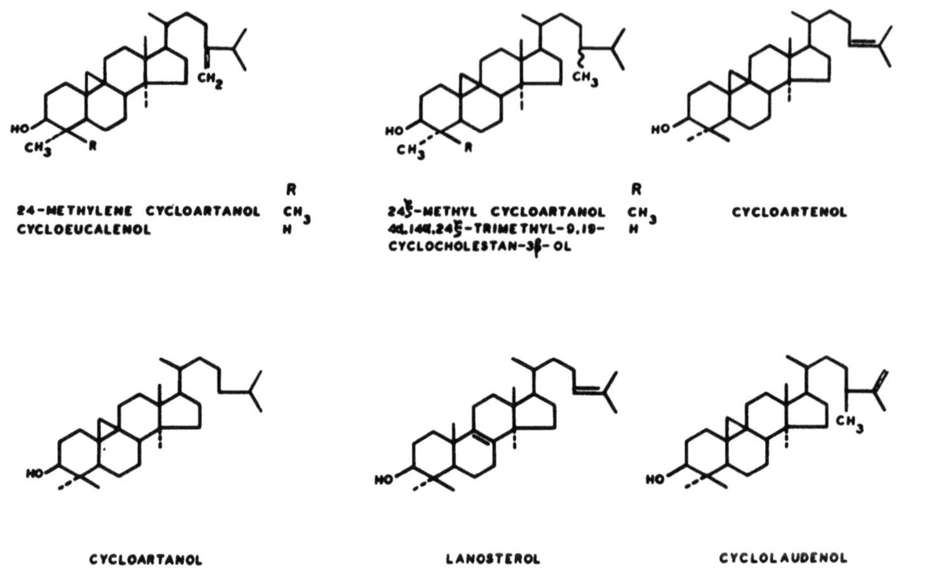

FIGURE I. TETRACYCLIC TRITERPENES

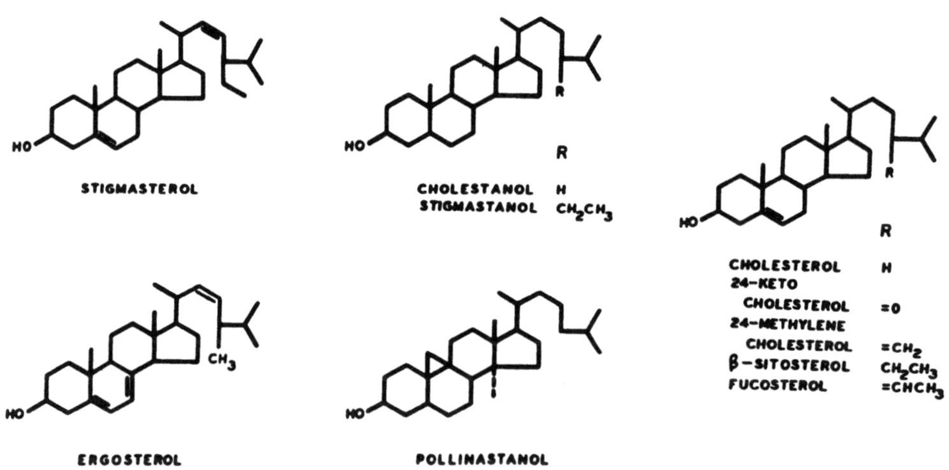

FIGURE II. STEROLS

point with changing acyl chain length. This is also the case with cholesteryl esters (18,19). The smooth curve relationship between transition temperature and chain length is not affected, however, again pointing to high purity (Figure III). The transitions described below are selected from those summarized in Tables I through III.

1. 24ξ-methyl cycloartanyl palmitate melted at 61-62° to a highly viscous, birefringent liquid. The color was purple with the polaroids crossed at 90°, but changed to green and rose as the eyepiece was rotated. At 63.5° a focal conic texture appeared, with the birefringence disappearing at 64° (cholesteric→ isotropic). The birefringence reappeared upon cooling to this same temperature (isotropic→ cholesteric). On the cooling cycle tiny batonnets appeared. The cholesteric mesophase could be cooled to 49° at which time agitation resulted in rapid crystallization.
2. The smectic mesophase of the hexanoate of 24-methylene cycloartanol and the corresponding ester of 24ξ-methyl cycloartanol display a beautiful mosaic type of birefringence similar to that illustrated for the type B smectic mesophase as described by Arnold and co-workers (3).
3. 24-Dihydro cycloeucalenyl hexanoate consisted of multicolored striated crystals when viewed through the polarizing microscope. It melted to an isotropic liquid at 92-93° and cooled to a bright birefringent liquid at 79° (isotropic→ cholesteric). When agitated, the mesophase changed to a deep red color. When viewed with the naked eye upon cooling from the isotropic liquid, the mesophase displayed brilliant colors varying from blue to deep red before solidification.
4. Several of the cholesteric mesophases described here respond rather dramatically to solvent vapors. This is apparently similar to the phenomenon mentioned by other workers (2,13,15). The palmitic acid ester of 24-dihydro lanosteryl palmitate forms a blue cholesteric mesophase. When a small amount of acetone is mixed with the isotropic melt, it cools to a blue mesophase that is stable at room temperature. The color of such mesophases has remained stable for several months. Cycloartenyl palmitate forms a similar stable mesophase. When heated, this system cools through a series of brilliant colors back to the stable blue color (Isotropic→red→orange→yellow→green→blue).

Triterpene Esters. Synthesis of the homologous series of esters of 24-methylene cycloartanol and its reduced form, 24ξ-methyl cycloartanol (Table I) further demonstrates the disruptive effect of the C-24 (31) double bond. While no esters of the unsaturated triterpene are cholesteric, several esters of the C-24 dihydro form display definite colors. These esters include the laurate, myristate and palmitate.

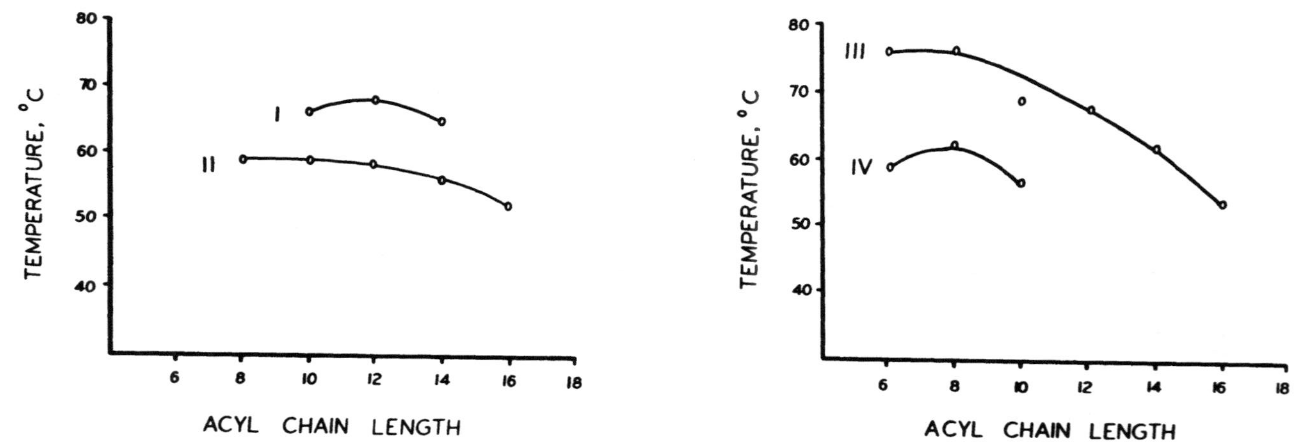

FIGURE III

CHOLESTERIC → ISOTROPIC TRANSITION TEMPERATURES OF ESTERS OF

I 24ξ-METHYL CYCLOARTANOL
II CYCLOARTENOL
III $4\alpha,14\alpha,24\xi$-TRIMETHYL-9,19-CYCLOCHOLESTAN-3β-OL
IV CYCLOEUCALENOL

The octanoate seemed to display a slight fugitive blue color and no color was observed with the decanoate. There should be no interruption in the cholesteric properties of a homologous series. Thus, if the laurate and octanoate are both cholesteric the decanoate should be cholesteric also. The decanoate was not colored, however, and since many milky white substances often appear blue to the naked eye, the octanoate has been designated as forming a smectic mesophase. The cholesteric isotropic transition temperatures of the three cholesteric esters are plotted as a function of acyl chain length in Figure III (I). Several esters of both the saturated and unsaturated triterpenes form a smectic mesophase. These include the butyrate and hexanoate esters of both triterpenes. In addition, neither the acetate, reduced acetate, laurate, myristate, stearate nor reduced stearate are mesomorphic. These results have also been discussed elsewhere (23).

To further investigate the disruptive effect of the C-24 (31) double bond and also the effect of a nuclear modification, the series of esters of cycloeucalenol were prepared (Figure I). This triterpene differs from 24-methylene cycloartanol in that it has only one methyl group at C-4 (4α-methyl). The properties of the esters of this triterpene and its reduced form (4α,14α,24ξ-trimethyl-9,19-cyclocholestan-3β-ol) are summarized in Table II. As noted in a preliminary communication (23), the palmitic acid ester of cycloeucalenol is not mesomorphic, the isotropic melt passing directly to the crystalline solid upon cooling. The C-24 dihydro ester forms a green cholesteric mesophase, further demonstrating the disruptive effect of the C-24 (31) double bond. The laurate, myristate, stearate and reduced stearate all form a smectic mesophase. In this regard the non-mesomorphic behavior of cycloeucalenyl palmitate must be an odd result. The laurate and myristate of the reduced triterpene also form a slightly blue mesophase, and are thus designated as cholesteric. With the shorter chain esters the situation is more complex than originally anticipated. The butyrate, hexanoate, octanoate and decanoate of the unsaturated triterpene are cholesteric, the colors varying from red (butyrate and hexanoate) to blue-green (decanoate). This is similar to the colors detected in the plane texture of the cholesteryl esters, the light scattered from the mesophase of the longer chain esters of cholesterol staying toward the blue end of the spectrum (19). Similarly, shorter chain esters (acetate, etc.) display colors toward the longer end of the spectrum. Although brilliant colors were observed with the butyrate ester of cycloeucalenol, it has not yet been satisfactorily crystallized. Its melting point and phase transition data are therefore not included in Table II. The hexanoate, octanoate and decanoate of the C-24 dihydro triterpene are also cholesteric. Neither the acetate nor the reduced acetate are mesomorphic. The cholesteric isotropic transition temperatures of esters of both the saturated and unsaturated triterpene are plotted in Figure III (III and IV). In analogy with those homologous series which have been studied, the transition temperatures steadily decrease with increasing acyl chain length (18,19). The transition temperature of the reduced decanoate is the only value out of line. The esters of cycloeucalenol differ markedly with those of 24-methylene cycloartanol where no esters of the unsaturated triterpene are cholesteric. The laurate, myristate, palmitate, stearate and reduced sterate of this triterpene are not even mesomorphic. Thus, the disruptive effect of the C-24 (31) double bond is regulated by both the nuclear structure and the fatty acyl chain length. It would be interesting to see if a similar role is played by the C-24 (31) double bond in esters of 24-methylene cholesterol (Figure II).

The structure of cycloartenol (Figure I) afforded an opportunity to study the effect of a C-24 (25) double bond while not altering the 9,19-cyclolanostan nuclear structure. Preparation of the homologous series of esters of this triterpene (Table III) demonstrate that the C-24 (25) double bond does not disrupt mesophase formation, at least when there is no substituent at C-24. The octanoate, decanoate, laurate, myristate and palmitate esters are cholesteric, the colors all being in the shorter end of the spectrum (Blue to Green). The hexanoic acid ester has not yet been available for analysis. These data are plotted in Figure III (II), and again demonstrate the decrease in transition temperature with increasing chain length. Although the complete series of esters of the C-24 dihydro triterpene have not yet been prepared, cycloartanyl palmitate forms a faint gray-green mesophase (m.p. 55-55.5^{0}, ch$\rightarrow$iso 72^{0}).

These studies indicate very complex structural requirements for cholesteric mesophase formation. One significant result is the disruptive influence of the C-24 (31) double bond, although this effect is augmented by other structural features, including fatty acyl chain length and nuclear structure. Dipole moments are extremely important with regard to the mesomorphic properties of many compounds (19). The moment associated with the 24-methylene bond is obviously not responsible for the inability of certain esters from forming a cholesteric mesophase since the shorter chain esters of cycloeucalenol are cholesteric. The disruptive effect must therefore be steric in nature. Since it is not known just how the side chain is arranged in the cholesteric mesophase with respect to both intra- and intermolecular interactions, there are several explanations for this steric effect. If the side chain is fully extended, the degrees of freedom lost by the presence of the double bond in the side chain could prevent it from assuming an orientation required for the maximal interaction of molecules. This steric effect, however, is only restricted to longer chain esters so the fatty acyl chain length must also be involved in this series of interactions. If the side chain were fully extended in the most energetically favored near-staggered configuration, it is difficult to observe any interaction of the C-31 hydrogen atoms with the nucleus. If this is the orientation the side chain does assume in the mesophase, any interaction must be intermolecular. It has been suggested that the side chain extends out of the molecular plane and is responsible for the helical structure of the cholesteric mesophase (15). In addition, the side chain has been shown by other workers to be important in mesophase formation (32). It is also possible that the side chain could extend from the molecular plane in such a way that the C-31 hydrogen atoms could interact with either the D-ring or the hydrogen atoms on C-21. In this conformation the side chain could not be fully extended. Such an interaction could be relieved by rotation of the C-24 (31) bond as in the 24ξ-methyl cycloartanol esters. This would in a sense relieve the interaction by the eclipsing of the hydrogen atoms. Rotation about the C-24 (31) double bond is of course not allowed. This would explain why no esters of 24-methylene cycloartanol are cholesteric. Since the shorter chain esters of cycloeucalenol are cholesteric, the nuclear structure must also effect mesophase formation, and intramolecular interactions must be of minor importance. The intermolecular interactions must therefore involve the side chain and nucleus. Further interpretation of these data must of course await relevant X-ray or other diffraction data. It would be interesting to see if any esters of a $\Delta^{25(26)}$ triterpene such as cyclolaudenol (Figure I) are cholesteric.

Molecular models indicate that the steric strain imposed by the 9,19-cyclopropane ring is approximately the same as that involving a C-8 (9) double bond and that the planarity of the B-ring is about the same in both cases. While cycloartenol contains a 9,19-cyclopropane ring, lanosterol (Figure I) has a C-19 methyl group and a C-8 (9) double bond. Other than this difference the structure and stereochemistry of these two triterpenes is the same. It thus seemed of interest to see if the palmitic acid ester of lanosterol is cholesteric, in analogy with the cholesteric properties of cycloartenyl palmitate. Pure lanosterol is difficult to obtain, the commercial product containing nearly an equal amount of

24-dihydro lanosterol. Careful purification of the palmitic acid esters of these two triterpenes, however, indicated 24-dihydro lanosteryl palmitate to form a cholesteric mesophase (m.p. 48-48.5°, ch→ iso 39°, blue mesophase). Although the lanosteryl palmitate was pure by all criterion available and crystallized nicely from a variety of solvents, no mesomorphic behavior could be detected with this ester. The isotropic liquid cooled upon standing to a glass. The melting behavior of this ester thus warrants further investigation.

It should be noted that all of the triterpenes thus studied have the same ring conformation (chair-boat-chair-boat-unfolded) and the same substituent stereochemistry (C-14 methyl, alpha; C-13 methyl, beta). It would be interesting to see if esters of tetracyclic triterpenes of different ring conformation or substituent stereochemistry could form a cholesteric mesophase (e.g. tirucallol, chair-chair-chair-unfolded). Theoretically if the overall planarity of a given nuclear structure is not changed, analogous esters should have similar mesomorphic properties. Although many structural features are important with respect to mesophase formation the side chain plays a unique role in determining the interaction of molecular species. From a strictly structural analysis esters of pentacyclic triterpenes would not be expected to form a cholesteric mesophase. These are triterpenes in which the side chain has cyclized to form an additional ring. We have found β-amyrin palmitate not to be mesomorphic (m.p. 76-77). The physical properties of various pentacyclic triterpenes are now being studied.

Sterols

The original work of Reinitzer (34) was concerned with cholesteryl benzoate and it is only fitting that the mesophase formed by such compounds be termed cholesteric. The liquid crystalline properties of many different cholesteryl esters have been studied and tables of these transition data are available (27). The most thorough study of the structural requirements for cholesteryl and related steryl esters to form a mesophase was made by Wiegand (37). While cholesteryl benzoate forms a cholesteric mesophase, epicholesteryl (3α-hydroxyl) benzoic acid ester is not mesomorphic, demonstrating the requirement for a 3β-acyl substituent. Likewise, coprostanyl (cis A/B ring juncture)benzoic acid ester is not mesomorphic. A trans A/B ring juncture therefore also appears to be necessary structural requirement. This is easily understood since the C-3 substituent projects out of the molecular plane in epicholesterol. In coprostanol, the A ring is not co-planar with the rest of the ring structure. These molecules are thus referred to as being 'kinked'. Wiegand also prepared the benzoic acid esters of cholestan-3β-ol, of the unsaturated sterols $\Delta^{5(6)}$, $\Delta^{7(8)}$, $\Delta^{8(9)}$, $\Delta^{8(14)}$, $\Delta^{14(15)}$-cholestan-3β-ol and of the diunsaturated forms $\Delta^{5,7}$, $\Delta^{6,8}$, $\Delta^{7,14}$, $\Delta^{8,24}$ and $\Delta^{14(15),24}$-cholestan-3β-ol. All of these esters were mesomorphic except those containing a C-14 double bond. Molecular models indicate that a double bond at the C-14 bridge head apparently tips the D-ring slightly, the side chain thus projecting from the molecular plane at a different angle. This slight difference must be just enough to disrupt mesophase formation. This is also a steric effect, and again demonstrates the important role played by the side chain in intramolecular interactions. The cholesteryl n-alkyl carbonates also exhibit cholesteric mesophases (12). The n-alkyl and α,ω-polymethylen bis(stigmasteryl) carbonates are not cholesteric (32). In comparison to cholesterol, stigmasterol has a Δ^{22} double bond and a C-24 methyl group. The inability of stigmasteryl esters to form a cholesteric mesophase again demonstrates the important role played by the alkyl side chain.

Evidently many different types of steryl esters can form a smectic mesophase. The cholesteryl ester series represents a well-studied class of such substances (4,5,18). Many of these esters form both a cholesteric and a smectic mesophase (polymorphism). In their original paper on the synthesis of 24-methylene cholesterol from cholenic acid, Riegel and Kaye noted the peculiar melting character of 24-keto cholesteryl acetate (Figure II). The crystalline compound melted at 127.5 to 128°, became turbid at 129-130° and finally melted at 131°. Although this double melting behavior may have been due to other factors, it is probable that it may have also been due to the formation of a smectic or cholesteric mesophase. Also, a recent report has attributed the double melting behavior of $\Delta^{5,7,22}$-cholestatrien-3β-ol (7,22-bisdehydrocholesterol) to the possible formation of a liquid crystalline phase (8). The crystalline sterol changed to an opaque liquid at 177-119° which became clear at 122-124°. This substance was not observed under a polarizing microscope and if it is mesomorphic (presumably smectic)it must represent a unique situation since such behavior is generally found only with steryl esters.

The collective data suggest that for a steryl ester to form a cholesteric mesophase, the sterol usually must meet at least three basic structural requirements: 1) trans A/B ring juncture, 2) 3β-acyl moiety, and 3) no double bond at C-14(15). Besides these studies there has been no systemmatic investigation of other general structural requirements. We have approached this problem by the synthesis of the palmitic acid esters of a number of different sterols.

Phase Transitions. As in the previous section, several representative phase transitions are listed below.

1. The transition temperatures for cholesteryl palmitate are in good agreement with those reported by other workers: m.p. 77-77.5°, cho→iso 82°, iso→cho 81°, cho→sm 77.5° (Literature, Reference 17: 70°, 83°, 80.5°, 78.5°, respectively).
2. Ergosterol palmitate melts (108.5-109°) to an isotropic liquid which forms a classic focal-conic texture when cooled to 102° (sm→iso). At 94° a crescent shaped front moves across the mesophase section as it crystallizes to a beautiful highly grained, birefringent solid.
3. The isotropic melt of 24-keto cholesteryl acetate cools to a brilliant birefringent liquid, which by a coverslip displacement changes to bright red. When heated and cooled in a glass tube, the red color is not observed until the melt is agitated with a glass rod.

Steryl Esters. The results of these studies are summarized in Table IV. When an ester has been previously described in the literature, the first values are the ones obtained by the authors. As tabulated above, the transition temperatures of cholesteryl palmitate agree well with those published elsewhere. Since the effect of nuclear unsaturation has been thoroughly studied by Wiegand, the effect of alkylation at C-24 in the steryl ester series has been investigated. In agreement with the results of Kuksis and Beveridge (28), neither β-sitosteryl palmitate nor stigmasteryl palmitate are mesomorphic. Stigmastanyl palmitate and fucosteryl palmitate are also not mesomorphic. Ergosterol and 24α-methyl cholestanyl palmitate, however, form well-defined smectic mesophases. With the palmitic acid esters, therefore, an ethyl or ethylidene group at C-24 is incompatible with mesophase formation. Again, this analysis is restricted to the palmitic acid esters since various stigmasteryl carbonates are smectic (32). Our preliminary work indicates that 24-keto cholesteryl acetate forms a brilliant red cholesteric mesophase. This sterol was synthesized from cholenic acid by the method of Riegel and Kaye (35). It is interesting that these earlier workers did not notice this color. This steryl ester demonstrates that the enhanced polarizability of the C-24 keto group does not effect mesophase formation. Although these data are of a preliminary nature, they demonstrate complex structural requirements for mesophase formation of steryl esters. Pollinastanol is the only 9,19-cyclopropane sterol that has been described (11). It would be interesting to study the possible mesomorphic behavior of esters of this sterol.

Table IV. Mesomorphic Properties of Various Steryl Esters

Plamitate Ester	Melting Point, °C	Transition Temperature, °C	Associated Color
Cholesterol	77-77.5	ch→iso 82	Violet
	78.4	ch→iso 80.5	Violet (18)
Ergosterol	1085-109.5	sm→iso 102	...
	106-108	...	... (31)
24α-Methyl Cholestanol	107-108	sm→iso 85	...
β-Sitosterol	91-94	...	...
	83.5-85.5	...	... (20)
	85.5	...	... (28)
Stigmasterol	101-102	...	...
	99.5	...	... (28)
Stigmastanol	102-103	...	...
Fucosterol	85	...	...

Naturally Occurring Liquid Crystals. Liquid crystals have unique optical properties and would be particularly well suited for biological systems by functioning in various sensory mechanisms. This subject has been reviewed in detail by other workers (16,36). The ubiquitous long chain esters of cholesterol have been isolated from mammalian sources (36). Ergosteryl palmitate, which forms a smectic mesophase (Table IV), has been isolated from Penicillium brevi compactum (31). A large percentage of the major sterols of the blue-tip oyster (Ostrea gigas) and of the pollen of Taraxacum dens leonis are esterified to long chain fatty acids (Knapp, et al, unpublished observations). These are only a few examples of steryl esters which have been isolated from natural sources. We have recently reported the isolation of cycloartenyl palmitate (Table III), a cholesteric liquid crystal, from banana peel and from the seeds of Strychnos nux-vomica (24, 25). To the best of our knowledge there have been no previous reports of such substances having been isolated from plants. The possibility that such substances are common in nature represents an exciting and challenging problem.

SUMMARY

Three new series of triterpene esters which are cholesteric have been described in detail. Mesophase formation by these esters has been shown to be dependent upon complex structural requirements. Both nuclear structure, fatty acyl chain length and side chain unsaturation are important structural features. In some cases, a C-24 (31) double bond can disrupt mesophase formation. In addition, the present study has demonstrated the possibility that esters of other triterpenes or sterols may form the cholesteric mesophase.

LITERATURE CITED

1. Adams, J. E., Haas, W., Wysocki, J., J. Chem. Phys. 50, 2458 (1969).

2. Adams, J. E., Haas, W., Wsyocki, J., Phys. Rev. Lett. 22, 92 (1969).

3. Arnold, H., Remus, D., Sackmann, H., Z. Phys. Chem. 222, 15 (1963).

4. Barrall, E. A. II, Sweeney, M. A., Mol Cryst. 5, 257 (1969).

5. Barrall, E. A. II, Porter, R. S., Johnson, J. F., J. Phys. Chem. 70, 385 (1966).

6. Barrall, E. A. II, Porter, R. S., Johnson, J. F., J. Phys. Chem. 71, 1224 (1967).

7. Castellano, J. A., Goldmacher, J. E., Barton, L. A., Kane, J. S., J. Org. Chem. 33, 3501 (1968).

8. Conner, R. L., Mallory, F. B., Landrey, J. R., Iyengar, C. W. L., J. Biol. Chem. 244, 2325 (1969).

9. Dave, J. S., Shimizu, M., J. Chem. Soc., 4617 (1954).

10. de Vries, H. L., Acta Cryst. 4, 219 (1951).

11. Devys, M., Barbier, M., Bulletin de la Société de chemie Biologique 49, 865 (1967).

12. Elser, W., Mol. Cryst. 2, 1 (1966).

13. Fergason, J. L., Scien. Amer. 211, 77 (1964).

14. Fergason, J. L., Mol. Cryst. 1, 293 (1966).

15. Fergason, J. L., Goldberg, N. N., Nadalin, R. J., Mol. Cryst. 1, 309 (1966).

16. Fergason, J. L., Brown, G. H., J. Amer. Oil Chem. Soc. 45, 120 (1968).

17. Friedel, G., Ann. Physique 18, 273 (1922).

18. Gray, G. W., J. Chem. Soc., 3733 (1956).

19. Gray, G. W., "Molecular Structure and the Properties of Liquid Crystals", Academic Press, New York, 1962.

20. King, F. E., Jurd, L., J. Chem. Soc., 1192 (1953).

21. Knapp, F. F., Nicholas, H. J., J. Org. Chem. 33, 3995 (1968).

22. Knapp, F. F., Nicholas, H. J., Phytochem. 8, 207 (1969).

23. Knapp, F. F., Nicholas, H. J., J. Org. Chem., In Press.

24. Knapp, F. F., Nicholas, H. J., submitted to Mol. Cryst. and Liq. Cryst.

25. Knapp, F. F., Nicholas, H. J., "Naturally Occurring Cholesteric Liquid Crystals", Abstract, VIth International Symposium on Natural Products (Terpenes and Steroids Mexico City, April, 1969.

26. Knapp, F. F., Nicholas, H. J., Phytochem., In Press.

27. Kast, W., Londolt-Bornstein, 6th edit., Vol. II, Part 2a, p. 266 (1959).

28. Kuksis, A., Beveridge, J. M. R., J. Org. Chem. 25, 1209 (1959).

29. Kuksis, A., McCarthur, M. J., Canad. J. Biochem. 40, 679 (1962).

30. Oseen, C., Trans. Far. Soc. 29, 883 (1933).

31. Oxford, A. E., Biochem. J. 27, 1176 (1933).

32. Pohlman, J. L. W., Mol. Cryst. 2, 15 (1966).

33. Porter, R. S., Barrall, E. M. II, Johnson, J. F., Acc. Chem. Res. 2, 53 (1969).

34. Reinitzer, F., Monatsch Chem. 9, 421 (1888).

35. Riegel, B., Kay, I. R., J. A. C. S. 66, 723 (1944).

36. Stewart, G. T., Mol. Cryst. 1, 563 (1966).

37. Wiegand, C., Zeit. Natur. 4b, 249 (1949).

ACKNOWLEDGEMENTS

We would like to thank Dr. Joseph P. Schroeder for determining many of the phase transition temperatures and for his helpful advice, and Mr. Joseph Kelley for his technical assistance.

NEMATIC MIXTURES AS STATIONARY LIQUID PHASES IN GAS-LIQUID CHROMATOGRAPHY (1)

J. P. Schroeder, D. C. Schroeder and M. Katsikas

Department of Chemistry, The University of North Carolina at Greensboro, Greensboro, N. C. 27412

Nematic liquid crystallinity (2,3) is exhibited by certain compounds having relatively rigid, polar, rod-shaped molecules which tend to be oriented with their long axes parallel due to mutual attractive forces. On heating such a compound, the crystalline solid melts to an anisotropic liquid (mesophase) in which neighboring molecules lie parallel to one another. At a higher temperature, disorientation of the ordered molecular arrangement occurs and there is a transition to isotropic liquid.

As solvents, nematic mesophases show a selective affinity toward linear, rod-shaped solute molecules relative to bulky, nonlinear molecules, presumably because the former fit more readily into the parallel nematic "lattice" (4). This has been demonstrated in the separation of close-boiling meta-para disubstituted benzene isomer pairs by gas-liquid partition chromatography (glpc) using nematic stationary liquid phases (6,7). The more linear para isomer is invariably eluted last.

The ability of nematic stationary phases to separate solutes on the basis of molecular shape is applicable, in principle, to many glpc problems. A practical limitation, however, is the restricted useful temperature range. Since the mesophase becomes less well ordered and, therefore, less selective as the transition to isotropic liquid is approached (6,7), the best operating temperatures are in the lower part of the nematic range. Maximum selectivity is limited by the lowest temperature to which the stationary phase can be cooled without crystallization, usually only a few degrees below the melting point for a pure compound. We have found that the use of mixtures is a convenient means of varying the position and breadth of the nematic range, increasing the versatility of this type of stationary phase.

Like other crystalline solids, a nematogenic compound (8) undergoes melting point depression on addition of a substance with which it is miscible. A mixed nematic mesophase originating at a lower temperature is the result. If the added substance is structurally dissimilar to the original compound, a small amount is sufficient to destroy liquid crystallinity by disrupting the mesomorphic "lattice" (5,9). If, however, it is structurally similar, nematic mesomorphism may persist to a high concentration of the second component and, in fact, may be exhibited by all compositions when the added compound is also nematogenic (10,11). Preliminary experiments with such a compatible binary system were encouraging. The components were the homologs 4,4'-dimethoxyazoxybenzene (**1a**) and 4, 4'-di-*n*-hexyloxyazoxybenzene (**1c**), both of which are nematogenic. It was found that the

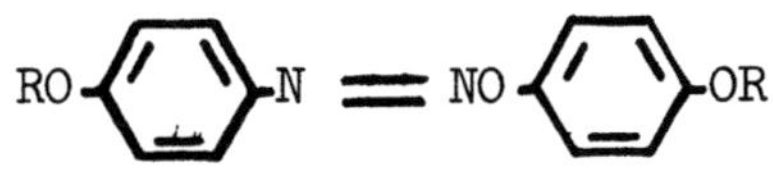

1a, R = CH_3
1b, R = C_2H_5
1c, R = *n*-C_6H_{13}

nematic mesophases of **1a** - **1c** mixtures were about as selective as those of the pure components (12). We then proceeded to the detailed studies of the **1a** - **1c** system and two others that are described below. For each system, selectivity was determined as a function of composition and temperature using ability to separate *m*- and *p*-xylene by glpc as the experimental criterion.

After this investigation was completed, we discovered that Kelker, Scheurle and Winterscheidt (13) had been working along similar lines. Where the two studies impinge, a comparison of the data will be presented.

DISCUSSION

The phase diagram of the **1a** - **1c** system (Figure 1) shows that all compositions are nematogenic. The nematic range of the eutectic composition is broader and originates at a lower temperature than that of either pure **1a** or **1c**. It also appears that addition of **1a** to **1c** or vice versa disrupts the nematic "lattice" very little since the mesophase-isotropic liquid transition curve is only slightly concave upward.

Glpc columns were prepared in which the stationary liquid phases were **1a**, **1c** and mixtures of the two containing 27 (eutectic), 34, 61 and 82 mole percent **1a**. The relative retention (α) of *p*-xylene based on *m*-xylene was determined at various temperatures in these columns (Figure 2). As isotropic liquids, **1a** and mixtures rich in **1a** gave lower α values than **1c** and mixtures rich in **1c**. This is

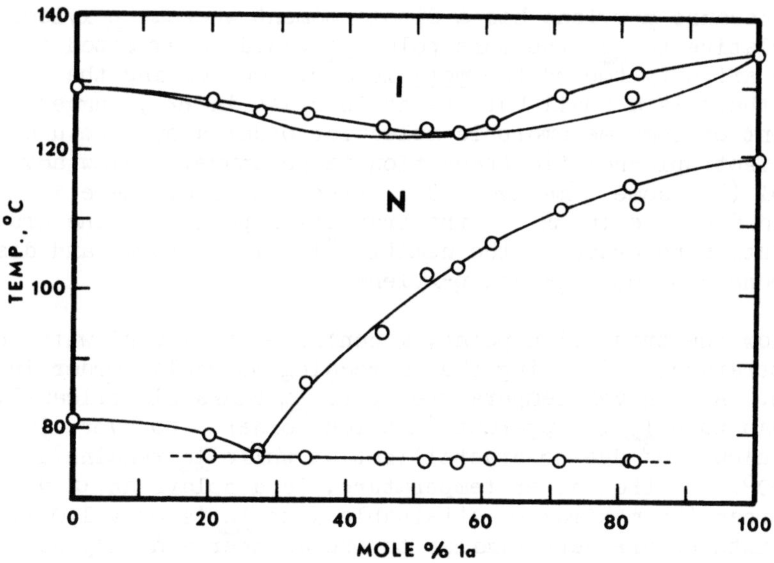

Figure 1. - Phase diagram of system 4,4'-dimethoxyazoxybenzene (1a) - 4,4'-di-n-hexyloxyazoxybenzene (1c). I = isotropic, N = nematic.

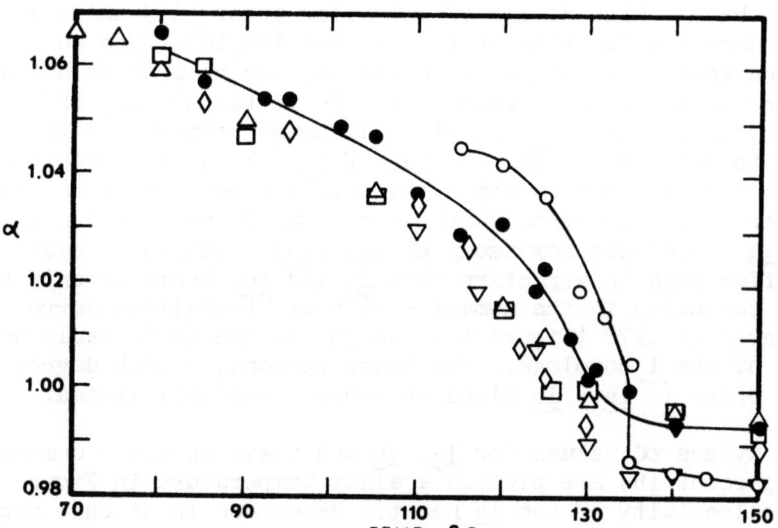

Figure 2. - Variation of the relative retention (α) of p-xylene (m-xylene = 1.00) with temperature using 1a, 1c and their mixtures as stationary liquid phases.

Mole °/o 1a:	0	27	34	61	82	100
Symbol:	●	△	□	◇	▽	○

probably because m-xylene has a dipole moment whereas p-xylene has none. Relative to 1c, the more polar 1a would be expected to exhibit a greater affinity toward the meta isomer. On cooling the isotropic liquids, there is a small but steady increase in α , suggesting the development of some mesomorphic molecular order above the usually assigned isotropic-nematic transition temperature, a view advanced by Tolstoi (14) some time ago. On further cooling, there is a much more rapid increase in α at the transition point and the order of isomer elution reverses as the nematic "lattice" forms and discriminates strongly between m- and p-xylene.

Below the transition point, α continues to ascend with decreasing temperature, reflecting the increasing molecular order in the mesophase. At a given temperature, pure 1a ranks significantly higher than pure 1c in apparent "lattice" order. However, 1a cannot be used below 115° due to crystallization while 1c remains liquid down to 80°. At the latter temperature, 1c displays an α value of 1.066 whereas the maximum α attainable with 1a is only 1.045. The 1a - 1c mixtures are very similar to one another and only slightly inferior to pure 1c in apparent "lattice" order. There appears to be little disruption of the nematic mesophase on adding 1a to 1c, somewhat more on addition of 1c to 1a.

These results support earlier evidence that mixed nematic mesophases can have a high degree of molecular order (15) and that this order is largely a function of temperature (15,16). The eutectic composition remains liquid and, therefore, can be used at a lower temperature than the pure components. As it turned out, the maximum selectivity of the 1a - 1c eutectic, although better than that of 1a, proved to be just equivalent to that of 1c. It seemed probable that a system could be found in which the eutectic would be more selective than either pure component. With this in mind, we turned next to a study of 1a - 4,4'-diethoxyazoxybenzene (1b) mixtures. These compounds differ less in structure than 1a and 1c, there is only the slightest concavity in the nematic-isotropic transition curve of their phase diagram (11,17) (Figure 3), and 1b has the most stable nematic mesophase of the 1 homologs. For these reasons, a high degree of molecular order in 1a - 1b mixed mesophases was anticipated.

The xylene α values for 1a, 1b and their eutectic composition (40 mole percent 1b) are plotted against temperature in Figure 4.. The high selectivity of the 1b nematic mesophase is at once apparent. Considerable order seems to persist well above the conventional nematic-isotropic transition point. Even at 181°, twelve degrees above this, α is greater than unity. At a given temperature, the eutectic composition's nematic mesophase is intermediate in molecular order between those of pure 1a and 1b. However, because it can be used at a lower temperature, it has the greatest maximum selectivity. Here, then, is an example of a mixed nematic stationary phase which performs better than either pure component. The shift in and broadening of the

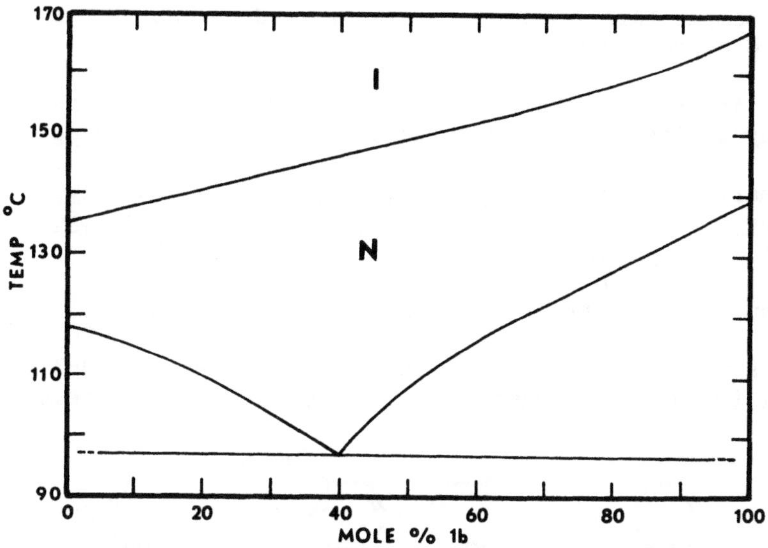

Figure 3. - Phase diagram of system 4,4'-dimethoxyazoxybenzene (1a) - 4,4'-diethoxyazoxybenzene (1b) (17). I = isotropic, N = nematic.

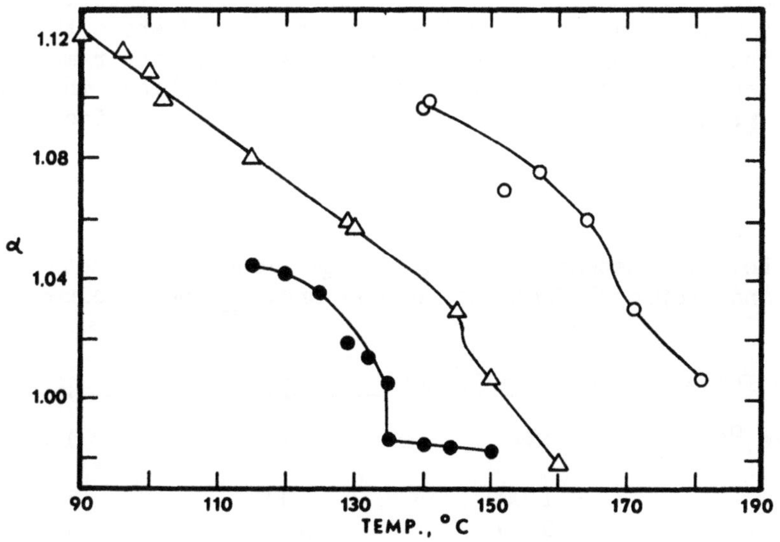

Figure 4. - Variation of the relative retention (α) of p-xylene (m-xylene = 1.00) with temperature using 1a, 1b and their eutectic composition as stationary liquid phases.

Mole °/o 1b:	0	40	100
Symbol:	●	△	○

operating temperature range which is effected by use of the mixture should also be emphasized.

This is the system which Kelker, Scheurle and Winterscheidt (13) had already studied. Their results and ours are in good agreement:

Stationary Phase	Temp., °C	α for p-xylene (m-xylene = 1.00) Ref. 13	This investigation
1a	119	1.04	1.04
1b	140	1.09	1.10
1a-1b eut.	99	1.10	1.11

The system 1a - 4,4'-bis(p-methoxybenzylideneamino)-3,3'dichlorobiphenyl (2) was examined next. The anil 2 has an extremely stable

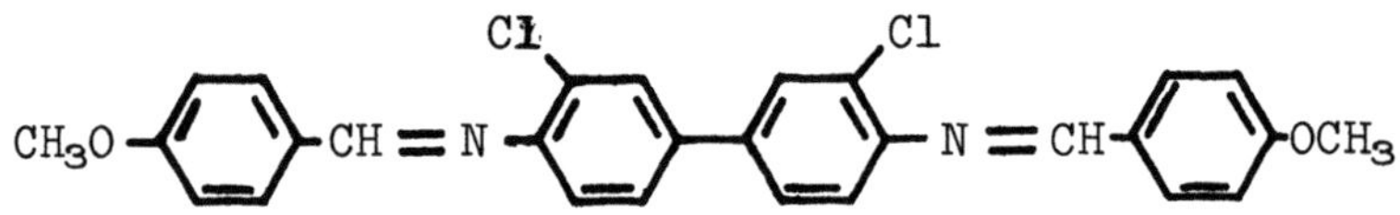

2

nematic mesophase that is highly selective toward solutes on the basis of molecular shape (18). It can be even more effective in this regard than 1b as shown by the data in Table I where the performances of the two compounds are compared just above their melting points. Results for a silicone stationary liquid phase, which discriminates primarily on the basis of boiling point, are included for comparison.

Table I. Relative Retentions of m- and p-Xylene and of Methyl, Ethyl and Isopropyl Benzoate in Silicone, 1b and 2 Columns

Stationary Phase	SE-30 Silicone	1b	2
Temperature, °C	104	141	160
m-Xylene	1.00	1.00	1.00
p-Xylene	0.99	1.10	1.16[a]
Methyl benzoate	1.00	1.00	1.00
Ethyl benzoate	1.64	1.23	1.11
Isopropyl benzoate	1.98	1.08	0.85

a. A value of 1.05 has been reported for almost identical experimental conditions (18). In consideration of the other data for 2, this appears to be inconsistently low.

The excellent selectivity of 2 is particularly well demonstrated by the data for the alkyl benzoates. In the silicone column, the order of elution is methyl-ethyl-isopropyl ester with a large interval between the first two in agreement with the respective boiling points: 199, 212 and 218.5°. In the 1b column, the retention times of methyl and ethyl benzoate are closer together and the isopropyl ester is eluted well before the lower boiling ethyl ester. Apparently, solute molecular shape, as it affects ability to fit readily into the nematic "lattice", has become an important factor. Methyl benzoate has the most linear molecule and, therefore, a long retention time in consideration of its boiling point. Conversely, isopropyl benzoate, with its bulky alkyl group, exhibits an anomalously short retention time. The same trends are observed for nematic 2 but to a more spectacular degree. Here, the difference between the retention times of methyl and ethyl benzoate is still smaller and the isopropyl ester is eluted even before the methyl ester.

The use of 1a - 2 mixtures takes advantage of the virtues and minimizes the faults of the pure components. 2 is a very selective solvent but high melting and heat sensitive (18). 1a is lower melting but less selective and has a narrow nematic range. The phase diagram for the system (Figure 5) shows that addition of 1a to 2 can depress the melting point 50° while retaining a stable, well ordered mixed mesophase (105-260° range). From the opposite standpoint, addition of a small amount of 2 to 1a broadens the nematic range significantly.

Xylene α values for 1a, 2 and their mixtures containing 3, 10 and 34.5 (eutectic) mole percent 2 are plotted against temperature in Figure 6. There are no data for isotropic 2 or the isotropic eutectic composition because 2 decomposes and 1a bleeds badly at the high temperatures required for these measurements. In agreement with Figure 5, the results show that addition of 1a to 2 depresses the melting point without disrupting the nematic molecular order appreciably. The eutectic composition supercools readily, remaining liquid for long periods at 60-70°. At these low temperatures, selectivity is almost as good as that of pure 2 just above its melting point. On the opposite end of the composition scale, addition of only three mole percent of 2 to 1a results in retention of nematic molecular order to a higher temperature and improved maximum selectivity at low temperature. With increasing amounts of 2, these trends are enhanced, culminating in the eutectic composition for which selectivity over the broad range 60-160° is markedly better than that attainable with pure 1a at any temperature.

This system is a good example of the practical merit of mixed nematic stationary phases. Although 2 is very selective, it is high melting and heat-sensitive, a combination which severely limits its usefulness. However, the 1a - 2 eutectic mixture permits operation at a much lower temperature without a great sacrifice in performance.

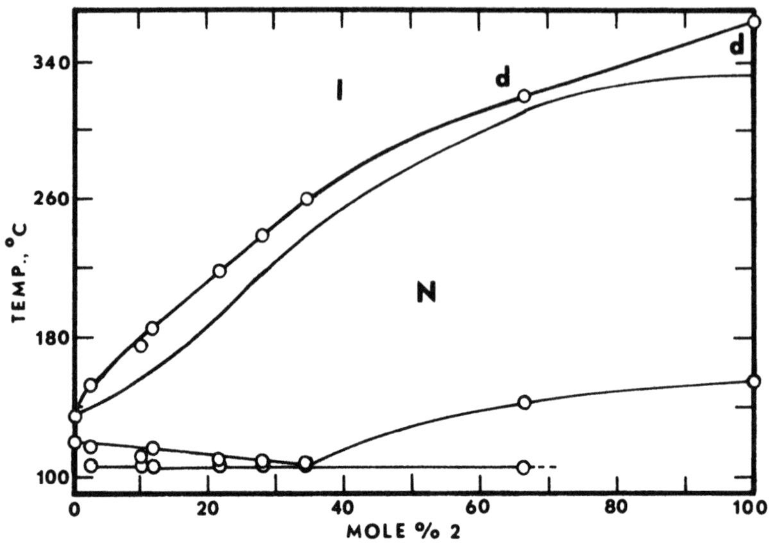

Figure 5. - Phase diagram of system 4,4'-dimethoxyazoxybenzene (1a) - 4,4'-bis(p-methoxybenzylideneamino)-3,3'-dichlorobiphenyl (2). I = isotropic, N = nematic, d = decomposition.

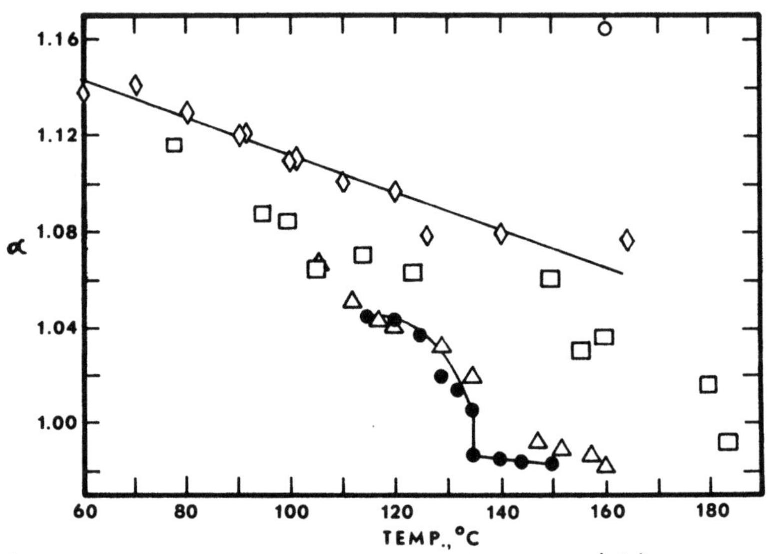

Figure 6. - Variation of the relative retention (α) of p-xylene (m-xylene = 1.00) with temperature using 1a, 2 and their mixtures as stationary liquid phases.

Mole $^{o}/o$ 2:	0	3	10	34.5	100
Symbol:	●	△	□	◇	○

Relative to 1a, the eutectic composition provides superior selectivity over a far broader temperature range.

Having demonstrated that these mixed nematic phases separate m- and p-xylene effectively, we tried them with other meta-para isomer pairs. In Table II, the relative retentions of p-ethyltoluene, p-dichlorobenzene and p-methylanisole based on the corresponding meta isomers are given for glpc columns containing 1a, 1b, 1c, 2 and the eutectic compositions of the three systems discussed above. In each case, the column temperature is in the lower part of the nematic range. Data for a silicone column are included for comparison. Note that all of the nematic liquids are highly selective but the eutectic mixture is invariably at least as selective as either pure component.

Table II. Relative Retentions of p-Ethyltoluene, p-Dichlorobenzene and p-Methylanisole (meta Isomer = 1.00) in 1a, 1b, 1c, 2, 1a - 1b, 1a - 1c, 1a - 2 and Silicone Columns

Stationary Liquid Phase	Temp., °C	p-$C_2H_5C_6H_4CH_3$	p-ClC_6H_4Cl	p-$CH_3OC_6H_4CH_3$
1a	120	1.09	1.17	1.12
1b	141	1.15	1.26	1.18
1a - 1b eut.	100	1.18[a]	1.28	1.23
1c	85	1.09	1.11	1.13
1a - 1c eut.	75	1.11	1.17	1.16
2	160	1.19	1.20	1.18
1a - 2 eut.	100	1.21	1.25	1.23
SE-30 silicone	100	1.01	1.04	1.02

a. Kelker et al. (13) found 1.18 at 99°.

In summary, this study shows that mixtures offer a promising approach to controlling the mesomorphic temperature ranges of nematic liquids. By adding a second substance to a compound which is nematic only at high temperatures, the lower limit of the nematic range can be depressed. Addition of a compound having a highly stable nematic mesophase to one with a less stable mesophase can lengthen the nematic range at both ends. A high degree of molecular order is retained in the mixed liquid crystal. Since the nematic mixtures tend to supercool readily, they can often be used at temperatures well below the melting point to take advantage of the accompanying further enhancement of molecular order.

The results also suggest that the ability of a nematic mesophase to differentiate between solutes on the basis of molecular shape could be used as a quantitative measure of its molecular order. However, if glpc were the experimental method, a correction would be needed for the temperature dependence of relative retention which is true of any stationary liquid phase (19). E. g., Wiseman (20) has shown that m- and p-Xylene are resolved at room temperature (separation factor 1.05) by dinonyl phthalate whereas there is no separation at 80°.

EXPERIMENTAL

Materials. - The 4,4'-dialkoxyazoxybenzenes (21) were recrystallized from benzene (1a and 1b) or 95 °/o ethanol (1c) and had the following transition temperatures:

Compound	Solid-nematic, °C	Nematic-isotropic, °C	Literature Values
1a	119.5	135	118.5, 135 (22)
1b	138.5	169.5	138.5, 168 (23)
1c	81	129	81, 127 (24)

4,4'-bis(p-Methoxybenzylideneamino)-3,3'-dichlorobiphenyl (2) was prepared as follows. 3,3'-Dichlorobenzidine dihydrochloride (technical grade) was treated with aqueous NaOH solution to give the free base in 95 °/o yield. Reaction of this with a slight excess of p-anisaldehyde in refluxing absolute ethanol using acetic acid as catalyst gave crude 2, mp 148-152°, in 97 °/o yield. Recrystallization from benzene provided the pure compound, mp 153-154.5°, nematic-isotropic transition point 363° d. (lit. (18) 154°, 334° d.).

The glpc solutes were commercial products which were not purified further. No major contaminants were detected in their chromatograms.

Procedure and Apparatus. - The mixtures used for constructing the phase diagrams were prepared by weighing the components accurately into a microbeaker, melting, stirring to give an intimate blend, and cooling with stirring until solidification occurred. Phase transitions were determined by observation of powdered samples in a calibrated Nalge-Axelrod hot stage melting point apparatus (12, 25).

The chromatograms were obtained with an Aerograph Hy-Fi gas chromatograph, model 600-D, using nitrogen as carrier gas and a flame ionization detector. A small sample size (0.5 μl of a 1 °/o solution in carbon disulfide) was used to minimize solute effects on the nematic stationary liquid phases. The glpc columns were made from 3.2 mm o.d., 1.8 mm i.d. annealed copper tubing and were 5 m in length. The column packings were prepared from Chromosorb W (60-80 mesh) as solid support and solutions of the stationary liquid phases in dichloromethane. All packings contained 15 $^{o}/_{o}$ liquid phase by weight.

Some of the chromatograms were determined at column temperatures below the melting point of the stationary liquid phase. When crystallization of the supercooled liquid occurred during such a run, it was readily apparent from a sharp drop in retention time.

Relative retentions were calculated from retention times corrected for column dead space. For this correction, the retention time of an unabsorbed gas (t_o) must be known. Since the flame ionization detector does not give an air peak, t_o was determined indirectly by extrapolation of retention time data for homologous hydrocarbons to zero carbon content. Attempts to calculate t_o for the nematic stationary liquid phases from n-alkane data by the method of Gold (26) gave different values depending on the molecular weight range of the alkanes which were used. This is undoubtedly associated with the selectivity of the mesophases toward linear molecules. Satisfactory results were obtained by including methane among the hydrocarbons, plotting retention time vs. number of carbon atoms per molecule, and extrapolating the smooth curve to zero carbon.

REFERENCES

(1) This work was supported by a grant-in-aid from the Research Council of the University of North Carolina at Greensboro.

(2) G. W. Gray, "Molecular Structure and the Properties of Liquid Crystals," Academic Press, Inc., New York - London, 1962.

(3) G. H. Brown and W. G. Shaw, Chem. Rev., 57, 1049 (1957).

(4) The term "lattice" as applied to liquid crystals, which was first employed by Dave and Dewar (5), has been criticized. Admittedly, the degree of order in a mesomorphic liquid does not approach that of a crystalline lattice. However, "lattice" is a most convenient short term for "anisotropic molecular arrangement" and, as long as it is enclosed in quotation marks, we see no harmful ambiguity in its use.

(5) J. S. Dave and M. J. S. Dewar, J. Chem. Soc., 4305 (1955).

(6) H. Kelker, Ber. Bunsenges. Phys. Chem., 67, 698 (1963); Z. Anal. Chem., 198, 254 (1963).

(7) M. J. S. Dewar and J. P. Schroeder, J. Am. Chem. Soc., 86, 5235 (1964).

(8) One which gives a nematic mesophase on melting (2).

(9) J. S. Dave and M. J. S. Dewar, J. Chem. Soc., 4616 (1954).

(10) A. C. de Kock, Z. Phys. Chem., 48, 129 (1904).

(11) A. Prins, ibid., 67, 689 (1909).

(12) M. J. S. Dewar, J. P. Schroeder and D. C. Schroeder, J. Org. Chem., 32, 1692 (1967).

(13) H. Kelker, B. Scheurle and H. Winterscheidt, Anal. Chim. Acta, 38, 17 (1967).
(14) N. A. Tolstoi, J. Exptl. Theoret. Phys. (USSR), 17, 724 (1947).
(15) D. H. Chen and G. R. Luckhurst, Trans. Faraday Soc., 65(3), 656 (1969).
(16) P. Diehl and C. L. Khetrapal, Mol. Physics, 14, 283 (1967).
(17) H. Arnold and H. Sackmann, Z. Phys. Chem., 213, 145 (1960).
(18) M. J. S. Dewar and J. P. Schroeder, J. Org. Chem., 30, 3485 (1965).
(19) A. B. Littlewood, "Gas Chromatography," Academic Press, Inc., New York - London, 1962, p. 78; S. Dal Nogare and R. S. Juvet, Jr., "Gas-Liquid Chromatography," Interscience Publishers, New York - London, 1962, p. 83.
(20) W. A. Wiseman, Nature, 185, 841 (1960).
(21) 1a, 1b and 1 c are commercially available from Frinton Laboratories.
(22) W. Davies and R. A. R. Down, J. Chem. Soc., 586 (1929); R. S. Porter and J. F. Johnson, J. Phys. Chem., 66, 1826 (1962).
(23) I. F. Homfray, J. Chem. Soc., 97, 1669 (1910).
(24) C. Weygand and R. Gabler, J. Prakt. Chem., 155, 332 (1940).
(25) J. P. Schroeder and D. C. Schroeder, J. Org. Chem., 33, 591 (1968).
(26) H. J. Gold, Anal. Chem., 34, 174 (1962).

SINGULAR SOLUTIONS IN LIQUID CRYSTAL THEORY

J. L. Ericksen

Mechanics Department, The Johns Hopkins University
Baltimore, Maryland

1. INTRODUCTION

Not infrequently, orientation patterns in liquid crystals are marred by imperfections, so it seems pertinent to attempt some general theoretical treatment of these, within the framework of continuum theory. We here present some thoughts concerning static theory for these.

We might reasonably consider three types of singularities. One is a surface, across which orientation changes abruptly, with well defined limiting orientations as we approach the surface from either side. Another involves a curve along which orientation is undefined, as in those called "disinclinations" by Frank [1]. In some cases, such curves may be edges of the aforementioned surfaces. Finally, there is the possibility of orientation being undefined at isolated points. Ericksen [2] gives a simple mathematical solution involving one such. With each, there is the problem of supplementing the governing differential equations with physically appropriate conditions. If we are to seriously study singular solutions, we must seek to better define the problems to be solved. At the same time, it can be instructive to explore simpler solutions which may or may not correspond to simply stated problems. We devote some space to both types of questions.

As is discussed by Ericksen [3], a surface bearing finite discontinuities will tend to move, provided the discontinuity is, in a well defined sense, weak. Thus we concentrate on the stronger types of discontinuity.

Simple disinclination solutions such as are discussed by

Frank [1] have two features worth pondering. One is that the energy density decays slowly with distance from the singular line. By itself, this leads to infinite energies in infinite regions. As is discussed below, requiring the energy density to decay rapidly enough to keep the energy finite leads to the notion that disinclinations tend to occur in certain groups. There is a similar divergence of energy at the singular line*. To remedy this, Oseen [5] suggested a modification of theory, allowing for compressibility. We explore an alternative, based on the notion that, in the immediate neighborhood of this line, there is a transformation to the isotropic liquid phase. It does not seem very obvious what is the remedy for the basic difficulty, so our treatment should be regarded as tentative. Formally, the analysis would apply to cases where the core is filled with a different isotropic fluid, a possibility suggested by Lehmann [6, pp. 47-48].

It seems premature to present a general viewpoint toward point imperfections until they are better explored. Of course, the notion of a phase transformation near the singular point could be invoked. As in the case of disinclinations, the energy is likely to decay slowly with distance from the singular point, and this is true of the solution given by Ericksen [2]. It is not very clear whether more rapid decay might obtain for groups of point imperfections.

Even after correction to remove divergent energy integrals, we might reasonably expect energies associated with most imperfections to be so high that such configurations will tend to be unstable. Since many are observed, we should not too hastily discard solutions which seem to involve instability.

To avoid excessive complication, we restrict our attention to simpler theories for liquid crystals of nematic type and ignore the compressibility effects mentioned above. Also, for example, we disregard the possible importance of surface energy at interfaces separating isotropic and liquid crystal phases.

2. GOVERNING EQUATIONS

We concern ourselves with liquid crystals, their orientation being described by a unit vector field $\underset{\sim}{d}$,

$$\underset{\sim}{d}\cdot\underset{\sim}{d} = 1 \ . \tag{2.1}$$

Important to the static theory is the stored energy per unit volume W , of the form

$$W = W(\underset{\sim}{d}, \nabla\underset{\sim}{d}) \ , \tag{2.2}$$

*Ericksen's [4] comments concerning this are befogged.

that commonly used being of the form, in Cartesian tensor notation,

$$2W = k_{11}(d_{k,k})^2 + k_{22}(\epsilon_{ijk}d_i d_{k,j})^2 + k_{33}d_{i,j}d_{i,k}d_j d_k - (k_{22} + k_{24})\left[(d_{k,k})^2 - d_{i,k}d_{k,i}\right] \geqq 0 , \tag{2.3}$$

the k's being material constants. With body forces and couples commonly considered, there is associated an energy per unit volume $\overline{W}$, of the form

$$\overline{W} = \overline{W}(\underset{\sim}{x}, \underset{\sim}{d}) . \tag{2.4}$$

The governing differential equations can be reduced to (2.1) and

$$\left(\frac{\partial U}{\partial d_{i,j}}\right)_{,j} - \frac{\partial U}{\partial d_i} = \lambda d_i , \tag{2.5}$$

where

$$U = W + \overline{W} \tag{2.6}$$

and λ is an unspecified function of position, a Lagrange multiplier associated with the constraint (2.1). With such a solution are associated surface forces and couples. The surface force per unit area $\underset{\sim}{F}$ is expressible in terms of the unit normal $\underset{\sim}{\nu}$ as follows:

$$F_i = t_{ji}\,\nu_j , \tag{2.7}$$

$$t_{ji} = (U + a)\,\delta_{ij} - \frac{\partial W}{\partial d_{k,j}} d_{k,i} , \tag{2.8}$$

a being an arbitrary constant. In part, the surface couple is the couple associated with this force, to which is added a couple per unit area $\underset{\sim}{L}$ of the form, which always acts perpendicular to $\underset{\sim}{d}$,

$$L_i = \epsilon_{ijk} d_j \frac{\partial W}{\partial d_{k,m}} \nu_m . \tag{2.9}$$

More detailed discussions of this theory, including symmetry considerations which restrict W, are given by Frank [1], Ericksen [7,8] and Leslie [9].

Consider a surface S across which $\underset{\sim}{d}$ and $\nabla\underset{\sim}{d}$ have finite discontinuities. Requiring equilibrium of forces and moments on volume elements containing part of S in their interior leads to the conditions

$$\left[F_i\right] = \left[L_i\right] = 0 , \tag{2.10}$$

where the square bracket denotes the jump in the quantity enclosed. The theory discussed treats $\underset{\sim}{d}$ and $-\underset{\sim}{d}$ as physically indistinguishable, so one can generate trivial examples where $\underset{\sim}{d}$ undergoes a simple reversal across a surface. This does not imply that all such discontinuities can be removed by reversing $\underset{\sim}{d}$ in subregions, there being counterexamples among the disinclination solutions.

Consider an interface separating isotropic liquid and liquid crystal phases. We adopt the simplest view, which is that the isotropic fluid exerts on the liquid crystal just those forces and moments deriving from its hydrostatic stress. Then on the liquid crystal side, we have

$$F_i = -p\, \nu_i , \tag{2.11}$$

$$L_i = 0 , \tag{2.12}$$

where ν_i is the unit normal to the interface, directed into the isotropic liquid, whose pressure is p. Violations might be explained by attributing some strength to the interfacial surface, but we here exclude this complication. Then (2.9) and (2.12) imply the existence of a scalar μ such that

$$\frac{\partial W}{\partial d_{k,m}} \nu_m = \mu\, d_k . \tag{2.13}$$

Using (2.1),

$$\frac{\partial W}{\partial d_{k,m}} d_{k,j}\, \nu_m = \mu\, d_k d_{k,j} = 0 .$$

With this, (2.7), (2.8) and (2.11), we find that

$$U + a = -p . \tag{2.14}$$

It is easily seen that (2.13) and (2.14) are necessary and sufficient conditions that (2.11) and (2.12) hold.

For simplicity, we henceforth assume that there are no external body forces or couples. Then, in (2.14), p should be constant, so

$$W = W_0 = \text{const.} \tag{2.15}$$

In places, we explore the notion that the liquid crystal will transform to the isotropic phase when W exceeds a critical value, W_0 then being interpretable as this critical value, simultaneously the maximum value of W in the liquid crystal regime. A typical calculation might go as follows: We somehow obtain a solution of the governing differential equations. Given the critical value W_0, we can locate the regions where W exceeds it, which should undergo phase transformation. By transforming, the liquid loses some ability to supply the variety of forces and couples which it could without transforming. In consequence, there must be some readjustment of orientation, in the neighborhood of the isotropic regions, as first estimated. Assuming static readjustment is possible, the shape and size of the isotropic regions, as well as orientation in the liquid crystal, will be somewhat different from those first estimated. If such readjustment is impossible, we might gain from static considerations some hint as to the nature of likely dynamical phenomena, but quantitative predictions seem hopeless. As is perhaps evident, it is generally not easy to correct for such readjustments or even to ascertain whether this is possible. I do not feel that it is impossible to make some headway with these problems, but we here concentrate on simpler analyses.

As is discussed by de Gennes [10] and Ericksen [11], the term multiplied by k_{24} in (2.3) has no effect in the equilibrium equations (2.5). Sometimes it influences the surface forces and couples considered above. If $\underset{\sim}{d}$ is normal to a family of surfaces, the second term in (2.3) has no effect in the equilibrium equations or on the surface actions. For the plane problems considered below neither modulus exerts any influence. The remaining two are subject to the inequalities

$$k_{11} \geqq 0 , \quad k_{33} \geqq 0 , \tag{2.16}$$

implied by the inequality in (2.3). We assume neither vanishes. Then

$$k \equiv k_{33}^{-1} (k_{11} - k_{33}) \tag{2.17}$$

satisfies

$$k > -1 . \tag{2.18}$$

3. PLANE SOLUTIONS

We consider plane solutions of the form

$$\underset{\sim}{d} = (\cos \theta, \sin \theta, 0) \qquad \theta = \theta (x_1, x_2) , \tag{3.1}$$

in which case

$$W = F(\theta , \nabla \theta) . \tag{3.2}$$

If (2.3) applies, we find that

$$2F = k_{33} \left\{ \theta_{,i} \, \theta_{,i} + k(\cos\theta \, \theta_{,2} - \sin\theta \, \theta_{,1})^2 \right\} , \tag{3.3}$$

where k is given by (2.17). Conditions considered above can be put in more convenient form, using relations of the type

$$\frac{\partial F}{\partial \theta_{,i}} = \frac{\partial W}{\partial d_{1,i}} (-\sin\theta) + \frac{\partial W}{\partial d_{2,i}} (\cos\theta) \tag{3.4}$$

$$\frac{\partial W}{\partial d_{1,3}} = \frac{\partial W}{\partial d_{2,3}} = \frac{\partial W}{\partial d_{3,1}} = \frac{\partial W}{\partial d_{3,2}} = \frac{\partial W}{\partial d_3} = 0 . \tag{3.5}$$

The latter relations, easily verified for (2.3), obtain quite generally for liquid crystals of nematic type because reflection of the x_3-axis, an admissable symmetry transformation, leaves $\underset{\sim}{d}$ and $\nabla \underset{\sim}{d}$ unaltered. In place of (2.9), we have, assuming $\nu_3 = 0$,

$$\left. \begin{aligned} L_1 &= L_2 = 0 , \\ L_3 &= \frac{\partial F}{\partial \theta_{,i}} \nu_i . \end{aligned} \right\} \tag{3.6}$$

$$\left. \begin{aligned} t_{ji} &= (F + a)\,\delta_{ij} - \frac{\partial F}{\partial \theta_{,j}} \theta_{,i} , \qquad i,j = 1,2 \\ t_{33} &= F + a , \\ t_{13} &= t_{23} = t_{31} = t_{32} = 0 . \end{aligned} \right\} \tag{3.7}$$

Further, eliminating λ in (2.5), we obtain

$$\frac{\partial F}{\partial \theta} - \left(\frac{\partial F}{\partial \theta_{,i}} \right)_{,i} = 0 . \tag{3.8}$$

Further, (2.10) gives, again assuming $\nu_3 = 0$, the jump conditions

$$\left[\frac{\partial F}{\partial \theta_{,i}} \right] \nu_i = 0 , \tag{3.9}$$

$$\left[F + a\right] \nu_i = \left[\theta_{,i}\right] \frac{\partial F}{\partial \theta_{,j}} \nu_j \tag{3.10}$$

Similarly, (2.13) and (2.14) yield the interfacial conditions

$$\frac{\partial F}{\partial \theta_{,i}} \nu_i = 0 , \tag{3.11}$$

$$F + a = -p = \text{const.} \tag{3.12}$$

It is interesting to note that, unless $L_3 = 0$ on both sides of a surface of discontinuity, the jump $\left[\theta_{,i}\right]$ is parallel to ν_i, implying that $[\theta]$ is constant over this surface. In the contrary case, $[F]$ is constant and L_3 vanishes. The latter case vaguely resembles the interfacial situation, where F is to be constant on surfaces where (3.11) holds. If the liquid crystal is uniformly oriented on one side, the correspondence is exact. For present purposes, we might bear in mind that analysis for the two is essentially the same.

When (3.3) applies, we can obtain relatively simple solutions as follows: refer the equations to polar coordinates,

$$x_1 = r\cos\psi , \qquad x_2 = r\sin\psi$$

and set

$$\phi = \theta - \psi ,$$

$$G = rF/k_{33} .$$

A routine calculation gives

$$2G = r\phi_r^2 + r^{-1}(1 + \phi_\psi)^2 + kr\left\{r^{-1}(1 + \phi_\psi)\cos\phi - \phi_r\sin\phi\right\}^2 , \tag{3.13}$$

(3.8) being replaced by

$$\frac{\partial}{\partial r}\frac{\partial G}{\partial \phi_r} + \frac{\partial}{\partial \psi}\frac{\partial G}{\partial \phi_\psi} - \frac{\partial G}{\partial \phi} = 0 . \tag{3.14}$$

Examination of (3.14) shows that there are solutions independent of r,

$$\phi = \phi(\psi) ,$$

governed by the ordinary differential equation

$$\phi'' + k\left\{(1+\phi')\cos^2\phi\right\}' + k(1+\phi')^2\sin\phi\cos\phi = 0. \qquad (3.15)$$

This has some solutions independent of k in addition to the trivial case $\phi = -\psi + \text{const.}$, corresponding to uniform orientation. These are

$$\phi \equiv 0 \quad \mathrm{mod}\ \pi/2 \qquad (3.16)$$

and

$$\phi = \psi + \text{const.} \qquad (3.17)$$

Also, (3.15), multiplied by ϕ', can be integrated once to give

$$\phi'^2(1 + k\cos^2\phi) - k\cos^2\phi = b = \text{const.} \qquad (3.18)$$

Because of the multiplication, this incorporates extraneous solutions with ϕ constant.

Generally, these solutions give θ as a multi-valued function of position, as is exemplified in (3.16) and (3.17). For these two, θ changes by a multiple of 2π as ψ varies from 0 to 2π. There are other cases where it changes by an odd multiple of π, giving rise to a rather trivial discontinuity where $\underset{\sim}{d}$ reverses itself on the ray $\psi = 0$, say, entirely compatible with (3.9) and (3.10). For the case $k = 0$, Frank [1] imposes the condition that the discontinuities be of this simple type, i.e. that $4b$ be the square of an integer. It is easily seen that this restriction is stronger than is required by the jump conditions. Still for the case $k = 0$, F reduces to a function of r alone. On any such circle, $L_3 = 0$. Thus in terms of ideas developed earlier, the shape of the isotropic core is circular. There is not space here to examine in detail solutions with $k \neq 0$. Obviously, F reduces to a function of r alone only in special cases, for example when (3.16) holds. In (3.17), we have an example of an orientation pattern completely independent of k. The curves of constant W are of the form

$$r^2 = c(1 + k\cos^2\phi), \qquad c = \text{const.}$$

With the restriction (2.18), these are closed curves encircling the origin. A calculation shows that, on these curves, $L_3 = 0$, so we can consistently regard one such as an interface bounding an isotropic core. Thus, even in this case, the value of k influences the shape of this core. Lehmann [6, pp. 47-48] suggested that the core is not really a cylinder, its diameter varying with height, orientation being non-planar near it.

For solutions given by (3.18), the energy density F is inversely proportional to r^2, the energy, integrated over an area containing the origin, is infinite. Even if we delete some area near $r = 0$, the energy per unit height, calculated for an infinite area, is infinite. We might thus expect a moderately large sample, containing one disinclination, to quickly find some way to change to a state of lower energy.

4. DISINCLINATION GROUPS

Still considering plane solutions, we consider the possibility of having solutions in the entire plane, involving a finite number of disinclinations, with energy decaying rapidly enough with distance from the group to permit the energy to converge. For this, we envisage excluding regions around the singular points to avoid infinite contributions from these. Also, with these, there may be associated curves of finite discontinuity, provided these are compatible with the proposed jump conditions.

For the case where $k = 0$ it is easy to construct solutions meeting these general requirements, for then θ satisfies a linear equation, Laplace's equation. The idea is to superpose elementary disinclination solutions such as are considered by Frank [1], in such a way that the singularities partially cancel at long distances. For example, we can take

$$\theta = \tan^{-1} \frac{x_2}{x_1 - d} - \tan^{-1} \frac{x_2}{x_1 + d} , \tag{4.1}$$

where $d > 0$ is a constant. With a suitable selection of branches of the arctangents, there is a finite discontinuity in θ of magnitude 2π, along the line segment on $x_2 = 0$ joining the singular points, θ being otherwise continuous except at the two singular points. At distances r from the origin, large compared with d, $\nabla\theta$ is then of order r^{-2}, the energy density of order r^{-4}, certainly fast enough to secure convergence. The corresponding vector $\underset{\sim}{d}$ is continuous everywhere but at the singular points. Thus, it seems energetically advantageous for disinclinations to occur in certain sets of this kind, where they serve to reduce the energy at long distances. From what little knowledge I have, this is consistent with experience, though I would be glad to be informed of evidence to the contrary. In terms of solutions characterized by integers, as are discussed by Frank [1], the important condition is that the sum of these integers vanish. There is consistency with the empirical laws of occurrence described by Friedel [11, p. 368], who also gives rules for combination and disassociation of disinclinations. Of course, the couple free circles associated with one will no longer be couple free and of constant energy, since one disinclination influences orientation near the other. Adding up simple disinclination solutions gives solutions in which L_3, integrated around circuits not passing through singular points, vanishes, though L_3 does not vanish at each point of such a curve. When $k = 0$, the energy density is proportional to

$$\theta_{,i}\, \theta_{,i} \, .$$

This implies that L_3 vanishes on the orthogonal trajectories of the curves θ = const., which are generally not curves on which W = const. Thus simple solutions involving a variety of curves on which interfacial conditions can be satisfied, must be regarded as rarities.

We now give up the assumption that $k = 0$ and turn to a more qualitative exploration. Suppose there are solutions having a finite number of singular points in the plane. If these decay sufficiently rapidly, the stress will approach the uniform hydrostatic pressure

$$t_{ij} = a\, \delta_{ij} \, .$$

Given this, and the critical value W_0, we can use (3.12), with $F = W_0$, to calculate the pressure p of the isotropic phase. With $W_0 > 0$ this will be less than the pressure at infinity. From observations concerning the effect of pressure on phase transition temperatures, such as are summarized by Brown & Shaw [13, p. 1130], it does not seem improbable that the assumed phase transformation might be caused by this local lowering of pressure. It is interesting to note that this pressure is always less than that at infinity and, for a given material and pressure at infinity, is a definite value, disregarding the possibility of jumps in a. It thus seems reasonable to associate with this phase an energy per unit height E_1, which is simply proportional to the area A occupied by the isotropic phase,

$$E_1 = \tilde{W} A \, , \qquad \tilde{W} = \text{const.} \tag{4.2}$$

The corresponding energy per unit height for the liquid crystal phase, E_2, is then given by

$$E_2 = \int_R W \, dx_1 dx_2 \, , \tag{4.3}$$

R being the entire plane, less that part occupied by the isotropic phase. We thus have the total energy per unit height

$$E = E_1 + E_2 \, . \tag{4.4}$$

With respect to its reference level, this clearly vanishes if the entire sample is uniformly oriented in the liquid crystal level, at the temperature considered.

To have $\widetilde{W} \neq W_0$ would seem to promote instability in the sense that the material could then attain lower energy by transforming more or less material to the isotropic phase. The argument is a bit weak, particularly since we envisage these configurations as being somewhat unstable. Tentatively, then

$$\widetilde{W} = W_0 \, . \tag{4.5}$$

This puts us in a position to make one plausible, though not entirely rigorous prediction: Suppose (3.3) applies, and we have some solution of the type discussed, with singular points at $\bar{\underset{\sim}{x}}_1, \ldots, \bar{x}_n$. We desire that the solution be defined, except at these points, even though it applies only in R , where

$$W \leqq W_0 \, .$$

Given this solution, $\theta = \theta(\underset{\sim}{x})$, set

$$\theta_\alpha(\underset{\sim}{x}) = \theta(\alpha \underset{\sim}{x}) \, , \qquad \alpha > 0 \, .$$

It is easily seen that, for constant α , θ_α satisfies the equilibrium equations (3.8) and, in obvious notation,

$$W_\alpha = \alpha^{-2} W \, .$$

The transformation $y = \alpha \underset{\sim}{x}$ maps R onto a region $\bar{R}_\alpha$, and we have

$$E_2 = \int_R W \, dx_1 dx_2 = \int_{\bar{R}_\alpha} W_\alpha \, dy_1 dy_2 \, .$$

The area A is similarly mapped onto a region of area

$$\alpha^2 A .$$

In $\bar{R}_\alpha$,

$$W_\alpha \leqq \alpha^{-2} W_0 . \tag{4.6}$$

Assume that $\alpha < 1$, so the singularities are moved closer together. Then, because of enhancement of energy implied by (4.6), some subregion of $\bar{R}_\alpha$, of area A_α should transform to the isotropic phase, and all parts which started isotropic should be mapped in to regions which remain isotropic. There will remain a region $R_\alpha \subset \bar{R}_\alpha$, which should remain in the isotropic phase. Note that the boundaries of the isotropic regions, as estimated, may be subject to surface couples, leading to a readjustment problem of the type discussed earlier. Ignoring this, we estimate the new energy as

$$E_\alpha = W_0(\alpha^2 A + A_\alpha) + \int_{R_\alpha} W_\alpha \, dy^1 dy^2 ,$$

so the change in energy is

$$\begin{aligned} E - E_\alpha &= W_0(1 - \alpha^2)A - W_0 A_\alpha + \int_{\bar{R}_\alpha} W_\alpha \, dy_1 dy_2 - \int_{R_\alpha} W_\alpha \, dy^1 dy^2 \\ &\geqq W_0(1 - \alpha^2)A > 0 . \end{aligned} \tag{4.7}$$

Thus, according to this estimate, the singularities tend to attract each other, granted that they will tend to move to reduce the total energy.

Clearly, there is considerable room for improvement in the theory here presented, and several improvements are incorporated in a forthcoming paper by Dafermos [14]. Among other things, he discusses calculation of forces exerted by a set of disinclinations on each other.

ACKNOWLEDGMENT

This work was supported by a grant from the National Science Foundation.

REFERENCES

1. F. C. Frank, On the theory of liquid crystals, Discuss. Faraday Soc. 25, 19-28 (1958).
2. J. L. Ericksen, General solutions in the hydrostatic theory of liquid crystals, Trans. Soc. Rheol. 11, 5-14 (1967).
3. J. L. Ericksen, Propagation of weak waves in liquid crystals of nematic type, J. Acoust. Soc. Am. 44, 444-446 (1968).
4. J. L. Ericksen, Continuum theory of liquid crystals, Applied Mech. Rev. 20, 1029-1032 (1967).
5. C. W. Oseen, The theory of liquid crystals, Trans. Faraday Soc. 29, 883-899 (1933).
6. O. Lehmann, Flüssige Kristalle und ihr scheinbares Leben, Verlag von Leopold Voss, Leipzig 1921.
7. J. L. Ericksen, Hydrostatic theory of liquid crystals, Arch. Rat'l. Mech. Anal. 9, 371-378 (1962).
8. J. L. Ericksen, Inequalities in liquid crystal theory, Physics Fluids 9, 1205-1207 (1966).
9. F. M. Leslie, Some constitutive equations for liquid crystals, Arch. Rat'l. Mech. Anal. 28, 265-283 (1968).
10. P.-G. de Gennes, Fluctuations d'orientation et diffusion Rayleigh dans un cristal nématique, Comptes Rendus 266(B), 15-17 (1968).
11. J. L. Ericksen, Nilpotent energies in liquid crystal theory, Arch. Rat'l. Mech. Anal. 10, 189-196 (1962).
12. G. Friedel, Les états mésomorphes de la matière, Ann. de Phys. 18 (9), 273-474 (1922).
13. G. H. Brown and W. G. Shaw, The mesomorphic state: liquid crystals, Chem. Rev. 57, 1049-1157 (1957).
14. C. M. Dafermos, Disinclinations in liquid crystals, forthcoming.

THEORY OF LIGHT SCATTERING BY NEMATICS

Orsay Liquid Crystal Group
Service de Physique des Solides
Faculté des Sciences
Orsay, France

ABSTRACT: The fluctuations of the molecular alignment in pure nematic systems give rise to a strong scattering of light. A study of the scattering intensity vs. angle allows for a determination of (a) the Frank elastic coefficients. (b) the sum of the two piezoelectric coefficients introduced by Meyer. The frequency distribution of the scattered light can be related to the Leslie friction coefficients.

In impure (slightly conducting) nematics, the piezoelectric charges induced by the fluctuations of interest are screened out when the Debye Huckel radius is smaller than the optical wavelength. Thus, for strongly piezoelectric nematics, the light scattering properties might be sensitive to very small amounts of conducting impurities.

I. FLUCTUATION MODES IN A PURE (INSULATING) NEMATIC

Light scattering by nematics is large (1). Early discussions of this effect were based on the somewhat vague concept of "swarms." We know now that scattering by a nematic single crystal is due to thermal fluctuations of the molecular alignment (2). For a given scattering geometry, i.e., for a given scattering vector $\underset{\sim}{q}$, the fluctuations of interest have a sinusoidal space dependence with wave vector $\underset{\sim}{q}$. There are two independent modes, as displayed on fig. 1. Mode 1 is a combination of bending and splay, in Frank's notation (3). Mode 2 is a combination of bending and twist.

If δ_1 and δ_2 are the angular amplitudes associated with

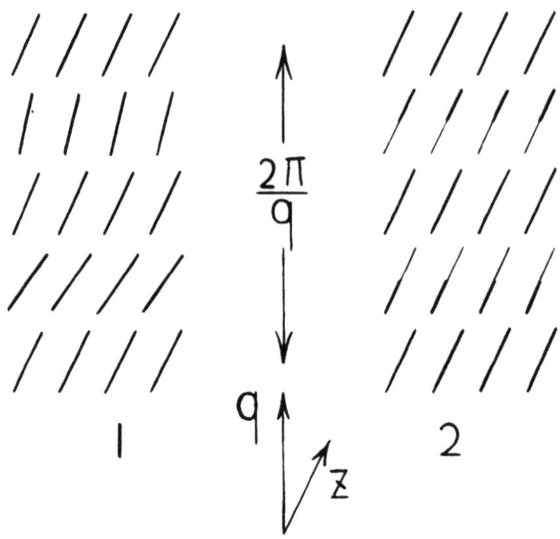

Fig. 1. The two fluctuation modes associated with a given wave-vector $\underset{\sim}{q}$ in a nematic single crystal. The direction of the molecules at rest is Oz. In mode 1 the molecules remain in the plane of the sheet. In mode 2 they are tilted with respect to this plane.

these two modes, the increase of free energy due to the associated distortions is of the form:

$$F = \frac{1}{2} K_1(\underset{\sim}{q}) |\delta_1|^2 + \frac{1}{2} K_2(\underset{\sim}{q}) |\delta_2|^2 \tag{1.1}$$

$$K_1(\underset{\sim}{q}) = K_{33} q_z^2 + K_{11} q_\perp^2 + \chi_a H^2 + \frac{(4\pi e_p q_z q_\perp)^2}{\varepsilon_{\parallel}^0 q_z^2 + \varepsilon_\perp^0 q_\perp^2} \tag{1.2}$$

$$K_2(\underset{\sim}{q}) = K_{33} q_z^2 + K_{22} q_\perp^2 + \chi_a H^2 \tag{1.3}$$

In these formulae, q_z and $q_\perp$ are the components of $\underset{\sim}{q}$ parallel and normal to the optical axis Oz. The constants K_{ii} are the Frank elasticity coefficients (3). χ_a is the anisotropy of the magnetic susceptibility, and is assumed to be positive. H is the magnetic field, applied along Oz to align the specimen. The terms involving the elastic constants and χ_a in (1.2) and (1.3) have already been discussed (2). The last term in eq. (1.2) is less familiar; it is a consequence of the piezoelectric effects recently introduced by Meyer (4). The distortions occurring in mode 1 create a sinusoidal charge distribution, and the term under discussion represents the electrostatic self energy of these charges. $e_p = e_{11} + e_{33}$ is the sum of the two Meyer coefficients, as defined in ref. (4). Finally, $\varepsilon^{\circ}_{\parallel}$ and $\varepsilon^{\circ}_{\perp}$ are the static dielectric constants of the nematic. It is important to note that the piezoelectric contribution in equ. (1.2) does not amount to a simple redefinition of the elastic constants: the angular dependence is more complex.

As shown in ref. (2), the scattering intensity for a given wave vector $\underset{\sim}{q}$ depends essentially on the thermal averages

$$< |\delta_i|^2 >$$

and, by suitable choices of polarisations, etc., these two averages may be extracted from the data. Their theoretical values may be derived from the equipartition theorem applied to eq. (1.1):

$$< |\delta_i|^2 > = \frac{kT}{K_i(\underset{\sim}{q})} \qquad (1.4)$$

Knowing only χ_a , and repeating the experiment with various values of q_z , $q_\perp$, and H, one may in principle obtain (using only relative measurements of intensity) the values of the elastic constants and of e_p . Unfortunately, a systematic study of this sort has not been undertaken yet, but the existing data -- for para-asoxyanisol, in 0 magnetic field -- (1), do agree with the general $1/q^2$ behavior predicted by eq. (1.4). It is not clear yet whether the piezoelectric terms are large enough to be detected in this system. (Also, they may sometimes be wiped out, by impurity effects, as we shall see in section 3).

2. DYNAMICS OF THE FLUCTUATIONS

With the present laser sources and beat methods, it is also possible to analyse the frequency distribution of the outgoing beam. Typical frequency widths, for the present problem, are in the kilocycle range. This provides us with crucial information on the dynamics of the fluctuations.

For the simplest case of non-piezoelectric nematics, the theoretical frequency distribution for modes 1 and 2 has recently been derived from the Ericksen-Leslie hydrodynamic equations (5). It is in fact possible to include piezoelectric effects in the analysis, and the results are as follows: for each mode (i=1 or 2) we expect a well defined relaxation rate, of the form:

$$\frac{1}{\tau_i(\underset{\sim}{q})} = \frac{K_i(\underset{\sim}{q})}{\eta_i} \qquad (2.1)$$

$K_i(\underset{\sim}{q})$ is defined in eq. (1.2)(1.3). η_i is an effective viscosity, related to the friction coefficients introduced by Leslie, and dependent on the direction of $\underset{\sim}{q}$ with respect to the nematic axis (6). The main assumptions underlying eq. (2.1) are that (a) the Leslie equations remain valid in piezoelectric materials; (b) a certain inequality

$$K\rho \ll \eta^2$$

be satisfied, where K is an average elastic constant, ρ the density, and η an average viscosity. (For materials such as p-asoxyanisol, the inequality is largely fulfilled.)

From the experimental frequency distributions one can extract the two relaxation rates $1/\tau_1$ and $1/\tau_2$; if $K_i(q)$ is known, one may then obtain η_i . This experiment, repeated for different q values, should ultimately lead to a measurement of all the Leslie friction coefficients. The method has been applied recently to p-asoxyanisol, and results will be presented in a companion paper (7). This optical technique is somewhat more powerful than conventional viscosity measurements, the latter being sensitive to orientational effects from the walls, disclination lines (3) from upstream, etc.

3. EFFECTS OF CONDUCTING IMPURITIES

Most commercial nematics display a weak but finite electrical conductivity (typically 10^{-9} ohms^{-1} cm^{-1}). Thus the piezoelectric charges predicted by the Meyer theory may be screened out if the Debye Huckel radius r_s is small enough. The distance between positive and negative charges in our fluctuation mode 1 is of order 1/q, and screening will be important if:

$$q r_s \ll 1 \qquad (3.1)$$

To estimate r_s , let us assume an assembly of n_o carriers per cm^3, each of charge e , and having the mobility of a Stokes sphere of radius R , moving in a fluid of viscosity η . Then the conductivity σ is given by:

$$\sigma = \frac{n_o e^2}{6\pi R \eta} \qquad (3.2)$$

Taking R = 3 Å, $\eta = 10^{-2}$ poise, $\sigma = 10^{-9}$ ohms^{-1} cm^{-1} (10^3 esu) we find $n_o = 2.10^{13}$. With an average dielectric constant of $\varepsilon \cong 2$ we get:

$$r_s \cong \left[\frac{kT\varepsilon}{4\pi n_o e^2} \right]^{1/2} \cong 3{,}500 \text{ Å} \qquad (3.3)$$

Again making an isotropy approximation, we may write for q :

$$q \cong \frac{4\pi}{\lambda} \sin(\theta/2) \qquad (3.4)$$

where λ is the optical wavelength and θ the scattering angle. We see that r_s is comparable to λ ; thus at small θ we will have $q r_s < 1$ (strong screening) while at large angles we will have $q r_s > 1$ (weak screening).

More precisely, the formulae of section 1 for the scattering intensity may be shown to remain valid, provided that we make in eq. (1.2) the substitutions:

$$\left.\begin{aligned} \varepsilon^{o}_{/\!/} &\to \varepsilon^{o}_{/\!/} + \frac{4\pi n_o e^2}{q^2} \\ \varepsilon^{o}_{\perp} &\to \varepsilon^{o}_{\perp} + \frac{4\pi n_o e^2}{q^2} \end{aligned}\right\} \qquad (3.5)$$

This amounts to included in the static dielectric constant of the undistorted nematic the contribution of the free carriers.

It is also possible to analyse the effect of these carriers on the dynamic properties discussed in section 2. (8). As expected, it is only the mode 1 which is coupled to the carriers. The new relaxation rate $(1/\tau_1)$ for this mode is given by the implicit equation

$$0 = q_z^2 \varepsilon^{o}_{/\!/} + q_\perp^2 \varepsilon^{o}_{\perp} + \frac{4\pi\sigma q^2}{Dq^2 - 1/\tau_1} \cdot \frac{(4\pi e_p q_z q_\perp)^2}{\eta_1 (u_{s1} - 1/\tau_1)} \qquad (3.6)$$

Here $D = \sigma kT/n_o e^2$ is the standard diffusion constant of the carriers, and u_{s1} is the relaxation rate in the absence of any piezoelectric effects. (For simplicity, we have taken D and σ to be isotropic.) Eq. (3.6) has two roots, one $(1/\tau_1')$ describing orientational relaxation, and the other $(1/\tau_1'')$ describing carrier diffusion. If the carriers and the nematic molecules are of comparable size, it is possible to see from equ. (3.2) and (2.1) that both roots are of the same order of magnitude.
We are mainly concerned here with $1/\tau_1'$, the exact value of which depends mainly on the quantity:

$$g(\underset{\sim}{q}) = \frac{4\pi\,(e_p q_z q_\perp)^2}{\sigma \eta_1 q^2} \qquad (3.7)$$

if $g \gg 1$, we recover eq. (2.1): free carrier effects are negligible. If $g \ll 1$, we have $1/\tau_1' = u_{s1}$ i.e., piezoelectric effects do not play a role in the dynamic properties.

At the time of writing (July 1969) we know of no experiments bearing on these carrier effects. But our results do suggest the following type of experiments:

(a) find a nematic with strong piezoelectric effects (i.e., where the last term is a significant contribution to $K_1(\underset{\sim}{q})$

in equ. (1.2);

(b) study the scattering of light by this material under various purity conditions, corresponding to a broad range of electrical conductivities (around 10^{-9} ohms^{-1} cm^{-1}). Our prediction is that both the intensity and the frequency width of the scattered light should then depend significantly on the conductivity.

REFERENCES:

(1) P. Chatelain, Acta Cryst. 1, 315, (1948).
(2) P. G. de Gennes, Compt. Rend. Acad. Sci. Paris 266, 15 (1968). See also: Proceedings of the 2nd Kent Conf. on Liquid Cryst. (to be published).
(3) F. C. Frank, Disc. Faraday Soc., 25, 1, (1958).
(4) R. B. Meyer, Phys. Rev. Lett. 22, 918, (1969).
(5) J. L. Erickson, Arch. Rat. Mech., 9, 371, (1962). F. M. Leslie, Quart. Journ. Mech. and Appl. Math. 19, 357, (1966).
(6) Orsay Liquid Crystal Group: to be published in Journ. Chem. Phys.
(7) Orsay Liquid Crystal Group: this symposium, p. 447.
(8) O. Parodi, to be published.

EFFECTS OF ELECTRIC FIELDS ON MIXTURES OF NEMATIC AND CHOLESTERIC LIQUID CRYSTALS*

E.F. Carr, J.H. Parker, and D.P. McLemore

Physics Department, University of Maine, Orono, Maine

ABSTRACT

This work involves nematic materials mixed with small amounts of cholesteryl acetate. It is assumed that the structures for these mixtures are similar to the cholesteric structure. The dielectric constant of p-methoxybenzylidene-p-cyanoaniline is greatest in a direction parallel to the long axes of the molecules; therefore, the structure for mixtures of this material with a cholesteric material can be changed to a nematic structure by applying external electric fields. NMR techniques and measurements of the dielectric loss at a microwave frequency are used to obtain information about the cholesteric-nematic phase transition as the external electric field and the concentration of cholesteryl acetate are varied. Although mixtures of p-[n-(p-methoxybenzylidene)-amino]-phenyl acetate and cholesteryl acetate cannot be changed to a nematic phase by employing electric fields because of the negative dielectric anisotropy, an ordering can be obtained with the screw axis of the helix parallel to a 300 kHz electric field.

INTRODUCTION

Mixtures involving a small amount of a liquid crystal of the cholesteric type dissolved in a nematic material have been known to exhibit a structure[1] similar to that of the cholesteric liquid crystal. Meyer[2] and Durand, Leger, Rondelez and Veyssie[3] have shown that these mixtures can be changed to nematic structures by applying external magnetic fields. These investigations supported the theory presented earlier by de Gennes[4] and Meyer[5].

The primary object of this work is to study the effects of

electric fields on the ordering in mixtures of cholesteric and nematic materials. Part of this investigation involves p-|n-(p-methoxybenzylidene)-amino|-phenylacetate (MBA), which exhibits a negative dielectric anisotropy at low frequencies, and CA. The nematic materials were chosen because of their large dielectric anisotropy. Carr[6] has shown that the behavior of MBA in electric fields is predictable at frequencies of a few hundred kHz. Although we are not aware of other work involving the effect of electric fields on nematic materials mixed with small amounts of cholesteric liquid crystals, work employing electric fields to study mixtures of cholesteric materials has been reported. Wysocki, Adams and Haas[7] have observed a field-induced phase transition in a mixture of cholesteryl chloride, nonanoate, and oleyl carbonate in an electric field of (3-4) X 10^5 V/cm. Baessler and Labes[8] have reported that a cholesteric-nematic phase transition can be induced by relatively weak fields in a mixture of cholesteryl chloride, and myristate, and that the threshold field varies inversely with the pitch of the helix as predicted[4,5].

The size of our samples is much larger than that used by most observers for related investigations, but is comparable to those used by Sackman, Meiboom, and Snyder[9], who investigated the effect of a magnetic field on cholesteric liquid crystals using an indirect NMR technique. In a later paper employing optical techniques Sackmann, Meiboom, Snyder, Meixner, and Dietz[10] discuss two cases that must be differentiated when discussing the behavior of cholesteric liquid crystals in a magnetic field. In the first case the long axis of the molecules tends to align parallel to the magnetic field, and no macroscopic alignment takes place unless the magnetic field is strong enough to unwind the helical structure. In the second case, the long axis of the molecules tends to align perpendicular to the magnetic field, and the axis of the helical structure aligns parallel to the magnetic field. This is similar to what is reported in this article except this work employs electric fields and is primarily concerned with the dielectric rather than the magnetic anisotropy.

EXPERIMENTAL

Two different experimental techniques were involved in this work. In one method measurements of the dielectric loss at a microwave frequency of 24 GHz were used to provide information about the ordering. This technique[11] involves a center plate inside the wave guide, but insulated from the walls so that an external electric field can be applied parallel to the microwave electric field in the guide. The second technique employed standard wide-line NMR equipment, except the sample holder contains parallel conducting plates used for applying an external electric field. This was discussed earlier by Carr, Hoar, and MacDonald[12]. All the NMR spectra were obtained at a relatively low magnetic field strength of approximately 1500 gauss. This required that the spectrometer operate at a very low r-f level which added noise to the spectra, but the authors believed this to be important because high magnetic fields would appreciably affect some results when studying electric effects. The relative intensities of lines within a given spectrum are significant. However, there is no correlation of intensity implied between different spectra because of routine adjustments in the NMR spectrometer.

For all the results reported in this article involving MBA a frequency of approximately 300 kHz was used for the externally applied electric fields. Electric fields at frequencies of approximately 5000 Hz were employed for most of the work involving MBC.

All the samples were obtained commercially and purified by recrystallization.

P-[N-(P-METHOXYBENZYLIDENE)-AMINO]-PHENYL ACETATE (MBA) AND CHOLESTERYL ACETATE (CA)

Since MBA exhibits a negative dielectric anisotropy at low frequencies and the mixtures are predominately MBA an ordering is preferred with the helix axis parallel to an externally applied electric field. This ordering is demonstrated in Fig. 1 for 1 mole % CA in MBA at a temperature of 90°C using the NMR spectrum of MBA. Fig. 1a shows a spectrum of pure MBA in a 300 kHz electric field of 4200 V/cm applied parallel to the magnetic field. This spectrum shows two inner and two outer side peaks symmetrically spaced about a center peak. This is the wide-line spectrum for the long axes of the molecules of MBA oriented perpendicular to the magnetic field. The details of this spectrum were discussed by Carr, Hoar and MacDonald[12].

Fig. 1b shows the NMR spectrum for 1 mole % CA in MBA in the absence of any externally applied electric field. The spectrum shows some evidence of the inner and outer side peaks with a spacing comparable to that shown in Fig. 1a. This implies a wall effect causing the long axes of the molecules to prefer an orientation parallel to the walls (the helix axis perpendicular to the walls). If wall effects are significant they should be indicated by the spectrum, because the conducting plates divided the sample holder into rectangular cells with dimensions of 0.24 cm X 0.8 cm, and the wide side of the cells was perpendicular to the magnetic field while the spectrum in Fig. 1b was obtained.

The NMR spectrum for 1 mole % CA in MBA in a 300 kHz electric field of 4200 V/cm applied parallel to the magnetic field is shown in Fig. 1c. The similarity of this spectrum with the one shown in Fig. 1a implies that the long axes of the molecules are perpendicu-

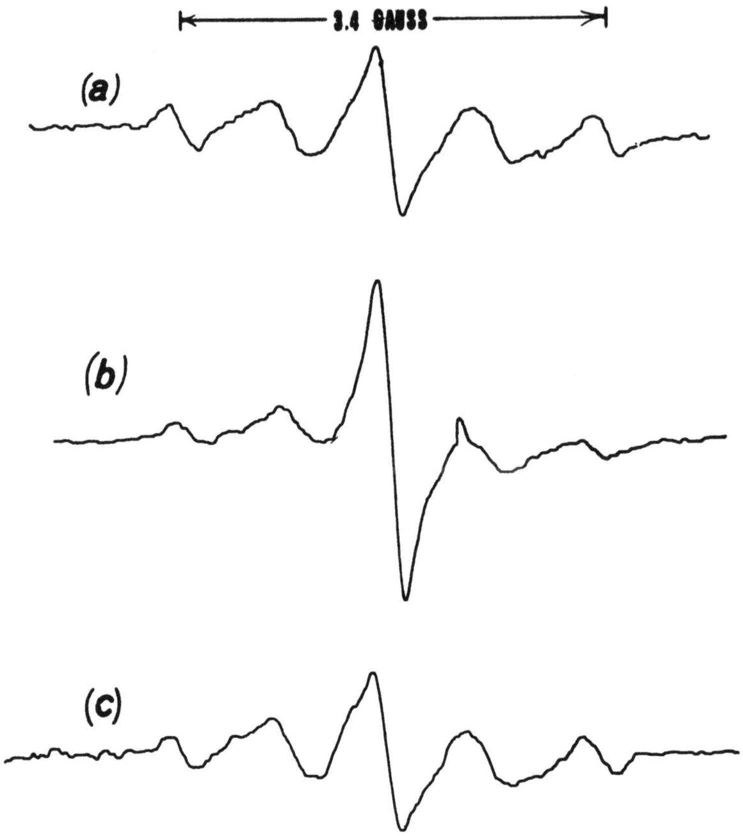

Fig. 1. NMR spectra at 6.4 Mc. and 90°C.
(a) P-[n-(p-methoxybenzylidene)-amino]-phenyl acetate (MBA) with a 300 kHz electric field of 4200 V/cm applied parallel to the external magnetic field; (b) Mixture of 99 mole % MBA and 1 mole % cholesteryl acetate (CA); (c) Mixture of 99 mole % MBA and 1 mole % CA in 300 kHz electric field of 4200 V/cm applied parallel to the external magnetic field.

lar to the magnetic field throughout the entire sample. If one assumes a helical structure then the axis of the helix is parallel to the electric field. Although the spectrum for a 90 degree rotation of the electric field was not analyzed in detail, it did show clearly that the sample did not exhibit a nematic structure.

Fig. 2 shows the effect of electric and magnetic fields on the ordering in MBA containing 1 mole % CA employing dielectric techniques. For given values of the magnetic field the dielectric loss is plotted as a function of the 300 kHz electric field applied parallel to the magnetic field. For any given value of the magnetic field, values for the electric field were reached beyond which the dielectric loss did not show an appreciable increase. This suggests that for high electric fields all regions of the sample are ordered with the helix axes parallel to the electric field, and this was assumed when plotting the results shown in Fig. 2. The best evidence for this assumption is shown in Fig. 1 which was previously discussed. In analyzing the data for the results shown in Fig. 2, a number of assumptions had to be made. The dielectric loss in the mixture was assumed not to be appreciably different from that of pure MBA for an ordering with the long molecular axes perpendicular to the microwave electric field[6]. Since only relative values of the dielectric loss were obtained with the apparatus employing electric fields, and an ordering with the long axes of the molecules parallel to the microwave electric field was not possible with the available equipment for this particular mixture, the effective length of the sample was not known as accurately as in previous work[6,9]. Because of some fluxuations in the sample, plots of the dielectric loss for two separate runs will not always be identical. The curves shown represent a typical set of results.

Because of the diamagnetic anisotropy an ordering in a magnetic field is preferred with the helix axis perpendicular to the field and this would correspond to a value of $\varepsilon'' = 0.42$ for the dielectric loss. The results shown in Fig. 2 indicate that in a field of 9000 gauss this is not the case, whereas, an electric field of 4000 V/cm, which is equivalent to a 9000 gauss magnetic field for producing molecular alignment[6] in pure MBA, is very effective in producing an ordering with the helix axis parallel to the electric field. The ordering in a 4000 V/cm field is nearly complete with a 6000 gauss field opposing it. The spacing between the curves for various values of the magnetic field is similar to that reported earlier[6] for pure MBA. A noticeable difference is a hysteresis effect which is indicated by the dotted curve for a field of 9000 gauss.

Fig. 3 illustrates the effect of an electric field on a mixture of MBA and CA containing 25 mole % CA. A visible inspection of this mixture showed green and red colors in the anisotropic

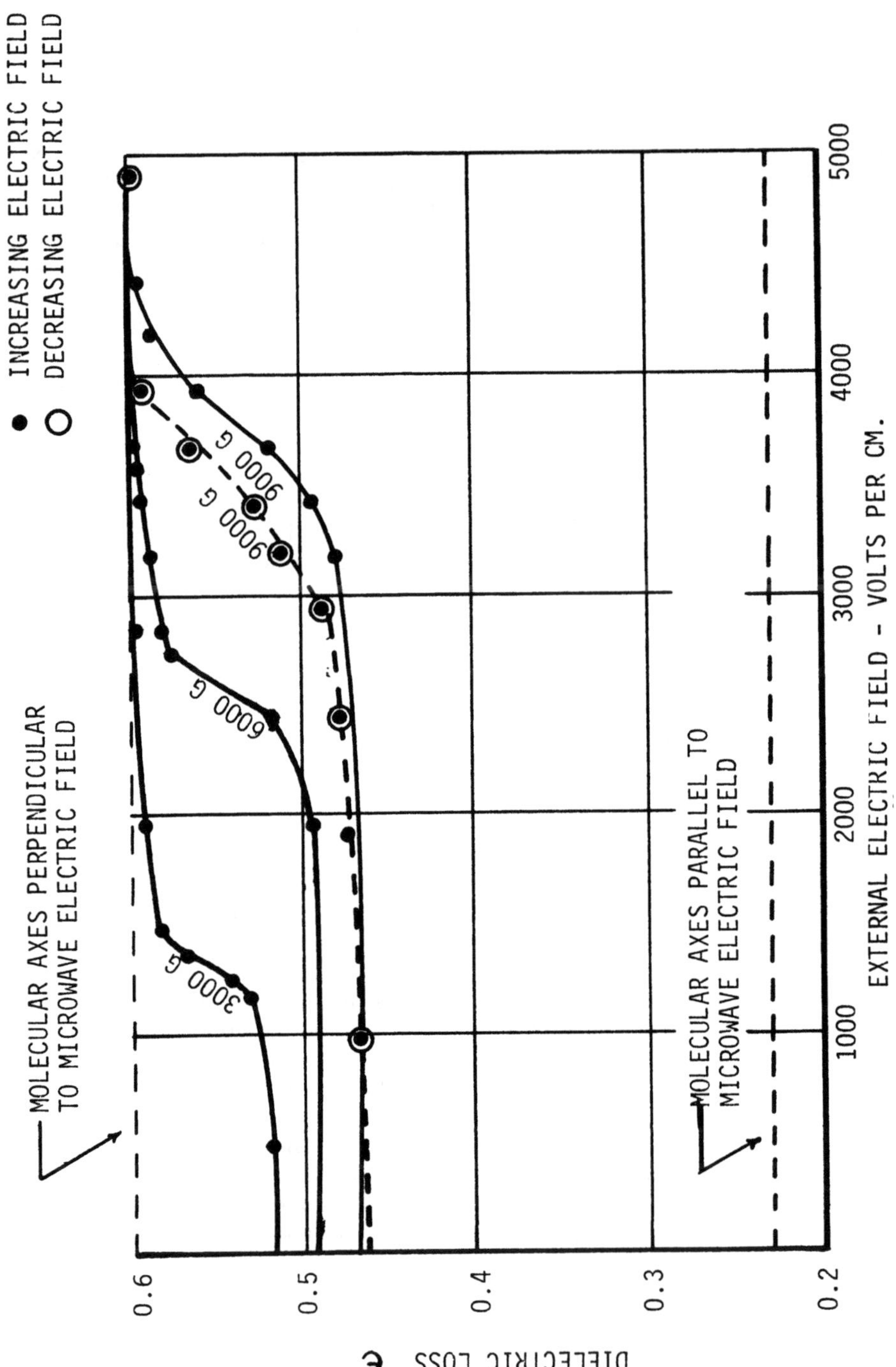

Fig. 2. Dielectric loss in 1 mole % mixture of CA in MBA at 24 GHz and 90° C as a function of a 300 kHz electric field applied to an external magnetic field.

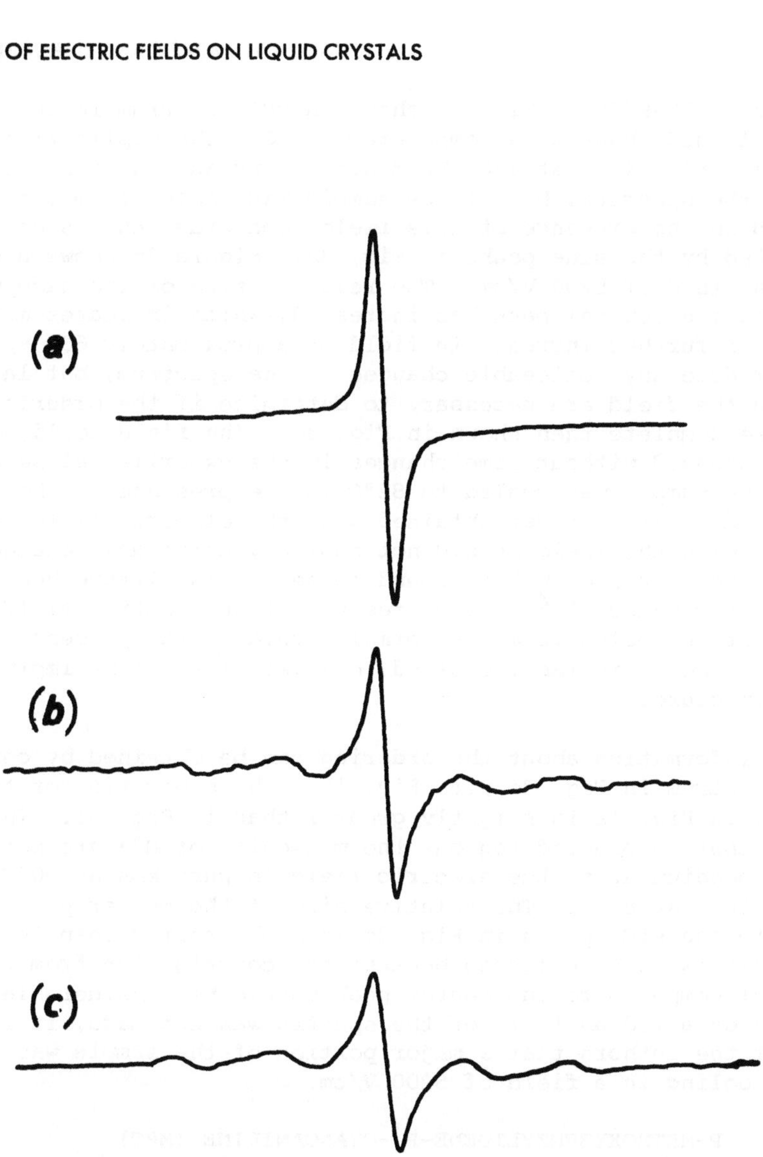

Fig. 3. NMR spectra at 6.4 Mc and 82° C in mixture of 75 mole % MBA and 25 mole % CA
(a) No electric field; (b) Cooled from normal liquid in 300 kHz electric field of 4200 V/cm; (c) Cooled from normal liquid in field of 5200 V/cm.

liquid phase (77-96°C). Fig. 3a shows an NMR spectrum in the anisotropic liquid phase at a temperature 82°C. The application of a 300 kHz electric field at 4200 V/cm did not produce a significant change in the spectrum, but if the sample was cooled from the normal liquid in the presence of this field some ordering was obtained as indicated by the side peaks in Fig. 3b. Figure 3c shows a spectrum for a field of 5200 V/cm. The relative size of the side peaks compared to the central peak has increased, which indicates more ordering. A further increase in field of approximately 600 V/cm did not produce any noticeable changes in the spectrum, but larger changes in the field are necessary to determine if the ordering can be more complete than shown in Fig. 3c. The field could not be further increased without some changes in the experimental setup. Although the sample was cooled to 82°C in the presence of the field, the spectrum in Fig. 3c was obtained with the electric field off. A spectrum with the field on did not show any noticeable changes other than an increase in background noise. This mixture behaves like a smectic material[14] in the presence of an electric field in that it must be cooled from its normal liquid in the presence of the field to be appreciably ordered; however, its colors imply a helical structure.

Some information about the ordering can be obtained by comparing the spectrum in Fig. 3c with Fig. 1a. The separation of the side peaks in Fig. 1a is slightly greater than in Fig. 3c. This indicates that in ordered regions the molecules of MBA are more nearly perpendicular to the electric field in pure MBA at 90°C than in the mixture at 82°C. The relative size of the center peak as compared to the side peaks in Fig. 3c is much greater than in Fig. 1a, but this is to be expected because the contribution from the cholesteryl compound to the center peak should be considerable. Although a detailed analysis of the spectra was not made, it is the opinion of the authors that a major portion of the sample was ordered by cooling in a field of 5200 V/cm.

P-METHOXYBENZYLIDENE-P'-CYANOANILINE (MBC) AND CHOLESTERYL ACETATE (CA).

Since MBC exhibits a positive dielectric anisotropy at low frequencies and the mixtures are predominately MBC, an ordering is preferred with the helix axis perpendicular to an externally applied electric field. An electric field should be capable of producing a cholesteric-nematic phase transition similar to that reported by others[7,8] using mixtures that were predominately of the cholesteric type. Because of the behavior of nematic materials in electric fields, much of this discussion will have to be qualitative.

The cholesteric-nematic phase transition is illustrated in Fig. 4 employing the wide-line NMR spectrum of MBC. The conductivity of the sample was such that the heating effect in the presence of high electric fields was appreciable. The NMR spectrum for pure MBC in a 4000 V/cm electric field, applied parallel to the magnetic field, was not identical to that with no electric field. It is felt that turbulence owing to the electric fields, as well as temperature gradients, affected the spectrum; therefore, the spectrum of pure MBC in an external field of approximately 4000 V/cm was used to illustrate the electric field effects in 2 mole % CA in MBC. This spectrum is shown in Fig. 4a. The main difference is that the relative size of the central peak is larger in the presence of the electric field, which suggests a lot of motion in small portions of the sample at any instant. The spectrum shows a triplet structure similar to that reported by Jain, Lee and Spence[13] for P-azoxyanisole. The central component which shows some structure arises from the protons in the methyl and amino groups, and the side peaks are the result of a dipole-dipole splitting of the ortho aromatic protons.

The spectrum for 2 mole % CA in MBC in the absence of an electric field is shown in Fig. 4b. This spectrum implies ordered regions that are randomly oriented, because the dipole-dipole interactions in MBC are more like the dipole-dipole interactions of a solid in polycrystalline form than in a liquid. Figures c, d, and e show spectra with electric fields of 2500, 3300 and 4200 V/cm respectively. The similarity of Figures 4a and 4c shows that at a field of 4200 V/cm the structure is nematic. As the electric field increases the side peaks, with a separation comparable to the pure MBC increase in size, which implies that regions exist in the sample for fields of 2500 and 3300 V/cm where the ordering is as complete as in the nematic phase. The length of these regions should be less than the pitch as shown in the schematic representation (Fig. 1) in Meyer's[2] article.

The cholesteric-nematic phase transition has been induced by electric fields in mixtures of MBC and CA with the percentage of CA varying from 0.25 to 2 mole % employing microwave dielectric techniques to measure the degree of ordering. The results for 2 mole % agree quite well with the results shown in Fig. 4. Because of some uncertainty in the behavior of MBC at high electric field strengths and some minor problems involving the design of the experimental setup for these particular materials, a good quantitive check with theory is not possible at this time. For 0.25 mole % of CA the helical structure was destroyed at approximately 6000 gauss. The corresponding critical electric field was approximately 500 V/cm. This was as expected since preliminary investigations using pure MBC had indicated that 1000 gauss was equivalent to an electric field of 81V/cm for producing molecular alignment. The critical field is difficult to determine exactly from plots of the dielectric loss, because the dielectric loss is a measure of the

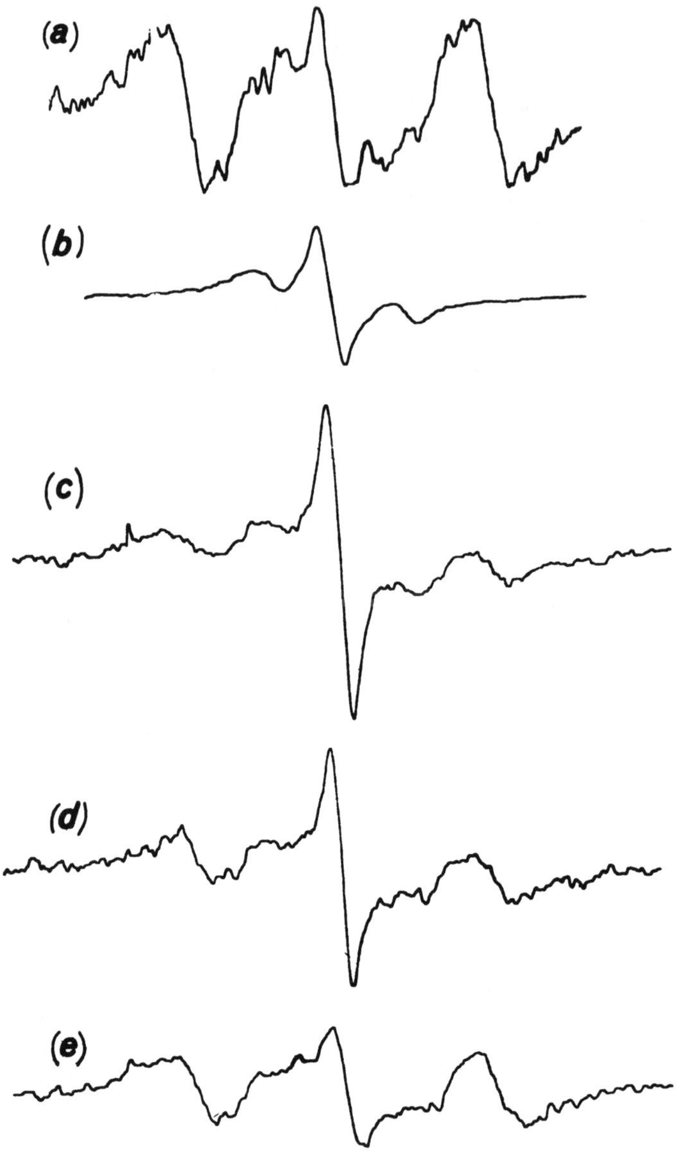

Fig. 4. NMR spectra at 6.4 Mc and 108°C
(a) P-methoxybenzylidene-p'-cyanoaniline (MBC) with 5000 Hz electric field parallel to the magnetic field; (b) 2 mole % cholesteryl acetate (CA) in MBC with no electric field; (c) 2 mole % CA in MBC with electric field of 2500 V/cm; (d) 2 mole % CA in MBC with electric field of 3300 V/cm; (e) 2 mole % CA in MBC with electric field of 4200 V/cm.

degree of ordering relative to that in the ordered nematic phase, and the ordering is not necessarily complete when the helical structure has been destroyed.

If the mixtures of MBC and CA behave like other mixtures containing nematic and cholesteric materials reported by Friedel[1] one would expect the pitch to be inversely proportional to the amount of CA. Since Baessler and Labes[8] have shown that the threshold electric field varies inversely with the pitch of the helix for mixtures of cholesteric materials, which is consistent with other observations[3,9] employing magnetic fields, one would expect the threshold field to be proportional to the amount of CA. The agreement is not as good as one would like, but it is the opinion of the authors that with improvements in the purity of the samples, and design of the experimental setup and a better understanding about the behavior of MBC in high fields, the agreement could be quite good.

CONCLUSIONS

The results of this work show that ordering in mixtures of nematic and cholesteric materials can be obtained for large samples by employing electric fields. A mixture in which the predominant component was a nematic material, with a negative low frequency dielectric anisotropy, can be ordered with its helix axis parallel to the electric field. The frequency of the applied field is critical. Although a frequency of 300 kHz was employed, the results could be obtained with lower frequencies. At low audio frequencies the results were different and are left for a later study. For mixtures of MBA and CA containing a few mole % CA, a field of a few thousand V/cm was very effective. For a mixture containing 25 mole % CA, which exhibited red and green colors, ordering was obtained only by cooling from the normal liquid in the presence of a high electric field. Although the field intensity was limited by the present experimental setup, it is the opinion of the authors that, if the sample is cooled in the presence of a field of at least 5000 V/cm, a major portion of the sample will become ordered with the helix axis parallel to the field. Some preliminary measurements with a mixture of p-azoxyanisole and CA indicated a behavior similar to the mixture of MBA and CA, but much higher fields are required because the low frequency dielectric constant in p-azoxyanisole is less anisotropic than in MBA.

For mixtures in which the nematic material exhibited a positive dielectric anisotropy, the helix axis prefers a direction perpendicular to the field. If a magnetic field is applied perpendicular to the electric field, the helix axis should prefer a direction perpendicular to both fields. Although our evidence to support this is not as good as the evidence for the ordering previously discussed, it is supported by the measurements involving microwave dielectric techniques. A magnetic field of 10,000 gauss and an electric field of at least comparable effectiveness in producing molecular alignment in the nematic material can be very effective for mixtures containing a few mole % CA. It is possible that by cooling from the normal liquid phase in the presence of crossed electric and magnetic fields a significant amount of ordering can be obtained in mixtures containing a large percentage of CA. Minor changes in the experimental procedure are now planned to check this possibility.

Electric fields can be very effective in producing a cholesteric-nematic phase transition for mixtures containing MBC. This is because the low frequency dielectric constant is highly anisotropic. An electric field of only 800 V/cm is as effective as a magnetic field of approximately 10,000 gauss. Although some experimental difficulties and a lack of knowledge about the behavior of MBC in high electric fields prevented a good quantitative check with theory[4,5], the results did indicate that good agreement is possible for mixtures containing a high as well as a low percentage of cholesteric material.

For investigations involving a helical structure, the mixtures of nematic and cholesteric materials should be quite useful. The experiments can cover a wide range of pitch, and the properties of one of the components of the mixture can be more easily obtained than the properties of most materials forming a helical structure. This is because the nematic material can be easily aligned in its pure state and the anisotropy associated with many of its properties can readily be determined.

ACKNOWLEDGMENTS

The authors wish to express their appreciation to Mr. Ronald Rippel and Mr. David Barker who helped with the measurements involving microwave techniques. They also wish to express their thanks to Mr. Peter Robinson for his valuable technical assistance.

FOOTNOTES AND REFERENCES

* Supported in part by U.S. Army Research Office - Durham Grant number DA-ARO-D-31-124-G1042

1. G. Friedel, Ann. Phys. (Paris) 18, 272 (1922); Compte Rend. 17b, 475 (1923).

2. R.B. Meyer, Bull. Am. Phys. Soc. 14, 73 (1969); Appl. Phys. Letters, 14, 208 (1969).

3. G. Durand, L. Leger, F. Rondelez, and M. Veyssie, Phys. Rev. Letters, 22, 227 (1969).

4. P.G. deGennes, Solid State Commun. 6, 163 (1968).

5. R.B. Meyer, Appl. Phys. Letters 12, 281 (1968).

6. E.F. Carr, Proceedings of the Second International Conference on Liquid Crystals, Kent, Ohio, August, 1968 (to be published)

7. J.J. Wysocki, J. Adams, and W. Haas, Phys. Rev. Letters 20, 1024 (1968).

8. H. Baessler and M.M. Labes, Phys. Rev. Letters 21, 1791 (1968).

9. E. Sackman, S. Meiboom, and L. Snyder, J. Am. Chem. Soc. 89, 5981 (1967).

10. E. Sackmann, S. Meiboom, L. Snyder, A. Meixner, and R. Dietz, J. Am. Chem. Soc. 90, 3567 (1968).

11. E.F. Carr, Advan. Chem. Series 63, 76 (1965).

12. E.F. Carr, E.A. Hoar, and W.T. MacDonald, J. Chem. Phys. 48, 2822 (1968).

13. P.L. Jain, J.C. Lee, and R.D. Spence, J. Chem. Phys. 23, 878 (1955).

14. E.F. Carr, J. Chem. Phys. 38, 1536 (1963).

SOME EXPERIMENTS ON ELECTRIC FIELD INDUCED STRUCTURAL CHANGES IN A MIXED LIQUID CRYSTAL SYSTEM

George H. Heilmeier, Louis A. Zanoni, Joel E. Goldmacher

RCA Laboratories, David Sarnoff Research Center

Princeton, New Jersey

SUMMARY

Electric field induced cholesteric to nematic structure changes have been investigated in mixtures of primary active amyl p-(4-cyano-benzylideneamino)cinnamate and the structurally similar nematics, cyano substituted benzylidene anilines. The high dielectric anisotropy of the latter materials plus their effect on the pitch of the cholesteric system enables the structure change to take place at fields as low as 2.5×10^4 V/cm. Studies of the transient behavior of the structure change show that the rise time over a limited range is proportional to exp αE^{-2}. This is suggestive of a nucleation dominated switching process. The relaxation back to the nematic state is independent of the cell thickness and depends only on the temperature and the percentage of nematic material present in the liquid.

The ability of an electric field to switch the bright yellow iridescence of the cholesteric compound is a potentially useful display effect. Reflective contrast ratios of greater than 10:1 can be achieved if a black background is used. This is required since the material is transparent in its nematic state.

INTRODUCTION

The possibility of changing the structure of a cholesteric liquid crystal into that of a nematic liquid crystal by the application of a sufficiently large electric or magnetic field was first observed experimentally by Sackmann et al.[1] (magnetic field) and Wysocki et al.[2] (electric field). The detection of the structural change due to the external magnetic field was made by means of NMR techniques while the changes induced by the electric field was observed optically. In both cases the conversion of a cholesteric system to a nematic is assumed to occur due to a breakdown of the helical structure characteristic of the cholesteric state, i.e., a finite pitch system transformed into a system of infinite pitch. When the field is removed the system reverts back to its helical structure.

Meyer[3] has shown that under certain conditions the critical electric field is given by

$$E_c = \frac{2\pi}{Z_o}\left(\frac{k_{22}}{\epsilon_p - \epsilon_t}\right)^{1/2} \tag{1}$$

where

E_c = critical electric field strength for cholesteric to nematic structural change to take place

Z_o = initial pitch of the helical structure

k_{22} = "twist" elastic constant

ϵ_p = dielectric constant parallel to the preferred molecular axis

ϵ_t = dielectric constant perpendicular to the preferred molecular axis

To estimate the magnitude of fields required for these effects in a typical cholesteric material one might assume that the values for the dielectric anisotropy and k_{22} are of the same order as those found for nematic p-azoxyanisole ($\Delta\epsilon \sim .1$ and $k_{22} \sim 10^{-6}$ dynes) and that the pitch is approximately 5000 Å. Under these assumptions, the magnitude of E_c is about 4×10^5 V/cm. Fields of this magnitude require either high voltages for thick samples or low voltages and critical thickness control to prevent localized breakdown in thin samples. In this paper we report some measurements on a particular cholesteric liquid crystal system which has a molecular structure which bears a similarity to a nematic liquid crystal. The effect of the electric field on mixtures of these structurally similar materials is of particular interest.

EXPERIMENTAL

The sandwich structure used in these experiments consisted of two pieces of tin oxide coated glass with a layer of liquid crystal material between the plates. The spacing was determined by mylar spacers and was varied from 6 μ to 25 μ. The cholesteric material was primary active amyl p-(4-cyanobenzylideneamino)cinnamate which exhibits liquid crystalline behavior between 92-110°C. This structure is shown in Figure 1a. When subjected to a DC electric field

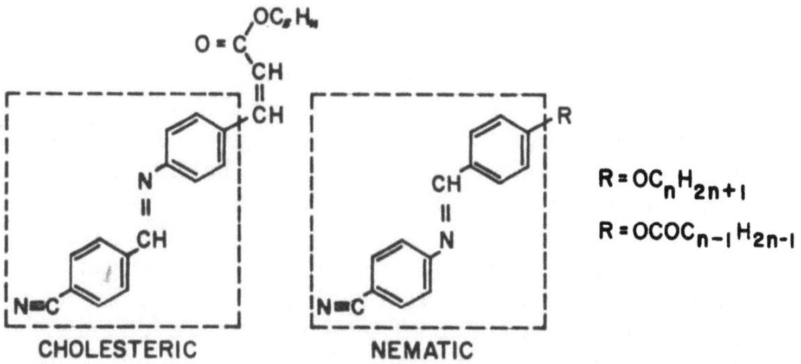

BASIC MOLECULAR STRUCTURES

FIG. 1

of $2x10^5$ V/cm at 95°C the material, which initially exhibited a bright yellow iridescence, became transparent (see Figure 2). The contrast ratio when a black background was used was greater than 10:1. Normally incident plane polarized light could be extinguished by a crossed analyzer but the material was not optically isotropic. Obliquely incident plane polarized light did not emerge from the sample plane polarized indicating that the material was active as a birefringent medium with its optic axis perpendicular to the film surface and parallel to the applied field. Further experiments using techniques outlined by Wysocki et al.[2] revealed that the material with field present was optically positive. From these experiments alone it is not possible to establish whether a smectic or nematic phase is present.

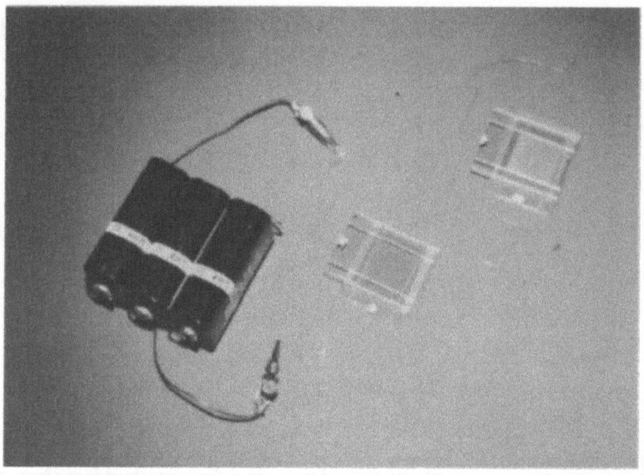

Fig. 2 SAMPLE APPEARANCE WITH FIELD ON AND OFF

The family of para substituted benzylidene anilines (shown in Figure 1b) are nematic materials with a marked structural similarity to primary active amyl p-(4-cyanobenzylideneamino)cinnamate. When equimolar mixtures of the three nematic materials shown were mixed with the cholesteric material, a striking decrease in the threshold field of the phase transition was noted. This behavior is shown in Figure 3. The characteristic iridescence of the cholesteric state was retained in the quiescent state for samples containing up to 70% nematic. The fact that a structurally similar nematic material can lower the threshold field for a structural change in a cholesteric liquid crystal strongly suggests that the structure change involved is of the cholesteric-nematic variety. The cyano substituted benzylidene anilines possess a strong electric dipole moment along their molecular axis. In fields of approximately 10^4 V/cm they align with their molecular axis and dipole moment along the direction of the applied field as evidenced by the fact that the sample is transparent and extinguishes light incident along the field direction when placed between crossed polarizers. This alignment is similar in appearance to that of the structurally similar primary active amyl p-(4-cyanobenzylideneamino)cinnamate which takes place at much higher fields. This similarity together with the threshold reduction in mixed systems, indicates that the field induced phase transition is of the cholesteric-nematic type and is due to the interaction of the electric field with the permanent electric dipole moment. The measured dielectric anisotropy for the nematic system is 14 ($\epsilon_p \sim 21$; $\epsilon_t \sim 7$).

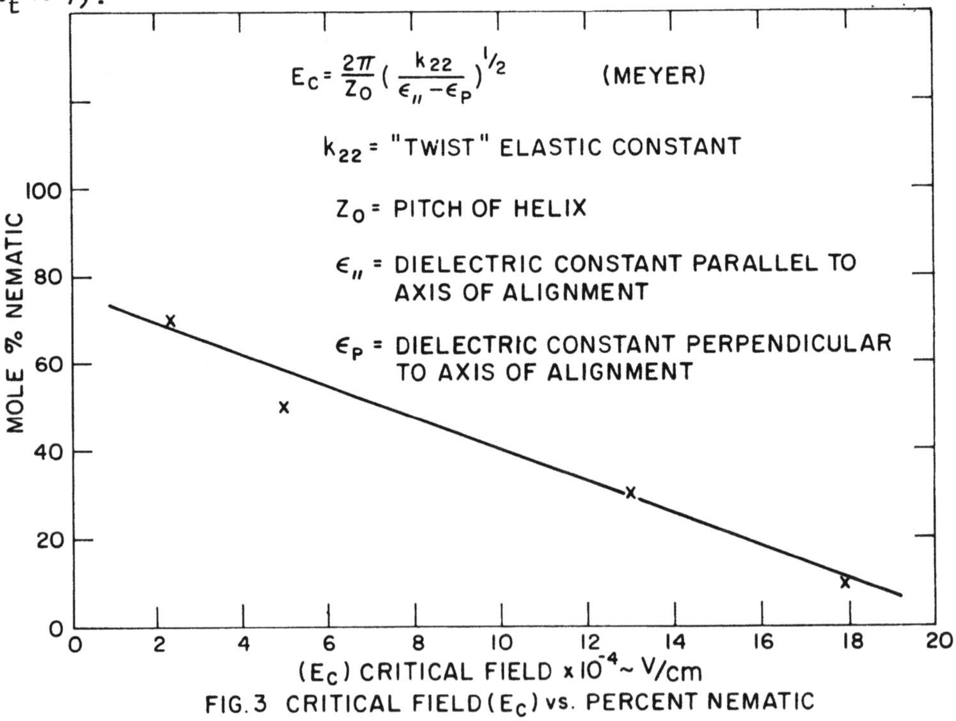

FIG. 3 CRITICAL FIELD (E_C) vs. PERCENT NEMATIC

Measurements of the transient behavior of the structure change were made by applying a voltage step and observing the clearing process by means of a photomultiplier which was fed to an oscilloscope. All data was taken at 25°C. The data for the rise time of a sample containing 70% nematic for several thicknesses is shown in Figure 4. Similar experiments were conducted for samples containing 50% and 30% nematic and this data is shown in Figures 5 and 6 respectively. The dependence of the rise time on field strength, exp αE^{-2}, is of particular interest. The relaxation time back to the nematic state seems to be independent of the sample thickness over the range measured depending only on the percent nematic in the mixture. This data is shown in Figure 7.

DISCUSSION OF RESULTS

The exponential dependence of the rise time of the cholesteric to nematic structural change on field strength is of particular interest because a similar behavior is noted for solid ferroelectric materials at low fields.[4] A possible explanation for this behavior in liquid crystal systems is to consider the switching process to be dominated by nucleation processes. Suppose that the rate of nucleation, dn/dt, of the nematic structure is given by

$$\frac{dn}{dt} \sim \exp^{-\Delta F/kT} \tag{2}$$

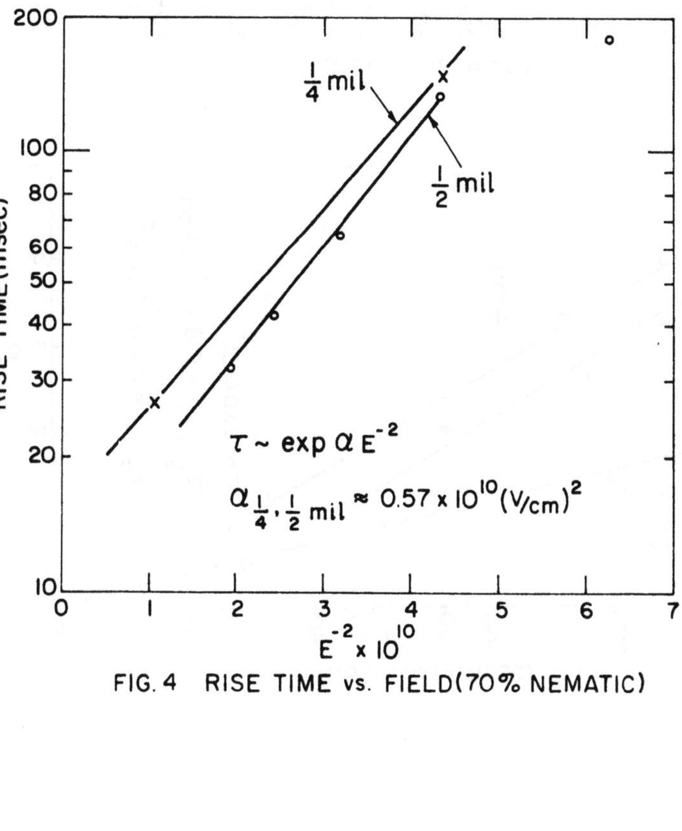

FIG. 4 RISE TIME vs. FIELD (70% NEMATIC)

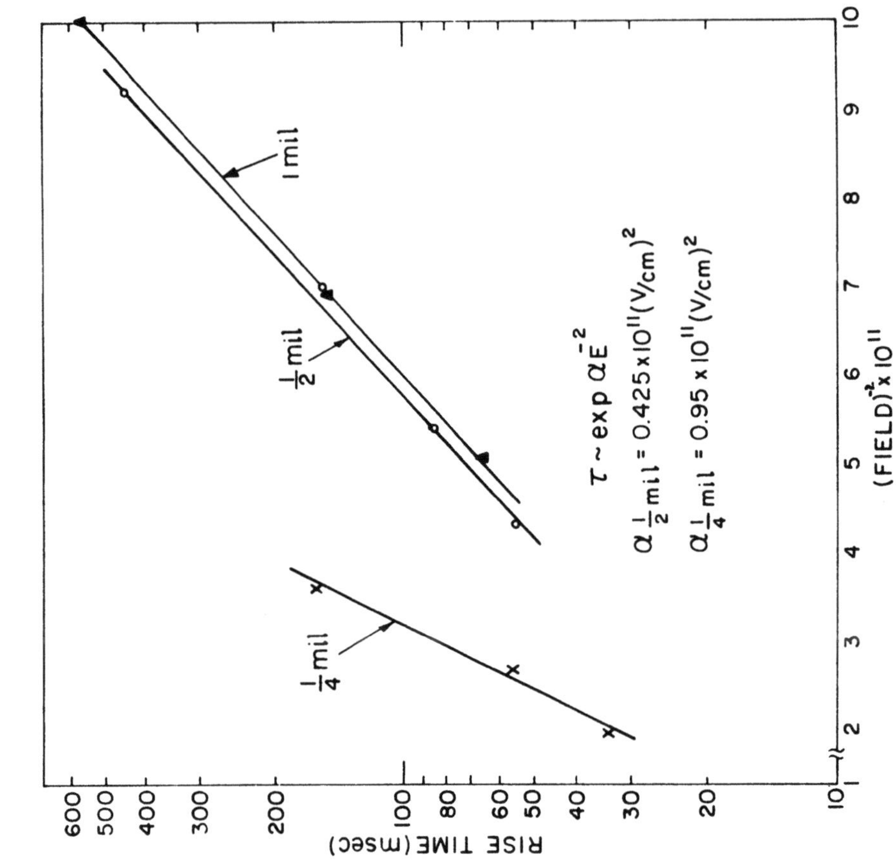

FIG.6 RISE TIME vs. FIELD(30% NEMATIC)

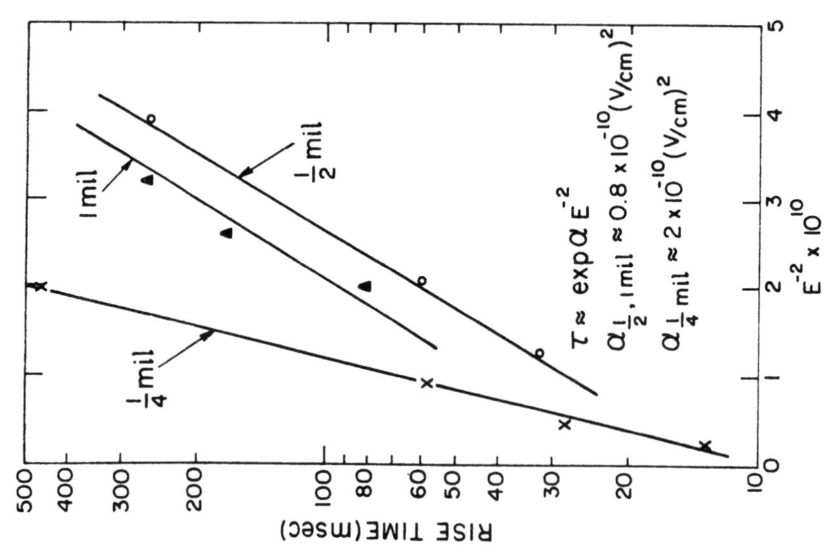

FIG. 5 RISE TIME vs. FIELD(50% NEMATIC)

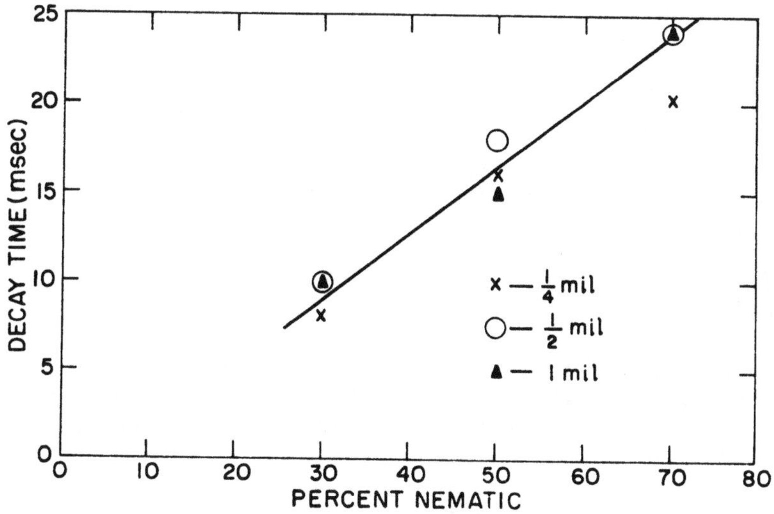

FIG.7 DECAY TIME vs. PERCENT NEMATIC

where ΔF = free energy of formation of a nucleus
kT = thermal energy

Now the free energy of formation of such a nucleus is just the sum of the polarization energy, $\mathcal{E}_p$, the surface energy, $\mathcal{E}_s$, and a depolarization energy, $\mathcal{E}_d$. For the sake of simplicity it is assumed that the nucleation is a needle-like center parallel to the applied field. In this case, the depolarization energy can be neglected hence

$$\Delta F = -\mathcal{E}_p + \mathcal{E}_s = -EPV + \sigma_s A \quad (3)$$

where

P = polarization per unit volume
E = field strength
V = volume of nucleus = πr^2
σ_s = surface energy per unit area
A = area of nucleation site = $2\pi r \ell$
ℓ = length of nucleation site
r = radius of nucleation site

The minimum free energy of formation is obtained by differentiating equation (3) with respect to r and equating to zero. The critical dimension is found to be

$$r = \frac{\sigma_s}{PE} \quad (4)$$

and hence the minimum energy is given by

$$\Delta F = \frac{\pi \ell \sigma_s^2}{PE} \quad (5)$$

The nucleation rate is then given by

$$\frac{dn}{dt} \sim \exp - \left(\frac{\pi \ell \sigma_s^{\ 2}}{EPkT} \right)$$

Since the switching speed is inversely proportional to the nucleation rate

$$\tau_{rise} \sim \exp \frac{\pi \ell \sigma_s^{\ 2}}{PkTE} = \exp \alpha E^{-1} \qquad (6)$$

where

$$\alpha = \frac{\pi \ell \sigma_s^{\ 2}}{PkT} \qquad (7)$$

In the case of solid ferroelectrics, the polarization is independent of the field. This is not likely to be the case here. If we assume that the alignment in the field can be treated by Boltzman statistics, the polarization is given by

$$P = \frac{N\mu^2}{3kT} E \qquad (8)$$

where

N = dipole density
μ = molecular dipole moment

Thus substituting equation (8) and (6) one obtains

$$\tau_{rise} \sim \exp \frac{3\pi \ell \sigma_s^{\ 2}}{N\mu^2 E^2} \qquad (9)$$

and

$$\alpha = \frac{3\pi \ell \sigma_s^{\ 2}}{N\mu^2} \qquad (10)$$

This behavior is a possible fit to that found experimentally (Figures 4, 5, 6). An additional confirmation is obtained by noting that in Figure 8, α is plotted as a function of the percent nematic and is seen to vary roughly as the inverse square of this parameter. If we assume that the surface energy decreases inversely with the nematic concentration, functional agreement with the behavior predicted by equation (10) is obtained. Since the data only extend over roughly 1.5 decades, one must view the functional agreement between theory and experiment with some reservations. Extension of the data to higher field strengths is not easy due to the problems associated with localized breakdown. It is with these reservations clearly in mind that we attempt to calculate some relevant parameters.

The surface energy can be computed from equation (10) if we make the assumption that the length of the nucleation site is equal to the pitch of the helix (~.5μ).

$$\sigma_s = \left(\frac{\alpha N\mu^2}{3\pi \ell} \right)^{1/2}$$

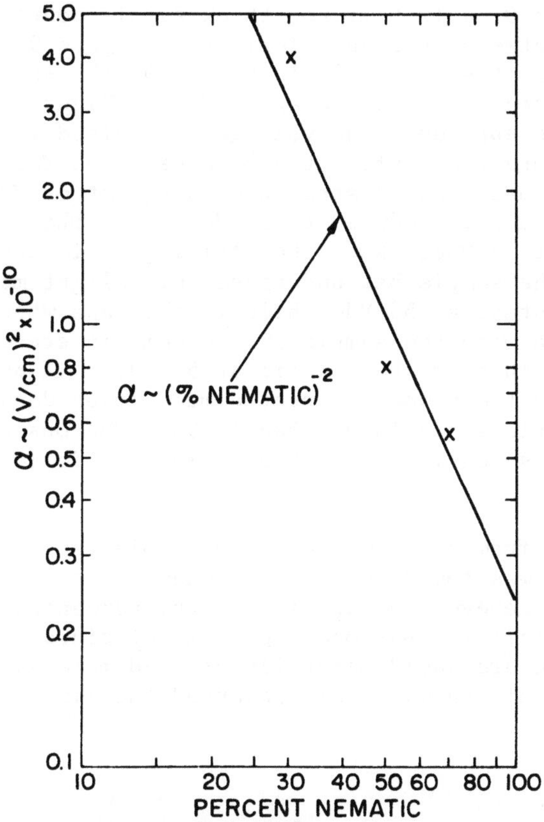

FIG. 8 α vs. PERCENT NEMATIC

Experimental values for the various parameters are as follows:
$N \sim 10^{22}\ cm^{-3}$ $\mu \sim 3.5$ debyes $\alpha \sim .57 \times 10^{10}\ (V/cm)^2$
This yields a value of 4×10^{-2} ergs/cm^2 for the surface energy. Typical wall radius of the nucleation site can be computed from equations (4) and (8)

$$r = \frac{3\sigma_s kT}{N\mu^2 E^2} \sim 250\ \text{Å}$$

From Figures 4, 5, and 6 it is noted that the value of α in most cases seems to be constant for sample thicknesses greater than 1/2 mil. Thinner samples seem to have higher values of α. This behavior suggests the presence of a surface layer which tends to inhibit the formation of nuclei. This behavior also has an interesting parallel in ferroelectricity and will be explored further in the future.

Additional support for the nucleation model of switching behavior is obtained from the observation of hysteresis in the switching behavior. In Figure 9, a 12.5 micron thick sample containing 50% nematic is viewed between crossed polarizers. Figure 9a shows the sample in its quiescent state while in Figure 9b the sample has undergone a structure change with 65 Vdc applied. In Figure 9c the sample is shown with 55 Vdc applied. The voltage was raised from zero to 55 Vdc and the picture taken after ten minutes at 55 Vdc. In Figure 9d the sample first underwent a structure change at 65 Vdc. The voltage was then gradually reduced to 55 Vdc. The photo was taken after ten minutes at 55 Vdc. Note the striking difference between 9c and 9d. In 9c the sample has undergone only slight modification from its quiescent state at 55 Vdc while in the case where the voltage has been lowered with the sample in the nematic state, the appearance is still that of the nematic at 55 Vdc. If nucleation processes are dominant one would expect a much lower density of nuclei for the nematic state in 9c than in 9d. The observed hysteresis behavior is thus consistent with a nucleation dominated switching process.

The relaxation from the nematic to the cholesteric structure after field removal was found to be independent of the sample thickness. It was, however, a function of the percentage of nematic material present. This is what one might expect since the return to the twisted structure would naturally proceed more slowly if a significant portion of the material preferred the linear structure.

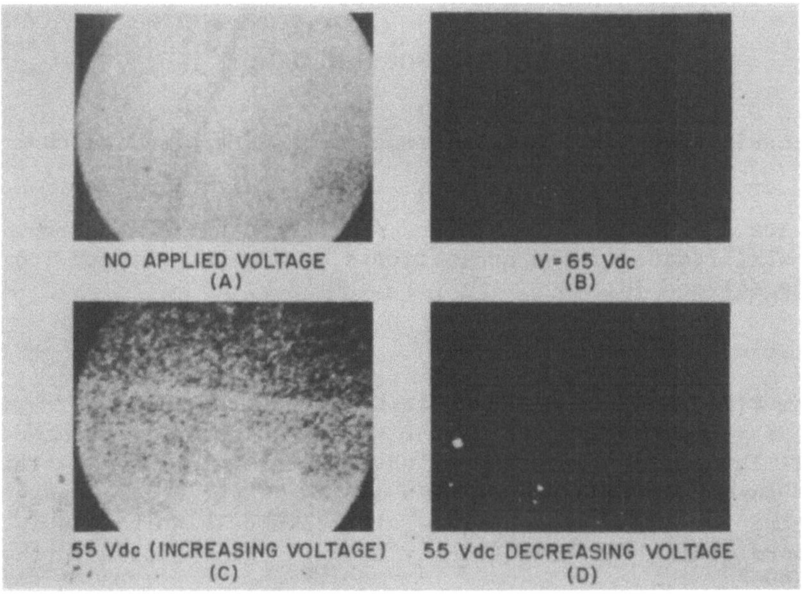

Fig. 9 HYSTERESIS EFFECTS IN FIELD INDUCED STRUCTURE CHANGES

Using equation (1), one can calculate the "twist" elastic constant since the critical field and dielectric anisotropy are known for this system.

$$k_{22} = \frac{E_c^2 Z_o \Delta\epsilon}{(2\pi)^2} \sim 3\text{-}4 \times 10^{-7} \text{ dynes}$$

where

$E_c \sim 4\text{-}5 \times 10^4$ V/cm (approximately 50% nematic)
$Z_o \sim 5000$ Å (yellow-green iridescence)
$\Delta\epsilon \sim 14$

This value for the twist elastic constant is practically the same as that found for p-azoxyanisole (4.3×10^{-7} dynes) and naturally varies with the percentage of nematic material.

CONCLUSIONS

Electric field induced cholesteric to nematic structure changes have been investigated in mixtures of primary active amyl p-(4-cyanobenzylideneamino)cinnamate and the structurally similar nematics, cyano substituted benzylidene anilines. The high dielectric anisotropy of the latter materials plus their effect on the pitch of the cholesteric system enables the structure change to take place at fields as low as 2.5×10^4 V/cm. Studies of the transient behavior of the structure change show that the rise time over a limited range is proportional to exp αE^{-2}. This is suggestive of a nucleation dominated switching process. The relaxation back to the nematic state is independent of the cell thickness and depends only on the temperature and the percentage of nematic material present in the liquid.

The ability of an electric field to switch the bright yellow iridescence of the cholesteric compound is a potentially useful display effect. Reflective contrast ratios of greater than 10:1 can be achieved if a black background is used. This is required since the material is transparent in its nematic state.

REFERENCES

1. E. Sackmann, S. Meiboom and L. C. Snyder, J. Am. Chem. Soc., 89, 5981 (1967).
2. J. J. Wysocki, J. Adams and W. Haas, Phys. Rev. Letters, 20, 1024 (1968).
3. R. B. Meyer, Appl. Phys. Letters, 12, 281 (1968).
4. W. J. Merz, Phys. Rev., 95, 690 (1954).

CAN A MODEL SYSTEM OF ROD-LIKE PARTICLES EXHIBIT BOTH A FLUID-FLUID AND A FLUID-SOLID PHASE TRANSITION?

Alexander Wulf and Andrew G. De Rocco

Department of Physics and Astronomy, University of Maryland, College Park, Maryland and Physical Sciences Laboratory, DCRT, National Institutes of Health, Bethesda, Maryland

INTRODUCTION

The existence of an isotropic-anisotropic phase transition in a system of long rigid rods has been noted by a number of workers. Apart from its intrinsic interest, this result has been considered to have application to at least the nematic-isotropic phase transition observed for nematic mesophases, and perhaps more generally to other mesophases as well.

Several criticisms can be made of this application to nematic mesophases: the relative density at which the transition occurs is much too low to represent liquid phases; the density change predicted for the transition is at best thirty or so times too large; no "reasonable" account is taken of the attractive forces since the transitions arise in systems with purely repulsive forces;[1] and, no evidence is ever presented that the transitions represent two fluid phases, one isotropic, the other anisotropic. In this paper we will deal primarily with the last of these considerations.

Earlier we have suggested that the principal role played by the attractive forces was in establishing the density of the fluid phases, and that the essential character of the phase transition could be understood in terms of the packing problem associated with the rigid particles - that is, with the repulsive forces. At that time we presented a mean-field model for the transition and indicated some of its effectiveness in characterizing the transition.[1] In a subsequent publication we shall give an extensive presentation of the results of this approach.[2]

Onsager first discovered that a system of long rigid rods showed a transition from an isotropic phase to a denser anisotropic phase.[3] His calculations were based on a second order virial expansion, a method which cannot be used to study the system at densities comparable to the close-packed density.

These density difficulties were partly overcome in the lattice models of Flory[5] and DiMarzio,[6] which gave the isotropic-anisotropic transition in semi-quantitative agreement with Onsager's theory. The treatment of Onsager was subsequently extended to the order of the seventh virial by Zwanzig for the particular case of rectangular parallelepipeds constrained to three mutually perpendicular directions, and in a certain limit involving particles of infinite length and zero cross-section.[4]

In this paper we shall deal with a model equivalent to that chosen by DiMarzio and by Zwanzig: a system of N long, rigid rectangular parallelepipeds allowed to point in only three mutually orthogonal directions. We show that in addition to the isotropic-anisotropic transition, a judicious application of DiMarzio's method implies the existence of another transition at higher density, which we interpret as the anisotropic fluid-solid transition. Thus the suggestion is offered, that systems of finite rigid particles are consistent with the existence of at least two fluid phases, one isotropic and the other anisotropic in some sense.

The present approach consists in dividing the Helmholtz free energy of N rods into two parts, $F_N = F_N^{\|} + \Delta_N$, where $F_N^{\|}$ represents the free energy of N parallel rods. It is then proven that $F_N^{\|}$ is identical to the free energy of an appropriate system of parallel cubes. The term $\Delta_N \equiv F_N - F_N^{\|}$ is treated (approximately) by the method of DiMarzio.

THEORY

It is physically evident that as the density, ρ, approaches the close packed density, ρ_o, the free energy approaches that of a system of parallel rods. The convergence should be good for this model since the rods cannot make small orientational fluctuations and the large fluctuations of 90° become very unlikely at high densities.

These considerations suggest that the division

$$
\begin{aligned}
F_N &= F_N^{\|} + \Delta_N \\
\Delta_N &\equiv F_N - F_N^{\|}
\end{aligned}
\tag{1}
$$

may be useful in studying the high density behavior of this system provided, of course, that accurate information about $F_N^{\|}$ can be obtained.

The following simple theorem establishes the equivalence of $F_N^{\|}$ and the free energy for a system of N parallel cubes. Since something is known about the equation of state of a system of cubes this proves to be a useful result.[7-9]

Theorem

$F_N^{\|}(V,T)$ equals the free energy of N parallel cubes in volume V and at temperature T, the volume of a single cube being equal to the volume of a single rectangular parallelpiped.

Proof

It is sufficient to show that the configurational integrals of the parallel rods and parallel cubes are equal. Suppose the rods are aligned along the z axis and have dimensions lxdxd.* The interaction potential between two rods, i and j, is

$$V_{ij} = v_d(x_j - x_i)v_d(y_j - y_i)v_\ell(z_j - z_i) \tag{2}$$

where

$$v_r(\zeta) = \begin{cases} 0 & |\zeta| > r \\ \infty & |\zeta| < r \end{cases} \tag{3}$$

With this notation we have

$$Q_N^{\|} = \frac{1}{N!}\int \cdots \int_V (d^3R)^N e^{-\beta \sum_{i<j} v_d(x_j - x_i)v_d(y_j - y_i)v_\ell(z_j - z_i)}$$

$$(d^3R)^N = \prod_{i=1}^{N} dx_i\, dy_i\, dz_i \tag{4}$$

$$\beta = (1/kT)$$

Making the following successive coordinate transformations

$$x_i = d \cdot x_i', \quad y_i = d \cdot y_i', \quad z_i = \ell \cdot z_i' \tag{i}$$

* It will be evident that the proof holds for parallelpipeds having three different edge lengths.

$$x_i' = x_i''/L, \quad y_i' = y_i''/L, \quad z_i' = z_i''/L \tag{ii}$$

we get

$$Q_N^{\|} = \frac{1}{N!}\int\cdots\int (d^3R'')^N \frac{(\ell d^2)^N}{L^{3N}} e^{-\beta\sum_{i<j} v_L(x_j''-x_i'')v_L(y_j''-y_i'')v_L(z_j''-z_i'')} \left(\frac{V}{\ell d^2}\cdot L^3\right) \tag{5}$$

But this can be seen to be the configurational integral for a system of N parallel cubes of volume $L^3 = \ell d^2$.

Q.E.D.

Next we turn to an evaluation of the term Δ_N. Following DiMarzio[6] (Eq.(7) of reference 6) we observe that the number of ways of packing N_i rigid rods in direction i (i=1,2,3) with N_o holes is

$$g(N_1,N_2,N_3,N_o) = \frac{\prod_{i=1}^{3}\left[N-(x-1)N_i\right]!}{N_o!\left(\prod_{i=1}^{3} N_i!\right)\left(N!\right)^2} \tag{6}$$

where the rods have dimensions (x,1,1), $x > 1$.

The configurational free energy is related to the logarithm of $g(N_i,N_o)$. Thus,

$$\begin{aligned}\ell n\, g(N_1,N_2,N_3,N_o) &= \sum_{i=1}^{3}[N-(x-1)N_i]\ell n[N-(x-1)N_i] \\ &- \sum_{i=1}^{3}[N-(x-1)N_i] \\ &- N_o\ell nN_o + N_o - \sum_{i=1}^{3} N_i\ell nN_i \\ &+ \sum_{i-1}^{3} N_i - 2(N\ell nN - N),\end{aligned} \tag{7}$$

where use has been made of the weak Stirling approximation.

For our purposes we find the following changes in notation to be convenient. Let

$$x \to \ell, \; N \to V(\text{volume of system}) \tag{i}$$

$$\sum_i N_i = N \text{ (total number of particles)} \tag{ii}$$

It follows then that $\ell N + N_o = V$, or $N_o = V - \ell N$. Further we define an orientational parameter s, $o \leqslant s \leqslant 1$, by the equations

$$N_1 = N_2 = sN, \quad N_3 = (1-2s)N \; . \tag{8}$$

With this notation Eq. (7) becomes

$$\begin{aligned} \ell n g(s,N,V) = \; & 2[V-(\ell-1)sN]\ell n[V-(\ell-1)sN] + \\ & [V-(\ell-1)(1-2s)N]\ell n[V-(\ell-1)(1-2s)N] - \\ & [V-\ell N]\ell n[V-\ell N] - 2V \,\ell n\, V - \\ & N[2s\ell n\, s + (1-2s)\ell n(1-2s) + \ell n\, N] \end{aligned} \tag{9}$$

It is easy enough to confirm that Eq. (9) is equivalent to the one which can be derived from Eq. (2) of reference 6.

The orientation parameter, s, is a function of the density. We select its value by the standard technique of maximizing the quantity g(s,N,V), thereby minimizing the free energy Δ_N. It should be pointed out that in the general case of arbitrary orientations this procedure leads to a difficult integral equation involving the orientational distribution function. The device of allowing only a finite set of orientations eliminates this difficulty, albeit at the expense of a certain physical reasonableness. However, the previous treatments of Zwanzig and DiMarzio give support to the notion that for the existence of such phase transitions as interests us here, the restriction to a finite (and limited) number of configurations is not of crucial importance.

The value of s is therefore determined by the solution of

$$\begin{aligned} & 2(\ell-1)N\ell n[V-(\ell-1)(1-2s)N] - 2(\ell-1)N\ell n[V-(\ell-1)sN] \\ & \quad - 2N\ell n \frac{s}{1-2s} = 0, \end{aligned} \tag{10}$$

which, upon introducing the reduced density

$$\rho^* \equiv \frac{\ell N}{V} = \frac{V_o}{V} = \frac{\rho}{\rho_o}$$

where V_o and ρ_o represent the close-packed volume and density respectively, becomes

$$\left[\frac{1-\frac{(\ell-1)}{\ell}(1-2s)\rho^*}{1-\frac{(\ell-1)}{\ell}s\rho^*}\right]^{\ell-1} = \frac{s}{1-2s} \tag{11}$$

The free energy Δ_N is given by

$$\frac{g(s,N,V)}{g(o,N,V)} \equiv \exp\left[-\beta\,\Delta_N\right], \tag{12}$$

and the pressure of the system by

$$\beta p = \beta p^{\#} + \left[\frac{\partial}{\partial V}(-\beta\Delta_N)\right]_{N,T} \tag{13}$$

where

$$\left[\frac{\partial}{\partial V}\left(-\frac{\Delta_N}{kT}\right)\right] = 2\,\ell n\left[1-\frac{(\ell-1)}{\ell}s\rho^*\right] + \ell n\left[\frac{1-\frac{(\ell-1)}{\ell}(1-2s)\rho^*}{1-\frac{(\ell-1)}{\ell}\rho^*}\right]. \tag{14}$$

Two remarks need to be made, one concerning Eq. (14) and one more generally concerned with the present approximation for Δ_N.

We expect that as $\rho^* \to o$ the equation of state should be of the form $\beta p \simeq \rho + 0(\rho^{*2})$, and expanding the logarithms for the case that $\rho^* \ll 1$ gives just this behavior. This serves to increase our confidence in the reasonableness of Eq. (14).

It should be noted, however, that the present approximation for Δ_N is not quite self-consistent at very high densities because Eq. (11) does not give s=o, exactly, at $\rho^*=1$. A better form close to $\rho^*=1$ would be

$$\left[\frac{1-(1-2s)\rho^*}{1-s\rho^*}\right]^{\ell} = \frac{s}{1-2s} \tag{15}$$

which, in any case, is quite close to Eq. (11) for $\ell > 10$ and leads to the same conclusions as does Eq. (11).

We turn now to an examination of the properties of this model in the region of high density, specifically to question whether it is possible to obtain evidence for a second transition at a density higher than that for the transition observed by DiMarzio[6] in general agreement with other workers.[1,3,4,5]

HIGH DENSITY BEHAVIOR; CALCULATIONS

It is the object of this section to demonstrate that the equation of state, Eq. (13), of a system of long rigid rods has a second van der Waals-like loop in the neighborhood of $\rho^* \simeq 0.7$, in addition to the low density loop corresponding to the isotropic-anisotropic transition.

We first make the obvious remark that our approximation for F_N still gives the isotropic-anisotropic transition. This is because in

$$F_N = F_N^{\|} + \Delta_N \simeq \mathcal{F}_N(s) + [F_N^{\|} - \mathcal{F}_N^{\|}]$$

where $\mathcal{F}_N$ represents the free energy for the DiMarzio model, only the leading term depends on the orientation parameter s. Thus the theory for the isotropic-anisotropic transition follows the outline given by DiMarzio, Section V of reference 6, with a small modification of the shape of the equation of state due to the term $(F_N^{\|} - \mathcal{F}_N^{\|})$, that is the increment arising from the difference in the free energy of a system of aligned cubes and the corresponding free energy for the aligned lattice in the DiMarzio approximation.

The basic observation relevant to the behavior of the system for high densities is that we expect a system of hard cubes, by analogy with that for hard spheres, to exhibit a fluid-solid phase transition in the neighborhood of $\rho^* \simeq 0.7$.

Explicit evidence for such a transition is not available, but from a variety of studies, evidence has accumulated to strongly support the inference.[7,9] We will assume the fluid-solid transition for hard cubes.

Thus to prove the existence of another loop in the neighborhood of $\rho^* \simeq 0.7$ it is sufficient to show that $[\partial(-\beta\Delta_N)/\partial V]$ is a decreasing function of ρ for $0.5_o \leqslant \rho \leqslant \rho_o$. This situation is indicated in Figure 1.

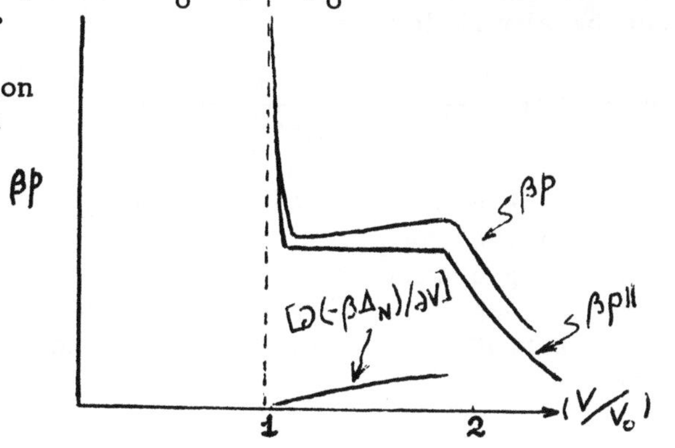

FIGURE 1. The equation of state for a system of rigid rods in the region near close-packed density.

A numerical solution of Eq. (11) suggests a useful simplification. For $\ell > 10$ and $\rho^* > 0.5$, $s \ll 0.001$ and a simple and excellent approximation for Eq. (11) is

$$s \simeq \left[1 - \frac{\ell-1}{\ell}\rho^*\right]^{\ell-1} \qquad (16)$$

It is easy enough to extend the calculation to the next order in s, and in doing so substantiate the assertion that Eq. (16) is correct to at least three significant figures.

The quantity of interest to us is $\rho^* s$ given by

$$\rho^* s \simeq \frac{\ell}{\ell-1}\left[1 - s^{1/\ell-1}\right]s + h(s^{1/\ell-1})s^2 \qquad (17)$$

where

$$h(s^{1/\ell-1}) \equiv \left[\frac{2\ell}{\ell-1} - \frac{\ell}{\ell-1}s^{1/\ell-1} - \frac{2\ell}{(\ell-1)^2}s^{1/\ell-1} - \frac{\ell}{\ell-1}s^{2/\ell-1}\right] \qquad (18)$$

and

$$h(s^{1/\ell-1}) = O(1).$$

Furthermore, it is only a matter of algebra to show that

$$\ell n\left[1 - \frac{\ell-1}{\ell}\rho^*\right] \simeq \frac{1}{\ell-1}\ell n s + \left[\frac{2}{\ell-1} - \frac{(\ell-1)}{\ell}\rho^*\right]s - 2\frac{\ell-1}{\ell}\rho^* s^{\ell-2/\ell-1} \qquad (19)$$

We are now ready to deal with $[\partial(-\beta\Delta_N)/\partial V]$. From Eq. (14), using Eq. (11) (for small s) and Eq. (19) we get

$$\frac{\partial}{\partial V}(-\beta\Delta_N) \simeq -2\frac{\ell-1}{\ell}\rho^* s + \ell n\left[1 - \frac{\ell-1}{\ell}(1-2s)\rho^*\right] - \ell n\left[1 - \frac{\ell-1}{\ell}\rho^*\right],$$

which can be simplified to

$$\frac{\partial}{\partial V}(-\beta\Delta_N) \simeq 2\frac{\ell-1}{\ell}\rho^* s\left(\frac{1}{s^{1/\ell-1}} - 1\right). \qquad (20)$$

It is clear from Eq. (20) that $[\partial(-\beta\Delta_N)/\partial V] \geq 0$, a result which is true for any $\ell \geq 1$ providing only that $s < 1$ (recall that in the anisotropic phase $N_3 \simeq N$ implying that $s \ll 1$). Furthermore, we see that $[\partial(-\beta\Delta_N)/\partial V]$ is monotone decreasing for sufficiently small s and for ρ^* sufficiently close to unity (i.e. decreasing with ρ^*).

Substituting Eq. (17) into Eq. (20) we obtain

$$\frac{\partial}{\partial V}(-\beta\Delta_N) \simeq 2s\left[\frac{1}{s^{1/\ell-1}} - 2 + s^{1/\ell-1}\right] +$$

$$s^2(1 - s^{1/\ell-1})\left[\frac{2\ell}{\ell+1}\cdot\frac{1}{s^{1/\ell-1}} - \frac{\ell}{\ell-1}\right.$$

$$\left. - \frac{2\ell}{(\ell-1)^2} - \frac{\ell}{\ell-1}\,s^{1/\ell-1}\right], \tag{21}$$

and what remains, only, is to show that

$$\frac{d}{ds}\left[\frac{\partial}{\partial V}(-\beta\Delta_N)\right] > 0$$

for $\rho^* \gtrsim 0.55$. The second term in Eq. (21) is of relative magnitude s compared with the first, and can be neglected for $s < 0.001$ (one easily checks that for $s < 0.001$ and $\ell > 10$, $\rho^* > 0.5$).

The derivative of the first term is

$$\frac{\ell-2}{\ell-1}\;\frac{1}{\sigma} + \frac{\ell}{\ell-1}\,\sigma - 2 \tag{22}$$

where we use the abbreviation $\sigma \equiv s^{1/\ell-1}$. Now for $\ell > 10$ and $\rho^* > 0.56, \sigma < 0.5$ and it is sufficient for our demonstration to show that

$$\frac{\ell-2}{\ell-1}\cdot\frac{1}{\sigma} + \frac{\ell}{\ell-1}\,\sigma - 2 > 0 \qquad (0 \leqslant \sigma < 1/2)$$

The minimum of this derivative occurs for $\sigma \simeq 1-1/\ell$ and is therefore increasing with decreasing σ in the range $0 \leqslant \sigma < 1/2$; and since at $\sigma = 1/2$ it is greater than zero, it satisfies this condition over the entire interval $0 \leqslant \sigma < 1/2$. We conclude, therefore, that

$$\frac{d}{ds}\left[\frac{\partial}{\partial V}(-\beta\Delta_N)\right] > 0 \text{ or } \frac{d}{d\rho}\left[\frac{\partial}{\partial V}(-\beta\Delta_N)\right] < 0$$

and with it the conclusion that the isotherm shows a second region of transition in the density interval of interest (i.e. $\rho^* \gtrsim 0.56$ and $\ell > 10$).

To check on the arguments given above we computed s from Eq. (16) for several values of ρ^* and these were substituted into the exact form, Eq. (11). Within the accuracy of 4-figure tables, at least, Eq. (11) was satisfied by values taken from Eq. (16). These values were then used to compute $[\partial(-\beta\Delta_N)/\partial V]$ and the results are exhibited in Table I. It can be seen that the incremental pressure decreases rapidly with increasing ρ^* in agreement with the general arguments presented previously.

TABLE 1. Values of the pressure, $[\partial(-\beta\Delta_N)/\partial V]$, for selected values of ρ^* and s. In this instance ℓ = 20.

ρ^*	s	$[\partial(-\beta\Delta_N)/\partial V]$
0.5	4.83×10^{-6}	4.13×10^{-6}
0.6	1.09×10^{-7}	1.65×10^{-7}
0.7	9.44×10^{-10}	2.48×10^{-9}
0.8	1.67×10^{-12}	8.1×10^{-12}
0.9	1.17×10^{-16}	$\backsim 10^{-15}$

FURTHER REMARKS

In this study we have said nothing about the relative density at which the anisotropic-isotropic transition occurs. It is of interest to remark, without the proof which will be forthcoming, that if one scales the density of the transition to a value suitable for real mesophase transitions, then the anticipated density change through the transition is O(1%). Furthermore, if the rods are made less rigid - in the simplest manner by representing the additional lateral flexibility as an effective decrease in the axial ratio-then the actual predicted density can be made to more nearly coincide with those observed for typical nematic mesophases.

None of the tentative conclusions made in this paper is modified in principle, if the term Δ_N is treated according to the lattice statistics of Flory[5], or, for that matter, with more recent and yet unpublished extensions.[10] The phenomenon seems inherent to a system of rigid rods, even when constrained to a limited part of configuration phase space.

It is worth mentioning that Δ_N can equally well be treated in a continuum fashion - that is allowing for the full range of configurational possibilities. In this case the technique of developing the free energy in terms of a multicomponent mixture is a suitable device[3], and the term Δ_N contains only those cluster diagrams where at least one particle is skew to the remainder. Clearly when all N particles are aligned we have the term $F_N^{\parallel}$. It is an interesting combinationial problem to exhibit those graphs which correspond to the difference - integrals appearing in the cluster development of $\mathcal{F}_N$.

It is clear that this argument would gain additional credence were a certain demonstration of the solid-fluid transition in a system of hard cubes to appear. Not surprisingly it is a basic question which remains unanswered.

ACKNOWLEDGEMENTS

One of us (A.W.) would like to thank the Instituto Venezolano de Investigaciones Cientificas (I.V.I.C.) for the award of a Predoctoral Fellowship, during which tenure this work was completed.

In addition we have benefitted from several discussions, notably with E. A. DiMarzio, W. G. Hoover, L. K. Runnels and R. W. Zwanzig.

REFERENCES

1. A. G. De Rocco, Paper No. 38, Second International Liquid Crystal Conference, Kent, Ohio, 1968.

2. A. G. De Rocco, and A. Wulf, to be published.

3. L. Onsager, Ann. N. Y. Acad. Sci. 51, 627 (1949).

4. R. Zwanzig, J. Chem. Phys. 39, 1714 (1963).

5. P. J. Flory, Proc. Roy. Soc. (London) A234, 73 (1956).

6. E. A. DiMarzio, J. Chem. Phys. 35, 658 (1961).

7. W. G. Hoover and A. G. De Rocco, J. Chem. Phys. 12, 3141 (1962)

8. W. G. Hoover and J. C. Poirier, J. Chem. Phys. 38, 327 (1963).

9. W. G. Hoover, J. Chem. Phys. 43, 371 (1965); W. G. Hoover, J. Chem. Phys. 44, 221 (1966); W. G. Hoover and F. H. Ree, J. Chem. Phys. 45, 3649 (1966); W. G. Hoover, Personal Communication (1969).

10. M. A. Cotter, Ph.D. Thesis, Georgetown University (1969).

HEAT GENERATION IN NEMATIC MESOPHASES SUBJECTED TO MAGNETIC FIELDS

Chang-Koo Yun and A. G. Fredrickson

Chemical Engineering Department

University of Minnesota, Minneapolis, Minn.

ABSTRACT

Moll and Ornstein and later Miesowicz and Jezewski observed that temperature changes were induced in nematic mesophases in the vicinity of solid interfaces by application or removal of magnetic fields. Such phenomena have been considered to be entropy effects associated with the ordering produced by the interface and the magnetic field. In the present work, these experiments have been conducted with two materials, p-azoxyanisole and p-n-decyloxy benzoic acid, but measurements of sample temperature were extended to much larger times and in larger volume that is relatively free of interfaces than those used by the earlier workers. It was found that the temperature rise upon application of a magnetic field was permanent rather than transient; the sample returned to the surroundings temperature only after the field was removed. Thus, a magnetic field causes a continuous generation of heat in a nematic mesophase, and this cannot be attributed simply to ordering effects.

Further experiments on this effect show that it is associated with nematic mesophases; thus, no temperature changes were observed in the isotropic phases of p-azoxyanisole or p-n-decyloxy benzoic acid or in the smectic phase of the latter. In addition, the magnitude of the heat generation rate is dependent on the past thermal history of the sample. Finally, the generation rate shows a saturation effect with respect to magnetic field strength. The implications of these findings are discussed and some possible sources for them are mentioned.

INTRODUCTION

The nematic mesophase is characterized by the cooperative orientation of its molecules when it is in the proximity of an interface or when it is subjected to an external electromagnetic field. It has been postulated that such cooperative orientation of molecules would involve much larger amounts of energy than is the case in isotropic liquids. In fact, it has been postulated that energy changes associated with changes of orientation ought to be detectable by macroscopic measurements.

There have been two previous attempts to measure the temperature variation due to orientational energy change. Moll and Ornstein (8) mounted a thin silver plate in nematic p-azoxyanisole and detected transient temperature changes of this plate of about 0.1°C when a magnetic field was applied or removed. Miesowicz and Jezewski (7) observed a transient temperature rise up to 1°C when they applied a magnetic field of 2,300 gauss to a 2 mm. thick sample of the same material. These workers attributed their observations to the work needed to orient molecules under the combined influence of the external field and the interface. Both sets of experiments were conducted under conditions where interfacial effects were undoubtedly very important.

The present work was started with the objective of repeating these measurements, but with samples of small surface-to-volume ratio so that interfacial effects could be minimized. It was also intended to measure the temperature variation as a function of magnetic field strength.

It was discovered at once——and unexpectedly——that the transient temperature rise upon application of a magnetic field was not followed by a decay back to the bath temperature; instead, the temperature approached an asymptotic value above the bath temperature and remained there as long as the field was applied. This forced the conclusion that the presence of a magnetic field caused a continuous generation of heat in the nematic mesophase. The experimental results and their possible explanations are presented in this paper.

EXPERIMENTS

Materials

p-azoxyanisole (PAA) and p-n-decyloxy benzoic acid (DBA) were used in the experiments. PAA was purchased from City Chemical Corp. of New York. It was dissolved in chloroform, filtered, recrystallized, dried, then recrystallized from ethanol three times. DBA was synthesized from p-hydroxy benzoic acid and n-decyl iodide according to the method of Herbert (3). The crude product was recrystallized once from glacial acetic acid and twice from ethanol.

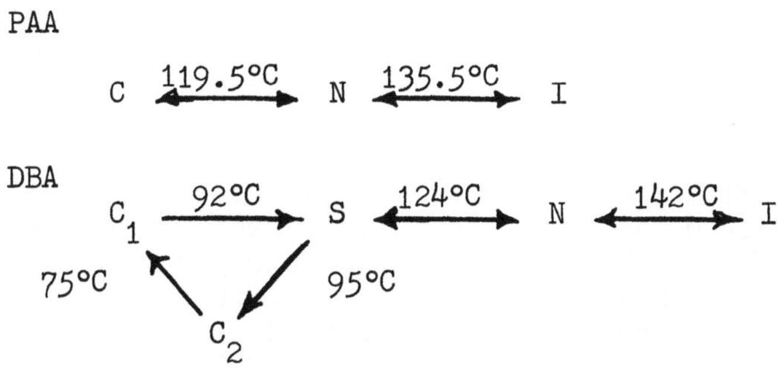

Both PAA and DBA were dried under vacuum. Phase transition temperatures were determined with a heated stage placed on a polarizing microscope. The samples used had the above transition temperatures, where C, S, N, and I are crystalline, smectic, nematic, and isotropic liquid phases, respectively.

Apparatus

As shown in Fig. 1 the sample was placed in a pyrex test tube 1.1 cm. in diameter and 10 cm. long. This was mounted in a bath through which constant temperature fluid circulated. In turn, the

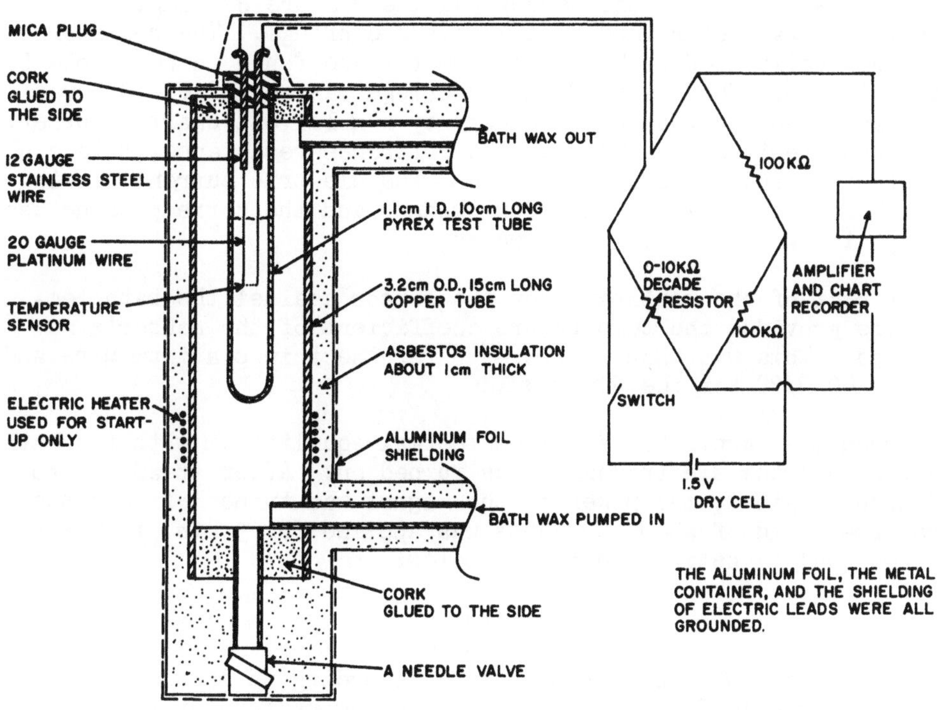

Fig. 1. Diagram of the experimental apparatus.

bath and its sample were placed between the pole faces of an electromagnet. The volume of the sample used was about 6 cm^3. The temperature of the sample in the bath fluctuated with an amplitude less than 0.02°C and with a frequency of about 1 cycle per 1.5 min.

Magnetic fields were provided by a water-cooled electromagnet, Alpha Scientific Model 7600, and a current-regulated power supply, Alpha Scientific Model 45-30. At the 2 inch gap width used, this magnet gave fields from -80 to 6500 gauss with the line drift less than 100 ppm in 8 hours. The field strength was determined and the magnet calibrated with a Dyna-Empire Model D855 gaussmeter and a 1000 gauss standard magnet. The gradient of the field strength in the space between the pole faces was less than 1%/cm..

Two different temperature sensors were used alternately. One was a fine platinum wire about 0.5 cm. long and 1.27 μ in diameter; this was highly responsive and also expected to give minimum interfacial effect. The other was a bead thermistor about 0.185 cm. in diameter, which was more stable than the platinum wire. The temperature sensor was suspended between the ends of the platinum probe. The upper half of the probe was made of stainless steel to minimize the heat conduction through the probe.

The resistance of the sensor was the measure of sample temperature; this was determined by a Wheatstone bridge. The resistors used were calibrated with a potentiometer and found to be stable to within $\pm$ 0.2%. The signal from the bridge was amplified 20,000 times and then recorded continuously by a millivolt chart recorder. The decade resistor was used as the reference resistance; this could be altered stepwise down to $\pm$ 0.1 Ω. The electric current flowing through the sensor was about 10 microamps and the corresponding heat dissipation was about 0.2 microwatts.

A plot of the resistance of the sensor against the bath temperature provided the temperature coefficient of the sensor's resistance. This was about + 1.3 Ω/°C for the thin platinum wire and about - 40 Ω/°C for the thermister.

When the sample had attained thermal equilibrium with the bath, the power to the electromagnet was turned on. After steady state had been attained, the power to the magnet was turned off and then a reverse field of about 80 gauss was applied for 5 min. to remove the residual magnetization of the poles.

Results

A typical (smoothed) temperature recording is shown in Fig. 2. When the magnetic field was applied, the temperature of the sample rose, reached a steady-state level in 5-10 min., and remained there indefinitely. When the field was removed, the temperature returned

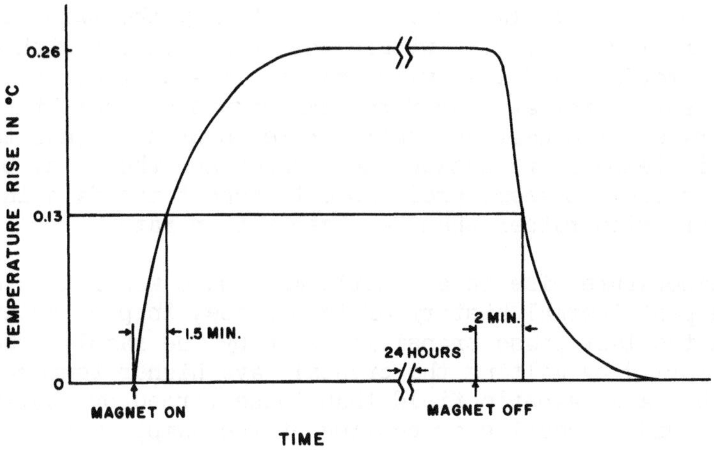

Fig. 2. A typical time course of the temperature rise in DBA. H = 3940 gauss, T = 126.0°C. Past history: C→N.

to that of the bath, in the reverse path. Evidently, the temperature rise must be due to a continuous heat generation in the sample, so long as the magnetic field is present.

The two temperature sensors used gave identical results even though the bead thermistor had a diameter about 1,500 times greater than the fine platinum wire. Hence, the heat generation is not an interfacial effect but is a phenomenon occurring in the bulk of the mesophase.

The measurements were reproducible to within about 4%. Even when measurements were made 2 months apart, during which time the sample had been held at temperatures within the nematic range, the differences did not exceed this limit.

The steady-state difference in temperature between the temperature sensor and the constant temperature bath can in principle be converted to the rate of heat generation per unit volume of the mesophase. Such a calculation is complicated by the twin facts that the thermal conductivity within the sample varies with proximity to an interface and also with magnetic field strength, and that the sensor is not at a uniform temperature so that it records only an average. The heat generation rate is proportional to the difference in temperature between the sensor and the bath, but for the reasons given, it seems preferable to report the data in terms of temperature rise rather than heat generation rate.

The temperature rise in a nematic mesophase was found to depend on the past thermal history of that phase; in particular, it depended on the last phase transition seen by the sample. Nematic mesophases formed by melting the crystal gave higher temperature rises for the same magnetic field than those formed by cooling the isotropic liquids. Heating or cooling of the sample within the nematic (N) and smectic (S) ranges, including changes across the S-N transition (in DBA), did not affect the temperature rise.

No temperature rise was detected in crystalline (C), smectic, and isotropic liquid (I) phases. The sample temperature ranges in which magnetic fields caused heating agreed with the phase transition temperatures determined from the heating stage method.

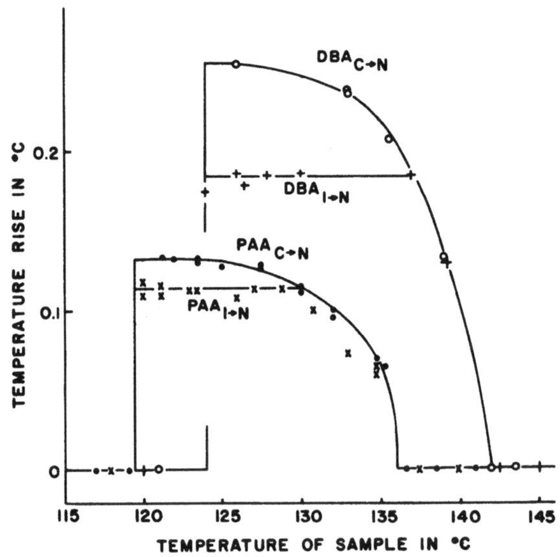

Fig. 3. The effects of sample temperature and past sample history on temperature rise. H = 6500 gauss.

In Fig. 3, the temperature rises of PAA and DBA are plotted against sample temperature. There is a temperature range $T_o < T < T_{N-I}$ for which the effect of past thermal history of the sample is not distinct. For the temperature range T_{C-N} (or T_{S-N}) $< T < T_o$, the temperature rise of the nematic phase formed by cooling the isotropic liquid is almost independent of sample temperature. The temperature T_o was about 130°C for PAA and about 137°C for DBA.

Fig. 4 shows the variation of temperature rise with applied magnetic field strength. In the case of PAA, the temperature rise increased with magnetic field strength, passed through a maximum value at about 800 gauss, and then decreased slowly to a (positive) limit. The largest field strength used was 6,500 gauss and the temperature rise seemed to have reached saturation at this value. DBA behaved in more or less the same manner, except that insufficient data were taken to see if its temperature rise also passes through a maximum before saturation is achieved. It should be noted that though past thermal history of the samples did affect the magnitude of the temperature rise, it did not affect the pattern revealed in Fig. 4.

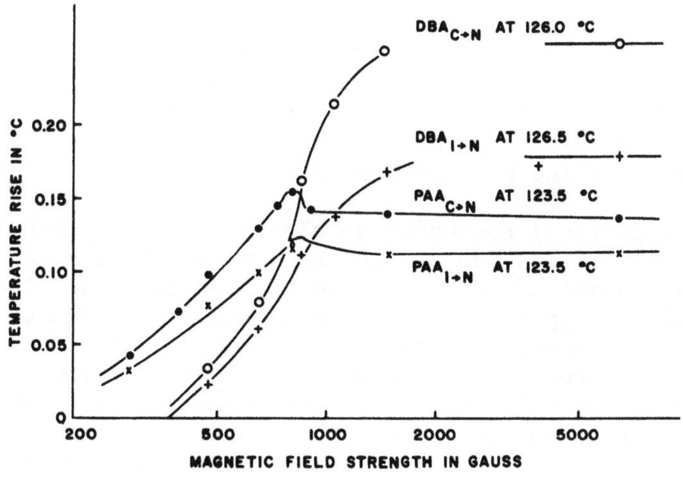

Fig. 4. The effect of magnetic field strength on temperature rise.

The foregoing results were so unexpected that it seemed at first that they must be due to some artifact. Hence, various experiments were tried to see if such an artifact could be detected. However, sending a discontinuous electric current through the sensor, reducing the height of the sample in the test tube to about two thirds of its original value, changing the orientation and polarity of the probe with respect to the magnetic field, further purification of the samples, and any combination of the foregoing expedients, did not cause any appreciable difference in the results. When the samples were passed through any of the transitions $N \rightarrow C$, $N \rightarrow S$, and $N \rightarrow I$ in a constant magnetic field up to 6,500 gauss, the removal of the field did not cause any temperature change, and this is as one would expect if the phenomenon is due to heat generation in the nematic mesophase.

A POSSIBLE EXPLANATION OF THE PHENOMENON

If a magnetic field be applied to a medium composed of diamagnetically anisotropic particles, there will be a tendency for each individual particle to orient itself with the direction in which its diamagnetic susceptibility is a maximum aligned in the direction of the field. If there were no thermal motions, all particles would be perfectly aligned in the foregoing fashion, and there would seem to be no possibility for a magnetic field to generate heat in the sample.

Of course, thermal motions are present, and so particles are continually being displaced from their position of equilibrium with respect to the field. As soon as this happens, the field exerts a restoring torque on the particle. The resulting rotational motion back towards the equilibrium position is resisted by frictional forces against which work must be done; this work may account for at least part of the heat generation.

If the particle possesses an electric dipole moment, $\underset{\sim}{p}_E$, some further possibilities for heating by a magnetic field come into play. Translation of such a dipole in a uniform magnetic field will cause equal but opposite Lorentz forces to be exerted at its ends. These have zero resultant, but their moments do not cancel and are equivalent to a couple of strength $\underset{\sim}{p}_E \times (\underset{\sim}{u} \times \underset{\sim}{H})$, where $\underset{\sim}{u}$ is the translational velocity and $\underset{\sim}{H}$ is the magnetic field intensity (with $\underset{\sim}{H}$ measured in gauss, the permeability constant has value unity and so need not be carried along in the equations). Rotation of such a dipole in a uniform magnetic field will cause equal Lorentz forces to be exerted at its ends. The moments of these forces cancel, but their resultant is a force of strength $(\underset{\sim}{\omega} \times \underset{\sim}{p}_E) \times \underset{\sim}{H}$, where $\underset{\sim}{\omega}$ is the angular velocity of the particle.

Hence, the force, $\underset{\sim}{g}$, acting on a particle in a uniform magnetic field is

$$\underset{\sim}{g} = (\underset{\sim}{\omega} \times \underset{\sim}{p}_E) \times \underset{\sim}{H} \tag{1}$$

whereas the torque, $\underset{\sim}{\ell}$, is

$$\underset{\sim}{\ell} = \underset{\sim}{p}_E \times (\underset{\sim}{u} \times \underset{\sim}{H}) + \underset{\sim}{p}_H \times \underset{\sim}{H} \tag{2}$$

where $\underset{\sim}{p}_H$ is the magnetic moment of the particle.

If the particle is surrounded by a viscous medium, the force and torque will quickly reach a local equilibrium with the frictional resistance of the medium. Under this condition, the translational velocity $\underset{\sim}{u}'$ which results <u>from the force due to the field</u> is given by

$$\underset{\sim}{u}' = \underset{\sim}{\alpha}_T \cdot \underset{\sim}{g} \tag{3}$$

whereas the angular velocity $\underset{\sim}{\omega}'$ which results <u>from the torque due to the field</u> is given by

$$\underset{\sim}{\omega}' = \underset{\sim}{\alpha}_R \cdot \underset{\sim}{\ell} \tag{4}$$

in which $\underset{\sim}{\alpha}_T$ and $\underset{\sim}{\alpha}_R$ are the translational and rotational mobility tensors of the particle. In writing these equations, it has been assumed that the coupling between rotational and translational motion could be neglected.

The rate at which the field does work against the frictional forces acting on the particle is just

$$\begin{aligned} &\underset{\sim}{u}' \cdot \underset{\sim}{g} + \underset{\sim}{\omega}' \cdot \underset{\sim}{\ell} \\ = \; &\underset{\sim}{\alpha}_T : \underset{\sim}{g}\underset{\sim}{g} + \underset{\sim}{\alpha}_R : \underset{\sim}{\ell}\underset{\sim}{\ell} \end{aligned}$$

so that the rate at which energy is transferred from the field to the medium through the particle is given by

$$-\frac{dW}{dt} = \underset{\sim}{\alpha}_T : \underset{\sim}{g}\underset{\sim}{g} + \underset{\sim}{\alpha}_R : \underset{\sim}{\ell}\underset{\sim}{\ell} \tag{5}$$

When a macroscopic thermal equilibrium is reached, the dissipation of electromagnetic energy will appear as a continuous heat generation of strength

$$\left\langle -\frac{dW}{dt} \right\rangle$$

per particle, where $\langle \; \rangle$ denotes the ensemble average over all possible orientations of the particle.

Evidently, the mechanism of heat generation by a magnetic field postulated in the foregoing will be operative not only in a nematic mesophase but also in the corresponding isotropic liquid and in the smectic mesophase, if there is one. Hence, the question arises as to why no heating effect is detectable in the isotropic liquid or in the smectic mesophase. We are unable to provide a satisfactory answer to this question. However, it appears that the answer—insofar as the isotropic fluid is concerned—must somehow involve the relative independence of molecular orientations in the isotropic liquid as compared to the situation in the nematic mesophase: in the nematic mesophase, a molecule is more or less constrained to be oriented in the direction of its neighbors; in the isotropic liquid, such constraints are nearly inoperative. As far as the smectic phase is concerned, its structure is too little known for us to offer any suggestion as to why no heating effect is observed therein.

APPLICATION OF THE SWARM MODEL

The so-called swarm model of the nematic mesophase can be combined with the foregoing considerations to yield an equation for the dependence of the heat generation rate on the magnetic field strength and the swarm size. The result is an equation containing but one adjustable parameter, and this gives a good fit of the experimental data. Moreover, the value of the parameter that fits the curve to the data is an agreement with other estimates of that parameter. Hence, though the theory is admittedly approximate, it appears useful to advance it anyway.

In the swarm model, molecules of the nematic mesophase are assumed to be grouped in swarms in which the molecules are oriented in more or less the same direction. Molecules within a swarm may undergo motion of translation and rotation about their own axes rather freely, but rotation about any axis normal to the axis of the molecule is assumed to be severely restricted by the tendency for alignment present in the swarm.

The swarms themselves are not uniformly oriented but rather are oriented according to some distribution law. Ornstein (9) calculated the distribution of swarm orientations in a magnetic field by the use of Boltzmann's principle. The probability density of the distribution of orientations according to his calculation is proportional to the factor

$$\exp\left\{\alpha \cos^2 \theta\right\}$$

where θ is the angle between the swarm axis and the magnetic field, and α is defined by

$$\alpha = \frac{NM\Delta\chi H^2}{2kT} \tag{6}$$

In the foregoing, N is the number of molecules in a swarm, M is the mass of a molecule, and $\Delta\chi$ is the difference between the axial and transverse diamagnetic susceptibilities of a molecule, per unit mass.

If now we assume that molecules within a swarm are more or less uniformly oriented—in the direction of the swarm axis, of course—then the rate of heat generation per unit volume, Q , will be given by

$$Q = n_s \left\langle - N\frac{dW}{dt} \right\rangle_s \tag{7}$$

where n_s is the number density of swarms, and $\langle\ \rangle_s$ denotes the average over the distribution of swarm orientations.

Consider now a single molecule within a swarm. Let $\underset{\sim}{d}$ be a unit vector pointing along the long axis of the molecule. Resolve the translational and rotational velocities of the molecule into components parallel and normal to $\underset{\sim}{d}$:

$$\underset{\sim}{u} = u_d \underset{\sim}{d} + u_e \underset{\sim}{e} \tag{8}$$

$$\underset{\sim}{\omega} = \omega_d \underset{\sim}{d} + \omega_f \underset{\sim}{f} \tag{9}$$

where $\underset{\sim}{e}$ and $\underset{\sim}{f}$ are unit vectors normal to $\underset{\sim}{d}$ but otherwise unrestricted. Similarly resolve the electric dipole moment:

$$\underset{\sim}{p}_E = p_{Ed} \underset{\sim}{d} + p_{Eh} \underset{\sim}{h} \tag{10}$$

where $\underset{\sim}{h}$ is normal to $\underset{\sim}{d}$. The directions $\underset{\sim}{d}$ and $\underset{\sim}{h}$ are shown in Fig. 5.

From Eq. (1), the force on the molecule can be written as

$$\underset{\sim}{g} = H\omega_f p_{Eh} (\underset{\sim}{f} \times \underset{\sim}{h}) \times \underset{\sim}{\hat{H}} + H\omega_d p_{Eh} (\underset{\sim}{d} \times \underset{\sim}{h}) \times \underset{\sim}{\hat{H}} + H\omega_f p_{Ed} (\underset{\sim}{f} \times \underset{\sim}{d}) \times \underset{\sim}{\hat{H}} \tag{11}$$

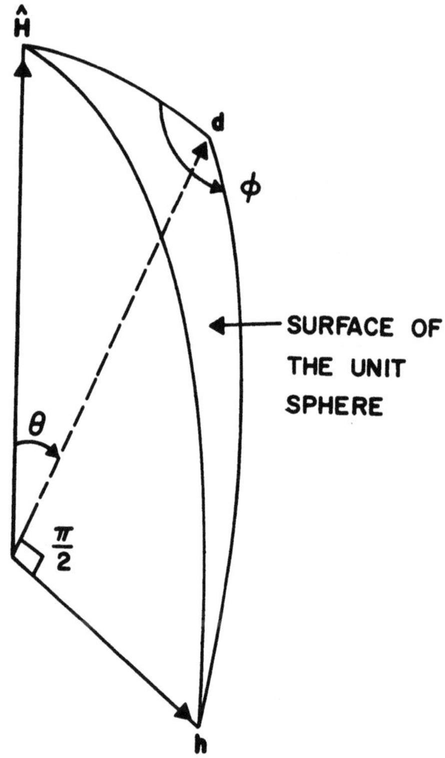

Fig. 5. Vectors and angles used in the theory.

where $\hat{\underset{\sim}{H}}$ is a unit vector in the direction of $\underset{\sim}{H}$. From Eq. (2), the torque is

$$\underset{\sim}{\ell} = Hu_d p_{Eh} \underset{\sim}{h} \times (\underset{\sim}{d} \times \hat{\underset{\sim}{H}}) + Hu_e p_{Ed} \underset{\sim}{d} \times (\underset{\sim}{e} \times \hat{\underset{\sim}{H}}) + Hu_d p_{Ed} \underset{\sim}{d} \times (\underset{\sim}{d} \times \hat{\underset{\sim}{H}}) +$$

$$Hu_e p_{Eh} \underset{\sim}{h} \times (\underset{\sim}{e} \times \hat{\underset{\sim}{H}}) + H^2 \Delta\chi M(\underset{\sim}{d} \cdot \underset{\sim}{H})(\underset{\sim}{d} \times \hat{\underset{\sim}{H}}) \tag{12}$$

where the last term on the right arises from the assumption that the molecule's magnetic susceptibility is transversely isotropic about its axis.

For a transversely isotropic molecule, the mobility tensors will be given by

$$\underset{\sim}{\alpha}_T = \alpha_T^t \underset{\sim}{I} + (\alpha_T^a - \alpha_T^t)\, \underset{\sim}{d}\underset{\sim}{d} \tag{13}$$

$$\underset{\sim}{\alpha}_R = \alpha_R^t \underset{\sim}{I} + (\alpha_R^a - \alpha_R^t)\, \underset{\sim\sim}{dd} \qquad (14)$$

where $\underset{\sim}{I}$ is the isotropic tensor and the superscripts t and a denote transverse and axial components, respectively.

We shall now assume that a molecule within a swarm can spin about its own axis and translate parallel to itself rather freely, but that oscillations about any axis normal to the molecular axis and transverse translational movements encounter a very large resistance. This means that α_T^t and α_R^t must be small compared to α_T^a and α_R^a, respectively, so that Eqs. (13) and (14) become approximately.

$$\underset{\sim}{\alpha}_T = \alpha_T^a \underset{\sim\sim}{dd} \qquad (15)$$

$$\underset{\sim}{\alpha}_R = \alpha_R^a \underset{\sim\sim}{dd} \qquad (16)$$

We can now calculate the rates of energy dissipation for a single molecule. For the contribution of the torques, we find

$$\underset{\sim}{\alpha}_R : \underset{\sim\sim}{\ell\ell} = \alpha_R^a (\underset{\sim}{d} \cdot \underset{\sim}{\ell})^2$$

$$= \alpha_R^a H^2 p_{Eh}^2 [u_d^2 (h \cdot \hat{H})^2 - 2u_d u_e (\underset{\sim}{h} \cdot \hat{H})(\underset{\sim}{d} \cdot \hat{H})(\underset{\sim}{h} \cdot \underset{\sim}{e}) + u_e^2 (\underset{\sim}{d} \cdot \hat{H})^2 (\underset{\sim}{h} \cdot \underset{\sim}{e})^2] \qquad (17)$$

It should be noted that under the assumption that α_R^t vanishes, the magnetic moment of the molecule contributes nothing to the heating. For the contribution of the force, we find

$$\underset{\sim}{\alpha}_T : \underset{\sim\sim}{gg} = \alpha_T^a (\underset{\sim}{d} \cdot \underset{\sim}{g})^2$$

$$= \alpha_T^2 H^2 [\omega_d^2 p_{Eh}^2 (\underset{\sim}{h} \cdot \hat{H})^2 - 2\omega_d \omega_f p_{Eh} p_{Ed} (\underset{\sim}{h} \cdot \hat{H})(\underset{\sim}{f} \cdot \hat{H}) + \omega_f^2 p_{Ed}^2 (\underset{\sim}{f} \cdot \hat{H})^2] \qquad (18)$$

In order to be consistent with the assumptions that α_T^t and α_R^t are small, we should also assume that u_e and ω_f vanish also. Hence, the expression for the heating rate per unit volume becomes

$$Q = n_s N H^2 p_{Eh}^2 \left\langle (\alpha_R^a u_d^2 + \alpha_T^a \omega_d^2)(\underset{\sim}{h} \cdot \hat{H})^2 \right\rangle_s \qquad (19)$$

With the help of the theorem of the equipartition of energy at an equilibrium state, we can take

$$u_d^2 = \frac{kT}{M} \tag{20}$$

$$\omega_d^2 = \frac{kT}{I_d} \tag{21}$$

where I_d is the moment of inertia of the molecule about its long axis. We shall assume that the molecule is a long, prolate ellipsoid, having semi-major axis of length a and semi-minor axis of length b (the length of the molecule is $L = 2a$). For such a molecule

$$I_d = \frac{ML^2}{10}(1 - \epsilon^2) \tag{22}$$

where ϵ is the eccentricity:

$$\epsilon = \sqrt{1 - \frac{b^2}{a^2}} \tag{23}$$

The mobilities may be found from the formulas given by Lamb (4) and Brenner (1); these are

$$\alpha_R^a = \frac{3}{4\pi\mu L^3} \cdot \frac{1}{(1-\epsilon^2)\epsilon^3}\left[2\epsilon - (1-\epsilon^2)\ln\frac{1+\epsilon}{1-\epsilon}\right] \tag{24}$$

$$\alpha_T^a = \frac{1}{8\pi\mu L} \cdot \frac{1}{\epsilon^3}\left[(\epsilon^2 + 1)\ln\frac{1+\epsilon}{1-\epsilon} - 2\epsilon\right] \tag{25}$$

where μ is the viscosity of the surroundings.

From the geometry shown in Fig. 5 , it follows that

$$\underset{\sim}{h} \cdot \hat{H} = \sin\theta \cos\phi \tag{26}$$

where θ and ϕ are the angles shown in the figure, the former being the angle between the swarm axis and the magnetic field.

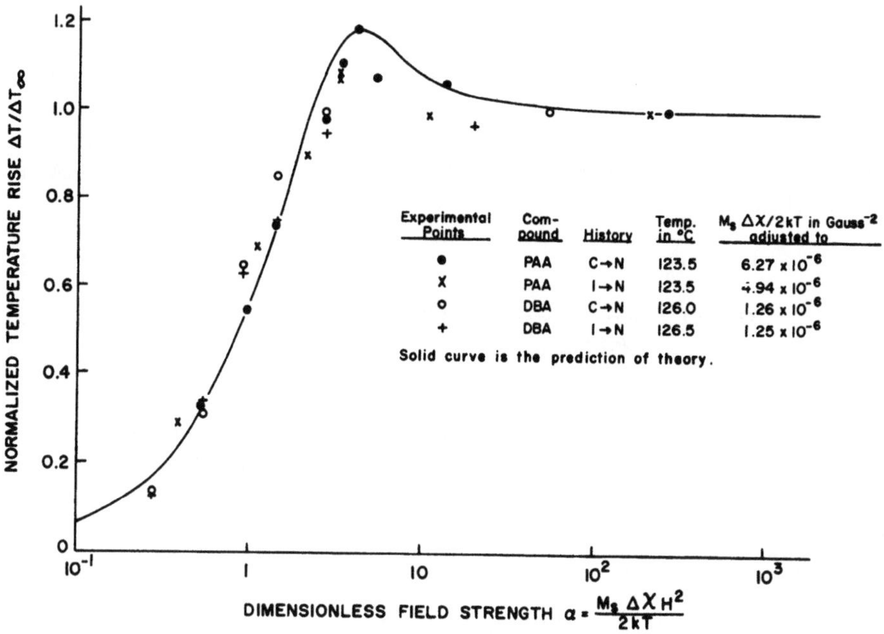

Fig. 6. Comparison of the experimental data on heat generation with the prediction of the theory.

Since ϕ and θ are assumed to be independent, we can write

$$\left\langle (\underset{\sim}{h} \cdot \hat{H})^2 \right\rangle_s = \left\langle \sin^2\theta \right\rangle_s \left\langle \cos^2\phi \right\rangle_s = \frac{1}{2} \left\langle \sin^2\theta \right\rangle_s \tag{27}$$

the result for $\left\langle \cos^2\phi \right\rangle_s$ following from the fact that the density function for the distribution of swarm orientations does not depend on ϕ .

Hence, the final expression for the heat generation rate is

$$Q = Q_\infty \left\langle \alpha \sin^2\theta \right\rangle_s \tag{28}$$

where α is the factor defined by Eq. (6) and Q_∞ , the heating rate at saturating field strength, is given by

$$Q_{\infty} = \frac{n_m (kTp_{Eh})^2}{MM_s \mu L^3 \Delta\chi} \varphi(\epsilon) \qquad (29)$$

in which M_s is the mass of a swarm (=NM), n_m is the number density of molecules ($=Nn_s$), and

$$\varphi(\epsilon) = \frac{1}{2\pi(1-\epsilon^2)\epsilon^3} [(4\epsilon^2+1)\ln\frac{1+\epsilon}{1-\epsilon} - 2\epsilon] \qquad (30)$$

The ensemble average in Eq. (28) is to be found using Ornstein's (9) density function for the distribution of swarm orientations. Since this is proportional to the factor $\exp(\alpha\cos^2\theta)$, and the density must be normalized, it follows that the complete density function is

$$G^{-1}(\alpha)e^{\alpha\cos^2\theta}$$

where

$$G(\alpha) = \int_0^{\pi} e^{\alpha\cos^2\theta}\, d\theta = \int_{-1}^{+1} e^{\alpha x^2}\, dx \qquad (31)$$

We have then

$$\begin{aligned} \langle \alpha \sin^2\theta \rangle_s &= \alpha(1 - \langle \cos^2\theta \rangle_s) \\ &= \alpha(1 - \frac{1}{G}\frac{dG}{d\alpha}) \\ &= \alpha(1 + \frac{1}{2\alpha} - \frac{e^{\alpha}}{\alpha G}) \\ &= \alpha + \frac{1}{2} - \frac{\sqrt{\alpha}}{2F(\sqrt{\alpha})} \end{aligned} \qquad (32)$$

where

$$F(x) = e^{-x^2}\int_0^x e^{\eta^2}\, d\eta \qquad \text{(Dawson's integral).}$$

We may note that

$$\lim_{\alpha \to \infty} \langle \alpha \sin^2\theta \rangle_s = 1 \tag{33}$$

so that Q_∞ is indeed the saturated heat generation rate.

The samples of nematic mesophase were held in a long cylindrical tube during the measurements. When a steady state is attained, the heat generation rate per unit volume will be related to the temperature rise at the center of the sample over that at the boundary by

$$Q = \frac{4\lambda}{R^2}(\Delta T)_{r=0} \tag{34}$$

in which λ is the thermal conductivity of the sample and R is its radius.

Assuming that λ is independent of field strength, we can write

$$\frac{\Delta T}{\Delta T_\infty} = \frac{Q}{Q_\infty} = \langle \alpha \sin^2\theta \rangle_s \tag{35}$$

where ΔT_∞ is the temperature rise at infinite field strength.

In Fig. 6, the measured values of $\Delta T/\Delta T_\infty$ are plotted against the dimensionless field strength $\alpha = M_s \Delta\chi/2kT$. The solid curve is the theoretical prediction from Eq. (32). One sees that the theory gives a good fit of data on both samples (PAA and DBA) and for both sets of thermal histories used. The data even seems to follow the "overshoot" predicted by the theory. The values of $M_s \Delta\chi/2kT$ required to fit the curve to the data are given in the figure.

In Fig. 6, the parameter $M_s \Delta\chi/2kT$ for PAA at 123.5°C was adjusted to 6.27×10^{-6} gauss^{-2}. Föex (2) reported the value of $\Delta\chi$ for crystalline PAA to be 2.42×10^{-7} cgs. units. From this, we can calculate the mass of a swarm to be

$$M_s = \left(\frac{M_s \Delta\chi}{2kT}\right) \cdot \frac{2kT}{\Delta\chi} = 2.84 \times 10^{-12} \text{gm.}$$

which is about ten times smaller than the value Massen, Poulis, and Spence (5) obtained from magnetic susceptibility measurements on nematic PAA. The linear dimension of a swarm may be estimated from the known density ($\rho = 1.15$ gm./cm.^{3}) to be

$$L_s \approx \left(\frac{M_s}{\rho}\right)^{1/3} = 1.35 \times 10^{-4}\ \text{cm.}$$

A further check on these results can be made from the maximum temperature rise. For PAA, this was $\Delta T_\infty = 0.132°C$ at 123.5°C. With $R = 0.55$ cm., $\lambda = 1.38 \times 10^4$ erg/°C.-cm.-sec. (Picot and Fredrickson (10)) we get from Eq. (34),

$$Q_\infty = 2.40 \times 10^4\ \text{erg/cm.}^3\text{-sec.}$$

A reasonable value of the axial ratio a/b for the PAA molecule is 5, so that $\epsilon = 0.98$ and $\varphi(\epsilon) = 85.7$. The electric dipole moment of PAA is about 2 Debyes. The viscosity of a swarm will be about 2.5 cp. (Miesowicz (6)). Hence, from Eq. (29) and the measured value of Q_∞, we get

$$M_s = 1.34 \times 10^{-13}\ \text{gm}$$

and so

$$L_s \approx 4.9 \times 10^{-5}\ \text{cm.}$$

Thus, the measurement of the maximum heating rate yields a swarm dimension about one-half as large as that found by fitting the theory to the temperature rise <u>vs</u>. field strength data. Considering the many approximations used, we believe that this is good agreement.

DISCUSSION AND CONCLUSIONS

Heat generation by magnetic fields is of no practical significance in most work with nematic mesophases. In such work, the surface-to-volume ratio is generally so large that the sample tends to be in near thermal equilibrium with its surroundings. Undoubtedly, this is the reason that the phenomenon has not been discovered earlier. However, heat generation is of potential practical importance, for if the mesophase were well insulated thermally from its surroundings, considerable temperature rises would be caused. In the limit of an adiabatic system, the temperature would rise to the N-I transition temperature.

A more important aspect of the heat generation effect is the evidence it provides for the structure and behavior of the nematic mesophase. We have shown how it is possible to explain the effect and to give a quantitative correlation of its strength with magnetic field strength from the swarm model of the nematic mesophase, with the modification that certain thermal motions of the molecules within a swarm are allowed.

The assumed restrictions on transverse translational motions and rotations about axes normal to the molecular axis follow logically enough from the swarm model and they are essential to the success of the calculations given above. If full freedom of motion is allowed to a molecule in a swarm, then terms in the expression for the heat generation rate arise that do not approach a limit as the field strength is increased without limit. Such terms contradict the experimental finding that the heat generation rate approaches a saturation value at large field strengths. Hence, it may be said that this experimental finding supports the view of the swarm model that molecules within a swarm are quite constrained to align in the same direction.

It is certainly true that the distortion theory of the nematic mesophase, and its descendant the continuum theory, also picture molecules of the nematic mesophase to be well aligned with each other. However, this theory, at least as far as we understand it, does not picture molecules as being arranged in groups or swarms, with neighboring swarms perhaps having differing direction of molecular orientation. If one accepts Eqs. (1), (2), and (5) as the basic explanation for the magnetic heating phenomenon, then it seems that the discrete nature of the swarm model is an important, if not essential, element of the application of the model to the data. For if we assume <u>all</u> molecules in the mesophase aligned in the same direction, with the only thermal motions permitted being spin about the molecular axis and translational parallel to that axis, then one can calculate from the foregoing analysis that there will be <u>no</u> heat generation in the magnetic field, since all molecules line up with the field and $\underset{\sim}{h}\cdot\underset{\sim}{H}$ vanishes.

One further consideration deserves mention. The experiments show that the rate of heat generation depends on the past thermal history of the sample. The swarm model can give an *a posteriori* explanation of this: the swarm size is presumed dependent on past thermal history; cf. the data given in Fig. 6 for $M_s \Delta\chi/2kT$. From this data one would conclude that the past history $C \rightarrow N$ gave a larger swarm size than the past history $I \rightarrow N$. Unfortunately, this explanation conflicts with observations on the saturation rate of heating. According to Eq. (29), Q_∞ should be inversely proportional to M_s, so that one would predict on the foregoing basis that the past history $C \rightarrow N$ would give a <u>lower</u> Q_∞ than would the past history $I \rightarrow N$. Of course, Fig. 3 shows that the opposite is true. Perhaps the explanation of this is that not all molecules in a nematic mesophase are associated with swarms and that the relative proportion so associated is smaller in case of the I→N past history.

ACKNOWLEDGEMENTS

This work was supported by the National Science Foundation, Grant GK 2900. We thank Dr. Jay Fisher for synthesizing the DBA used, Regents' Prof. A. O. C. Nier and Profs. S. R. B. Cooke and L. D. Schmidt for the loan of equipment, and Prof. J. S. Dahler for helpful discussions.

LITERATURE CITED

1. Brenner, H., Adv. Chem. Eng., **6**, 287-438 (1966).

2. Föex, G., Trans. Faraday Soc., **29**, 958-972 (1933).

3. Herbert, A. J., Trans. Faraday Soc., **63**, 555-560 (1967).

4. Lamb, H., "Hydrodynamics," pp. 604-605 (New York: Dover Publications, Inc., 1945).

5. Massen, C. H., Poulis, J. A., and Spence, R. D. pp. 72-75 in "Ordered Fluids and Liquid Crystals," ACS Monograph No. 63 (Washington, D.C.: American Chemical Society, 1967).

6. Miesowicz, M., Nature, **158**, 27 (1946).

7. Miesowicz, M., and Jezewski, M., Physik. Z., **36**, 107-109 (1935).

8. Moll, W. J. H., and Ornstein, L. S., Proc. Acad. Sci. Amsterdam, **21**, 259 (1919).

9. Ornstein, L. S., Z. Krist., **79**, 90-121 (1931).

10. Picot, J. J. C., and Fredrickson, A. G., I & EC Fund., **7**, 84-89 (1968).

NONBONDED INTERATOMIC POTENTIAL FUNCTIONS AND CRYSTAL STRUCTURE: NON HYDROGEN-BONDED ORGANIC MOLECULES

Dino R. Ferro and Jan Hermans, Jr.

Department of Biochemistry, University of North Carolina, Chapel Hill, N.C. 27514

Nonbonded interatomic potential energies are responsible in large part for the packing of solids and of highly organized macromolecules, such as enzymes, in solution. Considerable effort is being devoted to apply this principle to crystal structures of synthetic polymers[1-3] and the conformation of proteins.[5,6] With this application in mind, we have analyzed known crystal structures to determine those potential functions which are in best agreement with the packing of molecules in crystals.[7,8]

ANALYSIS OF STRUCTURES

We have proceeded as was first described by Williams for hydrocarbons.[7] Given the conformation of the molecules, the crystal structure is determined by a number of crystallographic parameters, p_k. Variable parameters may include the lengths of the three vectors which describe the unit cell (a, b and c), the angles between them (α, β and γ), three angles giving the orientation of the molecule (ϕ_x, ϕ_y, ϕ_z) and the three components of a translation vector (t_x, t_y, t_z). These are not all independently variable at the same time, but several will be determined by the symmetry of the structure. By making the following three assumptions, a set of equations is obtained which contain the interatomic energy parameters as unknowns: (1) It is assumed that the total lattice energy can be calculated as the sum of contributions from pairs of atoms, each the sum of a Lennard-Jones type term and an electrostatic term:

$$E_{ij} = -A\cdot r_{ij}^{-6} + B\cdot r_{ij}^{-12} + C\cdot r_{ij}^{-1}. \qquad (1)$$

(2) The lattice energy is put equal to the experimental heat of sublimation, when this is known and (3) one assumes that the experimental structure is the structure with minimum lattice energy, and that the derivatives of the energy with respect to each variable crystallographic parameter are equal to zero. The sums of r^{-6}, r^{-12}, r^{-1}, $\partial r^{-6}/\partial p_k$, $\partial r^{-12}/\partial p_k$, $\partial r^{-1}/\partial p_k$ for each type of atom pair can be obtained, and these are the coefficients in linear equations containing the unknown terms A_{ij}, B_{ij} and C_{ij}. Fitting of this set of equations by the least-squares method then provides the best parameters for each type of interaction.

Three difficulties which arise in the process of obtaining the parameters A and B from these linear equations were approached as follows: (1) As one considers different kinds of atoms, the number of unknown mixed interactions rises very quickly, while the number of crystal structures is limited. Hence, we have put all mixed coefficients equal to the geometric average of the coefficients for the participating atom types:

$$A_{ij} = \surd(A_{ii} \cdot A_{jj}) \quad \text{and} \quad B_{ij} = \surd(B_{ii} \cdot B_{jj}). \qquad (2)$$

(2) The set of equations which one obtains does not have a very well defined optimal solution. For this reason it is necessary to analyze many structures simultaneously, and to minimize the sum of the squares of the deviations for all equations. (3) Evaluation of the electrostatic term poses a special problem. We have set C_{ij} proportional to the product of the partial charges on the atoms, and since these charges depend more strongly on the structure of the entire molecule than on the type of the atom, the coefficients C differ from crystal to crystal, and one does not have enough equations to obtain meaningful values for all the independent variables, A, B and C. We have, therefore, calculated charge distributions a priori on the basis of other information. This is discussed in more detail below.

CALCULATION OF STRUCTURES OF MINIMUM ENERGY

In the second place, we have tested the reliability of the potential functions obtained in this way, by the reverse procedure of finding for each compound the crystal structure of minimum calculated energy, and comparing it with the experimental structure. Finally, we have extended this technique to a limited search for other packings of locally minimum energy, in order to determine if the experimental structure is truly energetically favored absolutely.

Given the A, B and C values, the energy and its derivatives are calculated, and subsequently the crystallographic parameters are modified according to Davidon's method in order to find the next structure to be considered.[9] This method has been applied by Gibson and Scheraga in energy minimization of proteins.[5] In the Davidon calculation, the energy and analytical partial first derivatives, $\partial E/\partial p_k$ are calculated and the search for a better structure is first made by the criterion of steepest descent. The differences in first derivatives for the two structures are used to obtain information about the partial second derivatives, $\partial^2 E/\partial p_i \partial p_j$, and after as many line searches as there are variables, a fairly accurate set of second derivatives is available. As a consequence, the method is a significant improvement over the steepect descent technique.

LIST OF STUDIED STRUCTURES

Aromatic hydrocarbons: Structures and heats of sublimation of benzene,[10] naphthalene,[11] anthracene,[12] phenanthrene,[13] pyrene,[14] chrysene,[15] triphenylene,[16] and ovalene.[17] All molecules were considered planar, and the best fitting plane was calculated. Ideal molecular symmetry was assumed for the molecules of naphthalene, anthracene, pyrene and ovalene. The C-H bond length was always assumed equal to 1.09 A. This kind of procedure was followed as well with many of the other molecules which we studied. Aliphatic hydrocarbons: Structures and heats of sublimation of n-octane,[18] n-pentane,[18] adamantane[19] and methane.[20] Molecules containing oxygen: p-dimethoxybenzene,[21] succinic anhydride,[22] maleic anhydride[23] (also heat of sublimation for both anhydrides), triketoindane,[24] dimeric cyclopentenone[25] and dibenzoylperoxide.[26] Molecules containing nitrogen: hexamethylenetetramine (also heat of sublimation),[27] pyrimidine,[28] pyrazine,[29] sym-triazine,[30] and sym-tetrazine.[31] Sulfur: Structure and heat of sublimation of cyclo-S_8 orthorhombic (α) sulfur[32] and structure of cyclo-S_6 rhombohedral (ρ) sulfur.[33]

POTENTIAL FUNCTIONS: HYDROCARBONS

In all work with the hydrocarbons we used a cutoff distance of 7 A for C..C, 6.5 A for C..H and 6 A for H..H pairs. Partial charges calculated by any method are relatively small (≃0.2 e.s.u.) and as a result, the energy contributed by these charges is quite small. We have, therefore, set all coefficients C to zero.

In fitting the equations with a single type of carbon atom, it was found that the sum of the squares of the errors was about 25% larger than when different constants for interaction between aliphatic carbons and aromatic carbons were used. Also, the hydrogen potential obtained with a single carbon type, was very shallow (cf. the observation of a shallow hydrogen potential when too many constraints are placed on the solution, made by Williams[7a]). We have, therefore, retained separate aromatic and aliphatic carbon potential functions. The best values obtained are given in Table I.

TABLE I. Calculated interatomic potential functions for carbon, hydrogen, oxygen, nitrogen and sulfur.[a]

Atom pair	A	B	$-E_o$	r_o	r_{VDW}
$C_{ar}\cdot C_{ar}$	600	1.17×10^6	.095	3.90	3.47
$C_{al}\cdot C_{al}$	460	1.33×10^6	.049	4.16	3.71
H...H	23	5.15×10^3	.033	3.36	2.99
$CH_i..CH_i$	2200	6.21×10^6	.194	4.22	3.76
O...O	580	2.58×10^5	.325	3.10	2.76
N...N	625	5.91×10^5	.165	3.52	3.13
S...S	2560	4.54×10^6	.361	3.91	3.48
N...N	330	7.65×10^5			
N'..N'	75	0 (assumed)			

a. E_o is the minimum energy in kcal/mole; r_o is the distance corresponding to the minimum energy, r_{VDW} the "van der Waals distance", where E=0, r_{max} the cutoff distance used in our calculations, all in A.

The solutions to the equations obtained with 6-12 type functions are not quite as good as those obtained by Williams using 6-exp type functions.[7] Most of the difference can probably be explained by the difference in shape between the 6-12 and 6-exp potential curves. The former are considerably steeper at distances where E≈0, which are the shortest interatomic distances observed. Hence, errors in the structures or in the function will give rise to larger errors in the derivatives of the energy.

The sums Σr^{-6}, etc. for carbon-carbon pairs in octane and pentane crystals, were used to calculate average methylene-methylene Lennard Jones coefficients. The values obtained (Table I) turn out to represent the equations for E and $\partial E/\partial p_k$ reasonably well. While this indicates that it may be possible, as a first approximation, to neglect the positions of methylene hydrogens in complex molecules, energy minimization of the octane crystal, using this function, gave a structure which was quite different from the experimental structure, and which, furthermore, did not return to the experimental structure when it was used as a starting point in a minimization with separate carbon and hydrogen atoms (see Tables II and III).

ENERGY MINIMIZATIONS: HYDROCARBONS

Using the experimental structures as starting points, structures of minimum energy were obtained. Experimental and calculated crystallographic parameters are given in Table II. The disagreement is of the same magnitude and often in the same direction as that found when 6-exp functions are used.[7] It was found that most unit cell lengths (a, b and c) become shortened from the experimental values and that the unit cell volume is generally smaller than that of the experimental structures.

The calculated structure for methane was much different from the experimental structure (almost identical results were obtained using Williams' 6-exp functions). When T^2 symmetry is retained and only the cell size is allowed to vary, the energy drops by 0.09 kcal/mole and the unit cell becomes larger. When one translation and one rotation are permitted (T^4 symmetry), the energy decreases by 0.5, to -2.36 kcal/mole and the unit cell shrinks. When tetragonal symmetry is assumed, cubic symmetry is not regained, but the minimum energy is only .07 kcal/mole below the minimum found when T^4 symmetry is maintained. In our opinion, this indicates that the methane structure differs, at least with respect to the position of the hydrogen atoms, from that reported. It is possible that the molecules are rotating freely or oscillating in the crystal, although this would be surprising, since the crystallography was reportedly performed below 20.4°K (at which temperature a transition is observed).[20a] The potential functions of Table I were calculated without use of the equations for methane.

The geometry of the octane molecule is such that a very large number of ways of tightly packing models of these molecules is possible, even if one restricts oneself to a triclinic unit cell, containing one molecule. This is illustrated by considering the surface of an octane molecule. The hydrogen atoms protrude, and between these protrusions there are indentations in which a

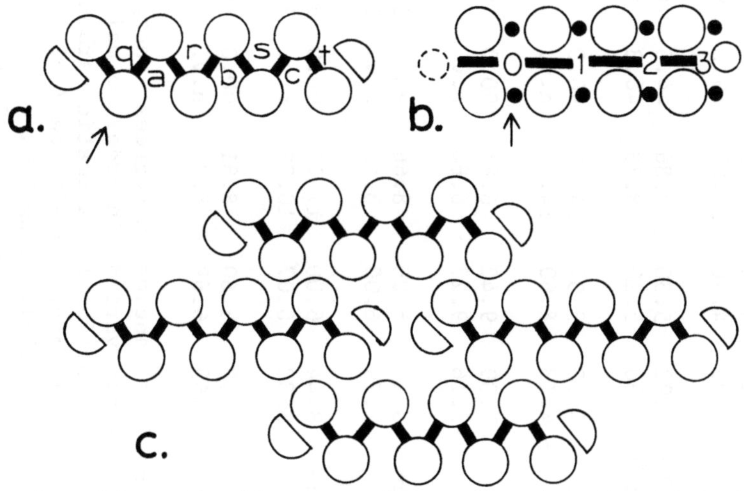

Figure 1. Schematic diagrams of octane molecules: (a) Side view, (b) top view, (c) Side view of one of many locking modes of packing characteristic of structures of locally minimum energy. ▬, carbon backbone; ○ , visible, ● , ◌, obscured hydrogens.

TABLE II. Experimental and calculated crystallographic parameters (hydrocarbons).[a]

Parameter:	a	b	c	α	β	γ	ϕ_x	ϕ_y	ϕ_z	t_y	E	Vol.
n-Octane[b,c]	4.16	4.75	11.00	94.8	84.5	105.1	-5.5	18.1	-14.3		-16.3	208
C&H	4.16	4.44	10.99	94.5	84.6	103.5	-4.5	18.1	-11.5	n.v.	-14.4	196
CH_i	4.01	4.46	12.09	89.0	98.0	116.0	0.0	1.0	-28.6		-18.5	191
Hexamethyl-	8.92	5.30	8.86	44.5	60.5	63.5	44.8	28.5	2.3		-17.8	248
benzene[d]	8.80	5.20	8.73	46.3	60.0	62.7	45.7	29.3	0.7	n.v.	-16.9	246
n-Pentane	4.10	9.04	14.70	90.0	90.0	90.0		0.0		-1.49	-20.0	545
	4.13	8.60	14.84	n.v.	n.v.	n.v.	n.v.	-1.0	n.v.	-1.42	-18.6	529
Benzene	7.39	9.42	6.81	90.0	90.0	90.0	0.0	0.0	0.0		-10.7	474
	7.16	9.27	6.84	n.v.	n.v.	n.v.	0.5	2.2	8.0	n.v.	-11.4	454
Naphthalene	8.23	6.00	8.66	90.0	122.9	90.0	49.4	60.9	25.7		-17.3	359
	7.79	6.00	8.42	n.v.	121.9	n.v.	46.7	63.3	27.6	n.v.	-17.7	336
Anthracene	8.44	6.00	11.12	90.0	125.6	90.0	37.5	58.1	14.4		-24.4	458
	8.33	6.07	10.72	n.v.	124.4	n.v.	31.2	58.4	11.0	n.v.	-24.0	447
Adamantane	6.60	6.60	8.81	90.0	90.0	90.0			0.0		-12.7	384
	6.69	6.69	8.33	n.v.	n.v.	n.v.	n.v.	n.v.	2.4	n.v.	-15.1	373

a. Lengths in A, angles in degrees, energies in kcal/mole, volumes in A^3; n.v. indicates that the parameter was not varied, in order to maintain experimental symmetry.
b. Using the function for aromatic carbon, the calculated structure is similar. However, the energy of the structure is -24 kcal/mole.
c. Second line (C&H), using C-C, C-H and H-H functions. Third line (CH_i), using methylene function.
d. Crystallographic data from L. O. Brockway, J. N. Robertson, J. Chem. Soc., 1324 (1939). Heat of subimation from S. Seki, H. Chihara, 1950, quoted by M. Frankosky, J. G. Aston, J. Phys. Chem. 69, 3126 (1965).

hydrogen atom of another molecule can fit. These positions are indicated with numbers and letters in Figure 1. If in a structure the hydrogen indicated with → of the next molecule (on top) covers, say, position 2, then we call this structure type 2. (Figure 1, b.) For side by side packing, placing the hydrogen atom indicated with → (Figure 1, a.) belonging to the next molecule behind the one drawn, say, in position b, produces packing type b. Movement to the left gives us additional arrangements, indicated as -1,... and -a,... . The experimental structure is a,0 in this notation. (This is also structure -a,0, but structure a,1 is not the same as structure -a,1.)

Any mode of stacking octane molecules side by side can be combined with any on top mode to give a two dimensional crystal. Given the side by side placement, the two packings obtained by adding a molecule on top in each position from -3 to 3 are no longer equivalent. Obviously, the best packing is obtained by moving the on top molecule away from the side where the side by side molecule protudes. This is confirmed in the structures of minimum energy obtained by us. The end to end packing of the sheets of molecules thus obtained, is largely determined at this point by requiring

TABLE III. Some n-Octane Structures of Locally Minimum Energy.[a]

Description	Energy	Volume
a,0[b]	-15.37 kcal/mole	190 A^3/mole
q,-1	-15.35	191
a,-1	-15.29	191
q,0[c]	-15.1	193
-a,1	-15.1	190
-a,-1	-14.8	195
a,2,-3	-13.8	200
c,-c,0	-13.8	200
s,-c,1	-13.8	199
-r,2,-3	-13.8	202
-r,1	-13.7	202
-b,-2,3	-13.1	206
t,-c,3,-3[d]	-10.7	233

a. These calculations were done with a preliminary set of functions, namely, A_{CC}=400, B_{CC}=1.45×10^6, A_{HH}=33, B_{HH}=4390.
b. This structure is obtained using the experimental crystal packing as a starting point.
c. This structure is reached from the minimum obtained when using the methylene function with the experimental structure as staring point.
d. This is the only structure of locally minimum energy with holes which we have found.

the closest possible end to end contact. In case of large offset, where the on top packing involves two top and two bottom molecules (e.g. Figure 1, c.), there is also the choice of having the top molecule straddle the same side, or different sides of the two bottom molecules. The latter appears to be more favorable in model building, and this finding is confirmed in the energy minimization of such structures. Locking modes of packing, in which the packing is not optimal and a hole is left at the end of each molecule, are conceivable. Several such structures were used as starting points in energy minimizations. It was found that the holes disappeared, except in one structure, which was obtained accidentally.

Given the large number of possible structures, we did not test all possible locking starting points in the energy minimization. The different structures of locally minimum energy which we have found are listed in Table III. It is seen that the experimental structure is lowest in energy among the listed structures. Cell constants and other parameters characterizing the experimental and the best calculated structure are given in Table II.

POTENTIAL FUNCTIONS: OXYGEN AND NITROGEN

Before we could calculate the attractive and repulsive coefficients of the Lennard-Jones expression from the linear equations in the unknowns, the partial charges had to be calculated, in order to obtain the electrostatic contribution to the energy and its derivatives. We successively used two different methods. The first was the CNDO method of Pople and Segal,[35] with parameters as given by these authors. This method appeared to give reasonable results for the molecular dipole moments (mostly too large). However, especially for the molecules containing nitrogen, the charge distributions were such that an extreme variety of electrostatic contributions to the crystal energy was calculated, from +0.4 to -14 kcal/mole. One expects that such large contributions must be reflected in a variation of the melting points of these compounds. However, the observed range of melting points of 50° for the azines is not in agreement with this expectation. As a further check, we have adjusted the parameters A_{OO}, B_{OO}, A_{NN} and B_{NN} to fit the equations for the energy and its derivatives, using the CNDO charges to calculate the electrostatic contributions, leaving as an additional adjustable parameter the ratio between E_{el} and the product of the charges. The "best" value of this ratio turned out very small, and this again indicates that the electrostatic term was not calculated correctly. Conclusive evidence can in principle be obtained if one knows the heats of sublimation of the azines, and we are presently attempting to determine these.

Having discarded the above results, we calculated the charge distribution with the MO-LCAO method of Del Re[36] and Pullman and

Pullman.[37] This method is more empirical and, therefore, represents the experimental dipole moments more exactly (for >C=O and -COO- we changed the charge parameters for π-electrons of Pullman in order to reproduce the experimental dipole moments of maleic anhydride, acetic acid and acetone). With the charges obtained in this manner (Table IV), the solution of the equations for energy and derivatives improved significantly, and the electrostatic contributions to the energy appear reasonable. The parameters A and B are given in Table I. Since the electrostatic energy varies as the reciprocal of the interatomic distance, we used a larger cutoff distance of 10 A for these molecules. The average error for the equations for the molecules containing oxygen is of the same order of magnitude as was observed with the hydrocarbons. However, with the molecules containing nitrogen, the errors are several times larger. We have, therefore, applied various stratagems to see if the error could be reduced. Of these we shall describe the two which cause a detectable improvement. The calculated values of the parameters of the Lennard-Jones expression are given in Table V.

(1) We have let A_{HN} and B_{HN} be independently varied, rather than fixed by equation 2. The rationale for this is that the electrostatic interaction between these atoms is largest and that one might have elements of hydrogen bonding here, even though the hydrogen atom is bonded to a carbon. As a result of freeing these coefficients, the error is reduced by 20%, and A_{NN} and B_{NN} become extremely small. This is acceptable, since there are no close N..N contacts in these crystals, but of course A_{CN} and B_{CN} are then also unreasonably small. We do not feel that the lower error is sufficient reason to prefer this set of parameters, which is not reproduced here.

(2) In the azine crystals (with the exception of tetrazine), there is a curious H..N approach, which appears to be too close, considering the values of A_{HN} and B_{HN}. (In fact, upon energy minimization, the H..N distance increases in these crystals.) It appeared to us that the hydrogen atom points between the nitrogen atom and the non bonding lone electron pair of the nitrogen atom. We, therefore, introduced an additional "atom", N', at the approximate center of the lone pair electron density, and adjusted the Lennard-Jones constants for this atom to obtain the best fit. If the equations for the structure of tetrazine are part of the group to be fitted, there is no improvement in the fit. However, without these equations, introduction of the N' atom halves the error. The resulting parameters are given in Table I. The fit to the equations for tetrazine alone is also improved by introducing the N' atom. However, the best fit is obtained with a <u>negative</u> value of $A_{HN'}$.

TABLE IV. Partial Charge Distributions Calculated by the MO-LCAO Method. Units are 10^{-10} e.s.u.

Atom type:	1	2	3	4	5	6	7	8
p-Dimethoxybenzene (I)	0.426	-0.380	-0.850	-0.112	0.264	0.256		
Succinic anhydride (II with four H atoms)	2.471	-0.227	-1.172	-2.054	0.198			
Maleic anhydride (II with two H atoms)	2.454	-0.065	-1.218	-2.040	0.260			
Triketoindane (III)	1.940	1.764	0.027	-0.165	-0.189	-1.942	-1.944	0.254
Cyclopentenone dimer (IV)	-0.286	-0.351	-0.162	-0.083	1.755	-1.977	0.188	0.180
Dibenzoylperoxide (V)	0.077	-0.219	2.417	-2.034	-0.635	0.254		
Pyrimidine (VI)	1.564	0.881	-0.193	-2.160	0.334	0.295	0.263	
Alloxan (VII)	-1.267	1.714	1.673	1.154	0.921	-1.672	-1.392	-1.066

Atom type:	C	N	H	S	O
s-Triazine	1.765	-2.101	0.336		
s-Tetrazine	1.654	-0.998	0.342		
Hexamethylenetetramine	0.100	-0.879	0.243		
Pyrazine	0.710	-2.018	0.299		
Dimethylsulfoxide	-0.225		0.250	0.930	-1.980

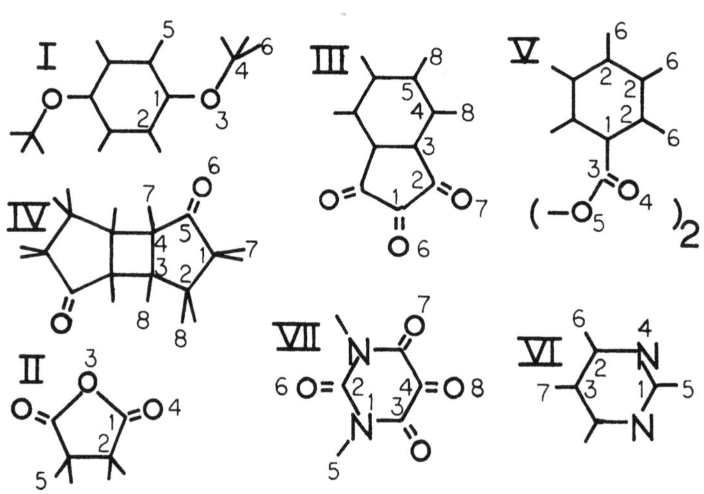

ENERGY MINIMIZATIONS: MOLECULES CONTAINING O AND N

Table V gives values of the crystallographic parameters as obtained by x-ray crystallography, and as calculated by energy minimization, using the experimental structure as a starting point. Clearly, the errors are larger than with the hydrocarbons, and in some cases unacceptable.

The minimization of the energy of dimethoxybenzene crystals results in considerable reorientation of the molecules. This effect persists when the dihedral angle of the $O-CH_3$ bond is variable. (The value of this angle is not known from experiment. In our calculation we added variable intramolecular terms and a threefold intrinsic potential with a barrier of 1 kcal/mole to the energy.) The triketoindane crystal presents unusually short C..O contacts: the structure appears to be held by a network of bridges of this kind. Bolton[24] has suggested that these "bonds" are purely electrostatic. After energy minimization we found a value of 2.93 A for the C..O distance. This is larger than the experimental 2.84 A, but much shorter than the van der Waals distance for a carbon-oxygen pair (3.15 A). Thus it is apparent that interatomic contacts can be very close if electrostatic interactions are favorable. Succinic and màleic anhydride have almost the same crystal structure. Our calculation reproduced the structure of the former reasonably closely but gave a rather large rearrangement of the latter. Use of the "larger" alipathic rather than aromatic carbon atom did not improve the agreement much.

The results for hexamethylenetetramine obtained without taking the lone electron pair into account are reasonable. Addition of the term $A_{HN'}$ makes the agreement somewhat poorer. In the case of triazine, the agreement is much improved by using the N' attractive center. For pyrimidine, the calculated crystal structure deviates significantly from the experimental. Addition of the term $A_{HN'}$ does not prevent the reorientation of the molecules to give a structure in which the molecular planes are all parallel and the special C-H.N arrangement is lost. This same arrangement is present in crystals of pyrazine. In the experimental structure the molecular plane makes an angle of -22° with the xy plane: all molecules centered on the yz plane are parallel, while the plane of adjacent molecules (at x = a/2) makes an angle of +22° with the xy plane. In the structure calculated by minimizing the energy, the molecule rotates about x by 21° so that all molecules now lie in the xy plane and are parallel. On the other hand, when the energy is minimized considering only a layer of molecules centered in the yz plane, the molecules assume an angle of -28° with xy. Apparently, interactions between adjacent layers cause the molecules to become parallel, especially the close N..H interaction. Use of the $A_{HN'}$ term causes the angle to be -31°,

TABLE V. Experimental and calculated structures. Cf. Table II.

	a	b	c	α	β	γ	ϕ_x	ϕ_y	ϕ_z	t_x	t_y	t_z	Vol.	-E
p-diCH_3O-benzene	7.29	6.30	16.55	90	90	90	129.0	-68.6	85.5				760	
	6.55	6.38	17.12	n.v.	n.v.	n.v.	111.3	-80.5	81.2	n.v.	n.v.	n.v.	716	18.3
Succinic anhydr.	6.96	11.71	11.73	90	90	90	122	16	70	0.70	1.45	2.04	441	
	6.68	11.73	5.30	n.v.	n.v.	n.v.	126	11	78	0.56	1.42	1.87	416	42.6
Maleic anhydride	7.18	11.23	5.39	90	90	90	114	14	63	0.76	1.32	1.96	435	
	6.33	11.84	5.10	n.v.	n.v.	n.v.	113	5.5	78	0.46	1.48	1.77	382	44.3
Triketoindane	7.06	(7.06)	28.77	90	90	90							1433	
	7.15	(7.15)	27.50	n.v.	n.v.	n.v.	n.v.	n.v.	n.v.	n.v.	n.v.	n.v.	1418	29.2
Cyclopentenone	6.78	7.23	8.67	90	98.9	90	0.0	0.0	0.0				420	
dimer	6.67	6.77	8.94	n.v.	101.6	n.v.	1.5	-5.4	-0.5	n.v.	n.v.	n.v.	389	22.3
Dibenzoyl	8.95	14.24	9.40	90	90	90	0.0	0.0	0.0	2.19	1.49	4.48	1198	
peroxide	8.62	14.29	8.80	n.v.	n.v.	n.v.	1.0	3.6	1.4	2.15	1.29	4.26	1084	69.9
Hexamethylene-	7.02	(7.02)	(7.02)	90	90	90							346	
tetramine	6.84	(6.84)	(6.84)	n.v.	n.v.	n.v.	n.v.	n.v.	n.v.	n.v.	n.v.	n.v.	320	17.8
s-Triazine	9.65	(9.65)	7.28	90	90	90							587	
	9.67	(9.67)	6.82	n.v.	n.v.	n.v.	n.v.	n.v.	n.v.	n.v.	n.v.	n.v.	552	27.6
Pyrimidine	11.70	9.49	3.81	90	90	90	179	-19.8	145.6	1.36	2.42		423	
	12.17	9.19	3.59	n.v.	n.v.	n.v.	180	0.0	142.5	1.44	2.29	n.v.	402	27.7

TABLE V. (continued)

	a	b	c	α	β	γ	ϕ_x	ϕ_y	ϕ_z	t_x	t_y	t_z	Vol.	-E
Pyrazine[a]	9.32	3.82	5.91	90	90	90	-22.5						210	
	8.62	3.63	6.48	n.v.	n.v.	n.v.	0.0	n.v.	n.v.	n.v.	n.v.	n.v.	202	13.6
	9.66	4.04	5.31				-31.1						207	11.8
s-Tetrazine	5.23	5.79	6.63	90	115.5	90	-150.3	-41.3	4.7				181	
	5.25	6.43	6.45	n.v.	127.3	n.v.	-133.0	-38.1	-0.3	n.v.	n.v.	n.v.	173	14.8
Orthorhombic	10.46	12.87	24.49	90	90	90			0.0			4.05	3297	24.5
sulfur	10.40	12.52	24.97	n.v.	n.v.	n.v.	n.v.	n.v.	2.2	n.v.	n.v.	4.05	3251	24.7
Rhombohedral	10.82	10.82	4.28	90	90	120			12.5				434	
sulfur	11.16	11.16	4.04	n.v.	n.v.	n.v.	n.v.	n.v.	10.9	n.v.	n.v.	n.v.	436	20.7
Dimethyl-	5.30	6.83	11.69	90	94.5	90	0.0	0.0	0.0	0.77	1.10	2.20	422	
sulfoxide	5.14	6.67	11.65	n.v.	97.3	n.v.	3.0	8.3	-0.5	0.76	0.92	2.22	396	28.1

[a] The third structure listed for pyrazine was calculated using a different potential function for nitrogen and the N' atom to represent the lone electron pair (Table I).

and the arrangement of the (entire) crystal is qualitatively unchanged. Unfortunately, the minimum energy of -11.8 kcal/mole calculated for this structure is 0.5 higher than the minimum energy calculated for a structure with parallel molecules, using in both cases a special term for the lone electron pair. The packing of tetrazine is more similar to that of benzene than to that of pyrazine, with adjacent molecules almost perpendicular. In the calculated structure, the molecules assume exactly normal positions. Use of the $A_{HN'}$ term makes very little difference, as expected. The crystal structure of alloxan, which does not contain hydrogen bonds in spite of the presence of N-H groups, is reproduced quite well in the energy minimization. This structure was not used in the evaluation of the energy functions.

RESULTS: SULFUR

The potential function for sulfur was obtained from the equations for two modifications of elemental sulfur (Table I). The functions of Table I were used to minimize the energy of the α and ρ crystal forms of sulfur and of dimethylsulfoxide crystals[38] (DMSO). The electrostatic energy term in the DMSO crystal was calculated assuming the point charges given in Table IV. These reproduce the experimental dipole moment of 3.96 D;[39] the dipole moment of the SO group is 3.0 D[40], and with the assumption of a positive charge of 0.25 on the hydrogens, the value of the CS bond dipole moment needed to obtain the total dipole moment is 0.9 D. (The average CS bond dipole moment calculated from data on diethyl, dimethyl and diphenyl sulfide is 1.26 D,[41] but also contains a contribution from the lone electron pair on the sulfur, here part of the SO bond.) Results are shown in Table V. The agreement with the experimental structures is reasonably good.

CONCLUSION

On the basis of crystal structures of organic molecules, we have obtained a set of 6-12 type interatomic potential functions, which, for the purpose of energy calculation and minimization, are certainly as reliable as any which have been suggested so far, or more so. We have tested the usefulness of these functions by minimizing the energy of the same crystals, using the experimental structure as a starting point. The deviation between calculated and experimental structures is small for the hydrocarbons, but occasionally quite unacceptably large for the molecules containing oxygen or nitrogen. Of all the assumptions which underlie this approach, two would appear to be most likely causes of this breakdown. These are: (1) The partial charge distribution on the molecules is not known precisely as required. Of the two methods which we used to calculate the charges, the more empirical MO-LCAO method, which is adjusted to give correct dipole moments, gives charge distributions which appear reasonable. However, we do not at the present wish to place too much confidence in these, either. (2) The assumed independence of the energy functions on the position of atoms other than the pair, especially atoms chemically bonded to one of the two interacting atoms, is very doubtful. In fact, improvement was obtained by relinquishing this assumption for nitrogen atoms. While the introduction of the lone pair attractive center did not produce good overall agreement either, the partial success of this stratagem indicates that the deviations from spherical symmetry in some of the functions are indeed significant. Further work is required to clarify this point.

The analysis of hydrogen bonded crystals remains to be done, and will clearly pose some new problems. Although the electrostatic terms will be even larger than in the crystals studied here, the difficulties need not be, since the hydrogen bonds will in many cases provide strong specific linkages between molecules, which will prevent the gross rearrangements produced by energy minimization, such as have been observed here in several instances. Since hydrogen bonds are thought to play precisely this rigidifying role in globular proteins, erroneous rearrangements of the calculated structures of such molecules produced by energy minimization need perhaps not at all be as extensive as was found for some crystals in this work.

This work was supported by research grants from the National Science Foundation (GB-5968), the National Institutes of Health (GM-12157) and from the U.N.C. Materials Research Center (Advanced Projects Agency, D.O.D., Contract SD-100). J. H. is a recipient of a Research Career Development Award of the National Institutes of Health, U.S. Public Health Service (Grant GM-22015).

REFERENCES

1. P. DeSantis, E. Giglio, A.M. Liquori, A. Ripamonti, J. Polymer Sci. A1, 1383 (1963); Nature 206, 456 (1965).
2. C.M. Venkatachalam, G.N. Ramachandran in Conformation of Biopolymers, G.N. Ramachandran, Ed., Academic Press, New York, N. Y., 1967, Vol. 1, p. 83.
3. S. J. Leach, G. Nemethy, H. A. Scheraga, Biopolymers, 4, 369 (1966), J. Phys. Chem. 70, 998 (1966).
4. H.A. Scheraga, S. J. Leach, R. A. Scott, G. Nemethy, Disc. Far. Soc. 40, 268 (1966).
5. K. D. Gibson, H. A. Scheraga, Proc. Natl. Acad. Sci., U.S., 58, 420 (1967).
6. R. A. Scott, G. Vanderkooi, R. W. Tuttle, P. M. Shames, H. A. Scheraga, ibid. 58, 2204 (1967).
7. D. E. Williams, (a) J. Chem. Phys. 45, 3770 (1966), (b) ibid. 47, 4680 (1967).

8. F. A. Momany, G. Vanderkooi, H. A. Scheraga, Proc. Natl. Acad. Sci., U.S., 61, 429 (1968)
9. As described by R. Fletcher, M.J.D. Powell, Comp. J. 6, 163 (1963).
10. (a) G. E. Bacon, N. A. Curry, S. A. Wilson, Proc. Roy. Soc. (London) A279, 98 (1964). (b) G. Milazzo, Ann. Chimica 46, 1105 (1956).
11. (a) D. W. J. Cruikshank, Acta Cryst. 10, 504 (1957). (b) R. S. Bradley, T. G. Cleasby, J. Chem. Soc. 1953,1690.
12. (a) R. Mason, Acta Cryst. 17, 547 (1967). (b) Ref. 11b.
13. (a) J. Trotter, Acta Cryst. 16, 605 (1963). (b) Ref. 11b.
14. (a) A. Camerman, J. Trotter, Acta Cryst. 18, 636 (1965). (b) H. Inokuchi, S. Shiba, T. Handa, H. A. Kamatsu, Bull.Chem. Soc. Japan 25, 299 (1952)
15. (a) D. M. Burns, J. Iball, Proc. Roy. Soc. (London) A257, 491 (1960), (b) H. Hoyer, W. Peperly, Z. Elektrochem. 62, 61 (1958)
16. (a) F. R. Ahmed, J. Trotter, Acta Cryst. 16, 503 (1963). (b) Ref. 14b.
17. (a) D. M. Donaldson, J. M. Robertson, Proc. Roy. Soc. (London) A220, 157 (1953). (b) Ref. 14b.
18. (a) N. Norman, H. Mathisen, Acta Chem. Scand. 15, 1747 (1961), 18, 353 (1964). (b) A. Bondi, J. Chem. Eng. Data 8, 371 (1963)
19. (a) C. E. Nordman, D. L. Schmitkons, Acta Cryst. 18, 764 (1965), J. Donohue, S. H. Goodman, Acta Cryst. 22, 352 (1967). (b) A. I. Kitaigorodskii, K. V. Mirskaya, Soviet Phys.-Crystallography 6, 408 (1962)
20. (a) A. Schallamach, Proc. Roy. Soc. (London) A171, 569 (1939), S. Kimel, A. Ron, D. F. Hornig, J. Chem. Phys. 40, 3351 (1964). (b) J. H. Colwell, E. K. Gill, J. A. Morrison, J. Chem. Phys. 40, 2041 (1964).
21. T. H. Goodwin, M. Przybylska, J. M. Robertson, Acta Cryst. 3, 279 (1959).
22. (a) M. Ehrenberg, Acta Cryst. 19, 698 (1965). (b) D. R. Stull, Ind. Eng. Chem. 39, 517 (1947).
23. (a) R. E. Marsh, E. Ubell, H. E. Wilcox, Acta Cryst. 15, 35 (1962). (b) L. O. Winstrom, L. Kulp, Ind. Eng. Chem. 41, 2584 (1949).
24. W. Bolton, Acta Cryst. 18, 5 (1965).
25. T. N. Margulis, Acta Cryst. 18, 742 (1965)
26. M. Sax, R. K. McMullan, Acta Cryst. 22, 281 (1962).
27. (a) R. Brill, H. G. Grimm, C. Hermann, C. Peters, Ann. Phys. [5], 34,435 (1931). (b) G. Klipping, I. N. Stranski, Z. Anorg. Allgem. Chem. 297, 23 (1958).
28. P. J. Wheatley, Acta Cryst. 13, 80 (1960).
29. P. J. Wheatley, Acta Cryst. 10, 181 (1957).
30. P. J. Wheatley, Acta Cryst. 8, 224 (1955).
31. E. Bertinotti, G. Giacomello, A. M. Liquori, Acta Cryst. 9, 510, (1956).

32. (a) A. S. Cooper, W. L. Bond, S. C. Abrahams, Acta Cryst. 14, 1008 (1961), A. Caron and J. Donohue, Acta Cryst. 18, 562 (1965). (b) G. B. Gurthrie, D. W. Scott, G. Waddington, J. Am. Chem. Soc. 76, 1448 (1954).
33. J. Donohue, A. Caron, E. Goldish, J. Am. Chem. Soc. 83, 3748 (1961).
34. A. I. Kitaigorodskii, Organic Chemical Crystallography, Consultants Bureau, New York, 1961. (translation of Organicheskaya Kristallokhimiya, Press of the U.S.S.R. Academy of Sciences, Moscow, 1955).
35. J. A. Pople, D. P. Satry, G. A. Segal, J. Chem. Phys. 43, S129 (1965), J. A. Pople, G. A. Segal, ibid. 43, S136 (1965).
36. G. Del Re, J. Chem. Soc. 4031 (1958).
37. B. Pullman, A. Pullman, Quantum Biochemistry, Interscience, New York, N. Y., 1963, H. Berthod, A. Pullman, J. Chimie Phys. 62, 942 (1965).
38. R. Thomas, C. B. Shoemaker, K. Eriks, Acta Cryst. 21, 12 (1966).
39. F. A. Cotton, R. Francis, J. Am. Chem. Soc. 82, 2986 (1960), H. Dreizler, G. Dendl, Z.Naturforsch. 19a, 512 (1964).
40. C. W. N. Cumper, S. Walker, Trans. Faraday Soc. 52, 193 (1956).
41. A. L. McLellan, Tables of Experimental Dipole Moments, W. H. Freeman, San Francisco, Cal., 1963.

THE INVESTIGATION OF LIPID-WATER SYSTEMS, PART 3. NUCLEAR MAGNETIC RESONANCE IN THE MONO-OCTANOIN-DEUTERIUM OXIDE SYSTEM

B. ELLIS[x], A.S.C. LAWRENCE[x], M.P. MC.DONALD[†], W.E. PEEL[†]

THE UNIVERSITY[x] AND THE POLYTECHNIC[†]

SHEFFIELD, ENGLAND

INTRODUCTION

The phase diagram for the 1-mono-octanoin (MG8) - water system has been reported by Larsson (1) and in common with the other long chain monoglycerides it shows a stable liquid crystal (l.c.) phase existing over a wide range of compositions (2).

The l.c. region is particularly interesting in the case of MG8 because the l.c./liquid phase boundary has an upper node similar to that of a compound with a congruent melting point. In the octylamine-water system there is a similarly shaped l.c. region with a maximum at the composition of an octylamine hexahydrate (3) and Ralston's phase diagram indicates that a solid hexahydrate also exists.

There are however few indications of specific hydrates in monoglyceride-water systems. Therefore, such features as the node in the l.c./liquid phase boundary probably indicate more labile hydrogen bonded aggregates (4) whose structure may be important to the functioning of monoglycerides as emulsifiers both in vivo and in industry.

Proton magnetic resonance methods have been used to study a number of lipid-water systems (6,5) in both solid and l.c. phases. It has been shown that further useful information can be obtained from the splittings of the deuteron resonance in the l.c. phases (7). These splittings arise when the average electric field gradient at the deuterium nuclei has a finite non-zero value and so indicate a residual anisotropy in the molecular motion of the deuterated species.

In this paper we report the phase diagram for MG8/D_2O and proton NMR measurements on the l.c. and solid phases. The quadrupole splitting of the deuteron resonance in the l.c. phase has been measured at a number of different compositions and temperatures.

EXPERIMENTAL

Racemic 1-mono-octanoin was prepared by Malkin's method (8) with slight modifications to the extraction procedure in order to reduce the loss of monoglyceride into the aqueous layer. After two recrystallizations from low boiling petroleum ether the material was found to be at least 99% pure by thin layer chromatography, (Mpt 38°C). The D_2O used was Koch Light 99.7% grade.

The MG8/D_2O samples were made by warming weighed mixtures to about 55°C, at which temperature most compositions form an isotropic solution, shaking briefly and then allowing them to cool.

The phase diagram was determined using the standard DTA module of a Du Pont Thermal Analyzer. Because of the supercooling which occurs in these systems it was not possible to determine transition temperatures on cooling runs. All samples were therefore cooled to -50°C in the DTA equipment and then allowed to warm up at 3°C min^{-1}.

The transition temperatures were taken to be the peaks of the curves on the thermograms (9). A rubber sleeve over the thermocouple lead and the top of the glass sample tube reduced the risk of water loss by evaporation during the run. All of the points on the phase diagram were mean values from at least two thermograms on MG8/D_2O mixtures differing in water content by not more than 5% between 0 and 50% (0·92 mole fraction) of D_2O.

Proton and deuteron magnetic resonance measurements were made at 60 MHz and 5·8 MHz respectively on the same 10mm samples using a JEOL3H-60 dual purpose spectrometer with variable temperature facilities at both frequencies.

Proton second moments (10) were calculated from at least six spectra of samples at -58°C. The line shapes of spectra of the 0·55 and 0·90 mole fraction samples were observed from this temperature up to their respective melting points. Separations between the maxima of the deuteron resonance spectra $\Delta\nu$ were measured at room temperature for mixtures between 0·55 and 0·92 mole fraction of D_2O and over the temperature range of the l.c.

phase for the samples containing 0·73 and 0·9 mole fraction of D_2O.

Difficulty was experienced in obtaining deuteron spectra of samples containing 0·55 mole fraction or smaller amounts of D_2O owing to the increasing broadness of the lines at these low concentrations.

RESULTS AND DISCUSSION

Phase Study

The phase diagram for the MG8/D_2O system is shown in Fig. 1 for the compositions between 0·22 and 0·92 mole fraction of D_2O.

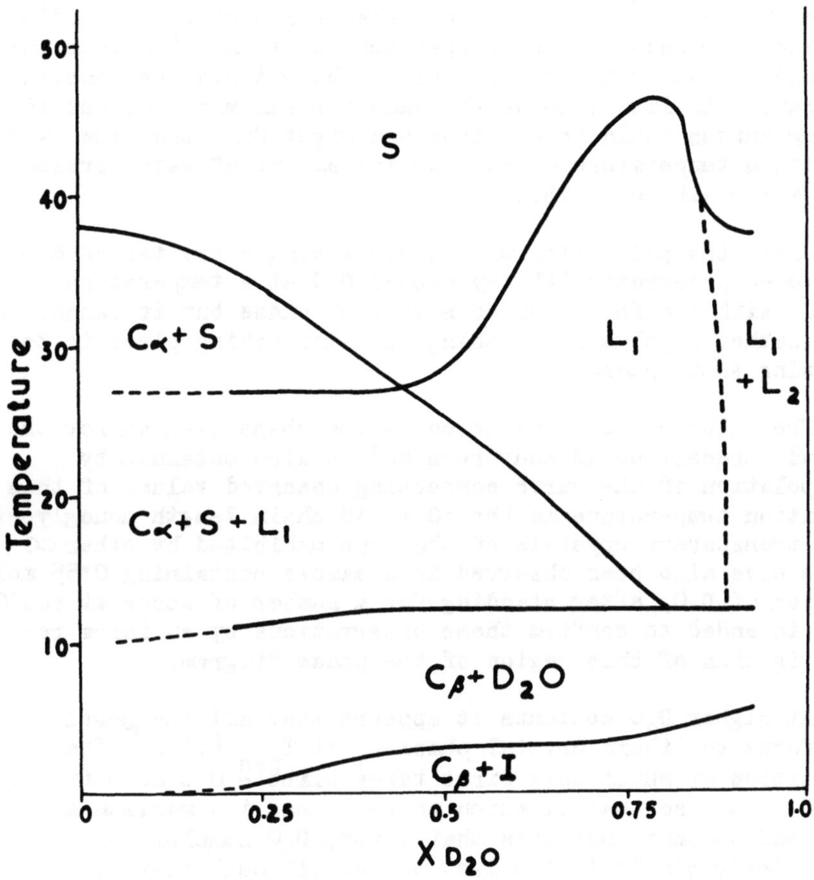

Fig. I. I-Mono-Octanoin - Deuterium Oxide phase diagram

It agrees in some respects with that of Larsson for MG8/H_2O down to 10°C (1). However, we do not observe the deeply plunging eutectic below 10°C shown in Larsson's phase diagram at 0·39 mole fraction of H_2O, but a much shallower eutectic at a temperature of 27°C and 0·45 mole fraction of D_2O.

The crystals C_α are thought to be an α-form of MG8 which may not be stable enough to exist in the anhydrous state. The α-form in long chain substances is characterized by its single short X-ray spacing (11) which is consistent with a hexagonal packing of the molecules within the layers of a bimolecular layer lattice. There is also evidence from X-ray (12), n.m.r.(13) and dielectric (14) data that rotation of the molecules about their longchain axes occurs in the α-form.

We consider that the changes occurring during a heating run on samples of low D_2O content are firstly the melting of D_2O ice between 0°C and 5°C and then the penetration of the liquid D_2O into the α-phase at the temperature of transition from the β-phase. Under the conditions of the DTA run the penetration temperature is likely to be the same for all mixtures but if cooling curves could be obtained one might find that the $\alpha \rightarrow \beta$ transition temperature depends on the amount of water present as in the n-alkanols (15).

Under the polarizing microscope a single crystal of MG8 is seen to be penetrated (16) by excess D_2O at a temperature of ca11°C. with the formation of some l.c. phase but it cannot be seen whether a polymorphic change is also taking place in the remaining solid phase.

The value of 10°C indicated by the phase diagram for the $\alpha \rightarrow \beta$ transition in anhydrous MG8 is also obtained by extrapolation of the curve connecting observed values of this transition temperature in the 10 to 18 chain length monoglycerides. Large transparent crystals of the type exhibited by other α-phases have also been observed in a sample containing 0·55 mole fraction of D_2O, after standing for a number of weeks at ca.20°C. It is intended to confirm these observations by an infra red investigation of this region of the phase diagram.

At higher D_2O contents it appears that all the β-solid transforms to liquid crystal phase L_1 at T_{pen} (16). The composition at which this first takes place - 0·8 mole fraction of D_2O - is also that at which phase L_1 has its maximum melting point and perhaps indicates that a MG8,$4D_2O$ complex is particularly stable in the l.c. phase, although there is no evidence of a solid tetrahydrate. In Larssons phase diagram the maximum melting point of the l.c. phase occurs at 0·75 mole fraction of H_2O.

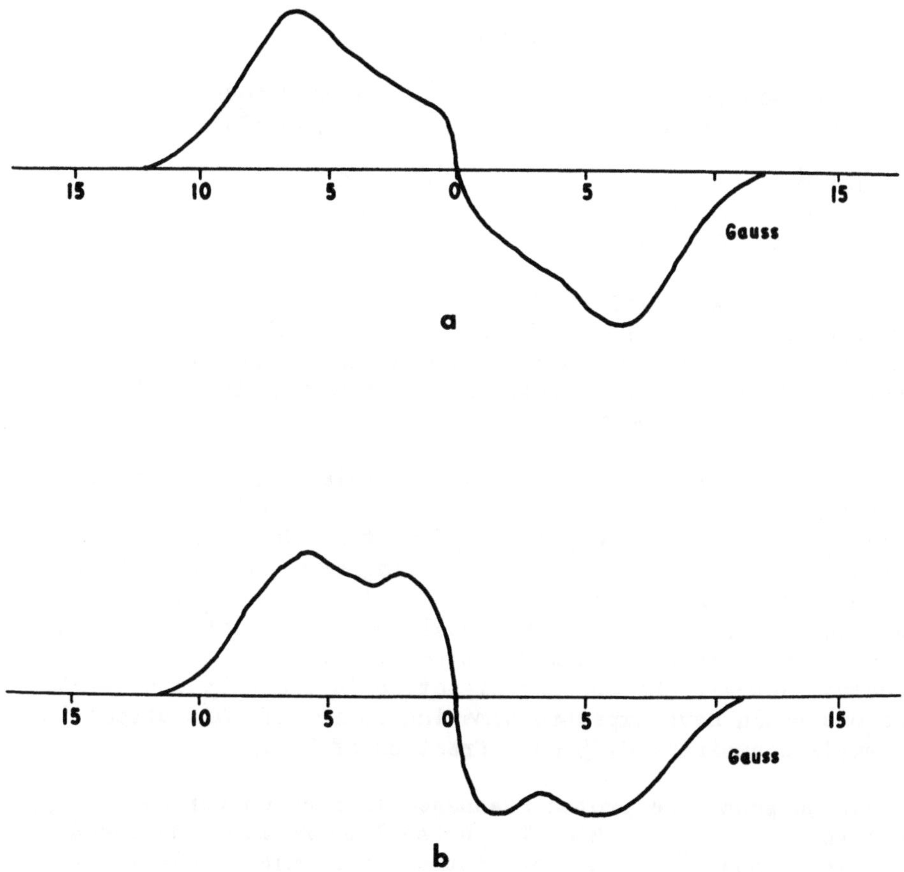

Fig. 2. Proton resonance spectra
a) MG8 + 0.55 mole fraction of D_2O at - 58°C
b) MG8 + 0.9 mole fraction of D_2O at - 58°C

Proton Magnetic Resonance

Figs. 2a and 2b show the shapes of the proton resonance lines at -58°C for samples of MG8 containing 0.55 and 0.9 mole fraction of D_2O respectively. The line shape of anhydrous MG8 is identical to Fig 2a. The line widths of the broad and narrow components in Fig 2b are 11.4 and 3.9 gauss respectively. Second moment values at -58°C are shown in Table 1.

Table 1

Sample Composition (mole fraction)	Second Moment ($gauss^2$)
pure MG8	16.4 ± 0.6
0.45 MG8 + 0.55 D_2O	16.5 ± 0.3
0.10 MG8 + 0.90 D_2O	14.8 ± 0.4

The line shapes and widths are similar to those observed previously in glycerides (17,5) but in the sample containing 0.9 mole fraction of D_2O there is a pronounced increase in amplitude of the narrower component accompanying the small reduction in second moment.

Chapman et al. (17) suggested that this narrower line in the spectrum of long chain glycerides was due to the end methyl groups and possibly the glycerol -CH proton. The exchange of glyceride -OH protons with deuterons, leaving very few in adjacent positions or in close proximity to the glycerol $-CH_2$ protons, should cause a reduction in the total proton dipolar broadening and hence an increase in the relative amplitude of the narrow component. However the above explanation is probably incomplete since one would have expected some indication of this effect in the sample containing 0.55 mole fraction of D_2O.

Fig 3a shows the proton resonance line shape for MG8 + 0.55 mole fraction of D_2O at 5°C. The broad line remains unchanged and there is now a very narrow central line which indicates a liquid like freedom of molecular or segmental reorientation. In pure MG8 this very narrow line begins to appear at temperatures above -33°C and gradually grows in intensity up to the melting point.

In samples containing more than 0.45 mole fraction of D_2O the broad resonance line disappears at the temperature of transition to the l.c. phase and a line shape of the type shown in Fig. 3b appears. In these spectra the broader component now has a line width of 1.3 gauss (peak to peak). It is similar to the 1.0 gauss line width observed in the l.c. phase of MG12 + H_2O at higher

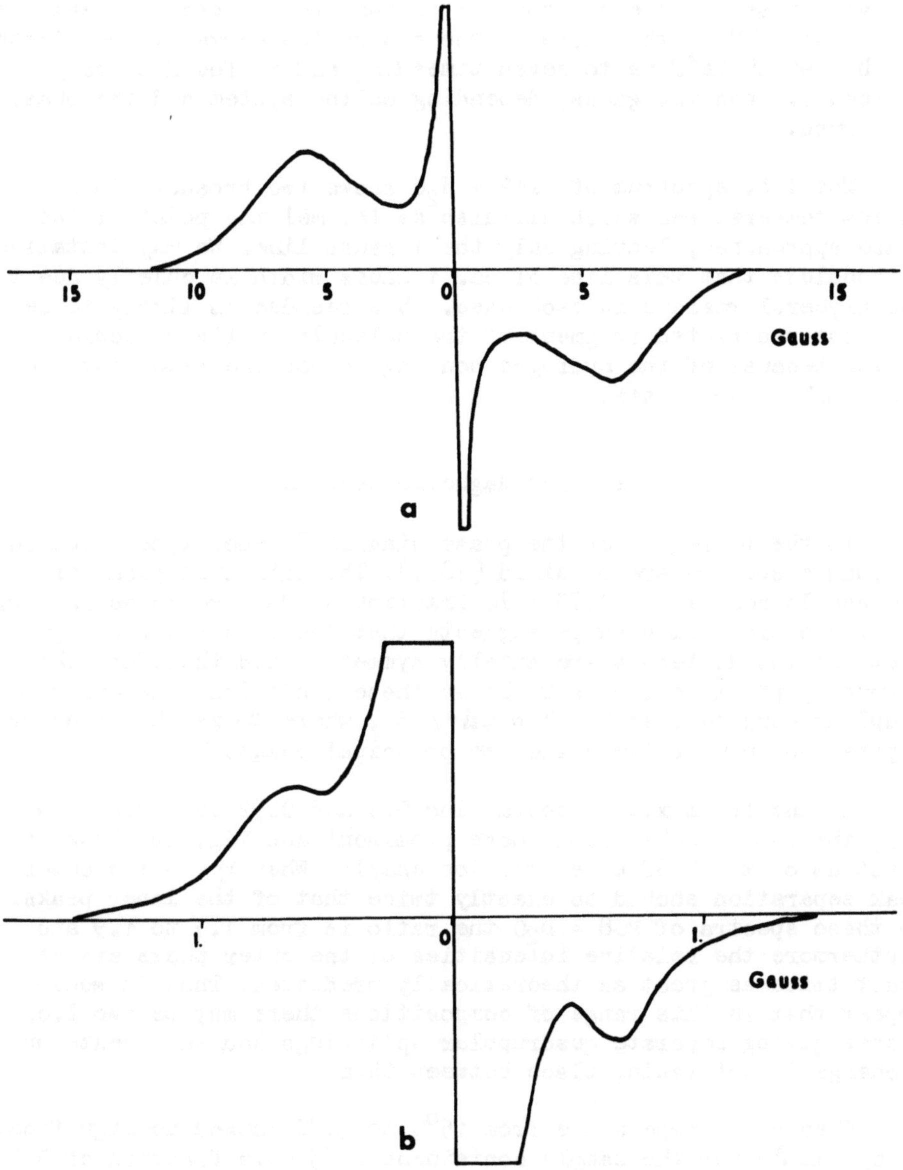

Fig. 5. Proton resonance spectra
a) MG8 + 0.55 mole fraction of D_2O at $5^{\circ}C$.
b) MG8 + 0.90 mole fraction of D_2O at $26^{\circ}C$.

temperatures (5), but is broader than $\Delta H_{\frac{1}{2}}$ recorded by Lawson and Flautt in the smectic phases of other amphiphile-D_2O systems (6,7). However it seems more appropriate to compare our peak to peak line widths with $\Delta H_{\frac{1}{8}}$, the width of the absorption curve at one eighth height, which is five to seven times $\Delta H_{\frac{1}{2}}$ and is found to vary between 0.4 and 1.4 gauss, depending on the system and the phase concerned.

The l.c. spectrum of MG12 + H_2O shows two broader lines at low temperatures which diminish as the melting point of the phase approaches, leaving only the 1 gauss line. We may tentatively conclude that this line of ca. 1 gauss width is chiefly due to the glycerol residue in each case. This residue is likely to be the most restricted fragment of the molecule in its molecular motion because of the hydrogen bonding across the water layer of the lamellar aggregate.

Deuteron Magnetic Resonance

In the L_1 region of the phase diagram "powder type" deuteron resonance spectra are obtained (18,7). The principal peaks of the sample containing 0.73 mole fraction of D_2O are shown in Fig. 4a. The overall line shape suggests that the electric field gradients at the deuterons are axially symmetric and therefore the asymmetry parameter, $\eta = 0$. Under these conditions the quadrupole coupling constant, $e^2qQ / h = 4\Delta\nu / 3$, where $\Delta\nu$ is the frequency separation in Hz between the two principal peaks.

In the two mixtures containing 0.9 and 0.92 mole fraction of D_2O, the outer peaks become more prominent and Fig. 4b shows the spectrum of the 0.92 mole fraction sample. When $\eta = 0$ the outer peak separation should be exactly twice that of the inner peaks. In these spectra of MG8 + D_2O the ratio is from 1.7 to 1.9 and furthermore the relative intensities of the outer peaks are at least twice as great as theoretically predicted. Thus it would appear that in this range of compositions there may be two l.c. phases giving separate quadrupolar splittings and that deuteron exchange is not taking place between them.

Change of temperature from 15°C to 40°C caused no significant change in $\Delta\nu$ for the sample containing 0.73 mole fraction of D_2O (L_1). Migchelson and Berendsen (19) found a similar lack of temperature dependence of the splitting in deuterohydrated collagen.

Below 12°C in the L_1 region the lines broaden and become unmeasurable, but in the $L_1 + L_2$ region a central line appears when the temperature reaches 18oC and increases in intensity as the temperature decreases. We are at present making a more

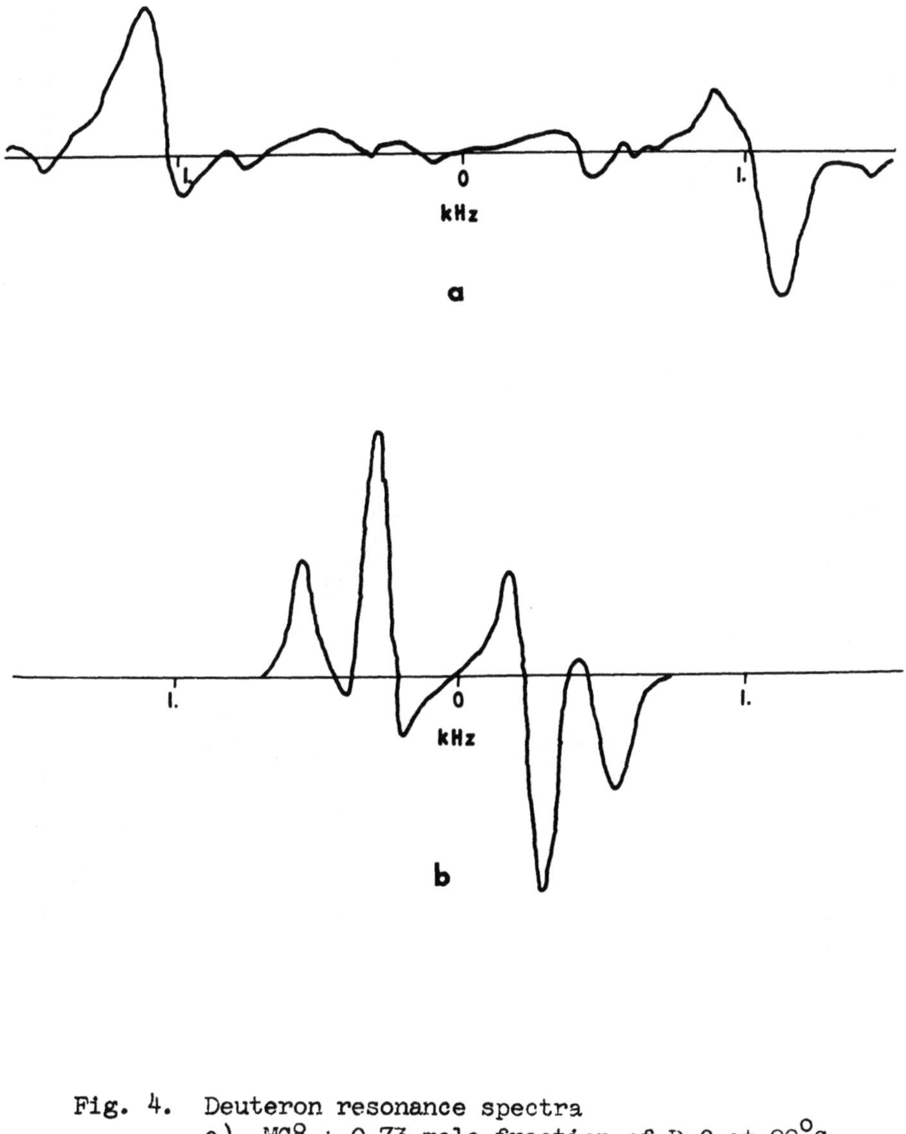

Fig. 4. Deuteron resonance spectra
a) MG8 + 0.73 mole fraction of D_2O at 22°C.
b) MG8 + 0.92 mole fraction of D_2O at 22°C.

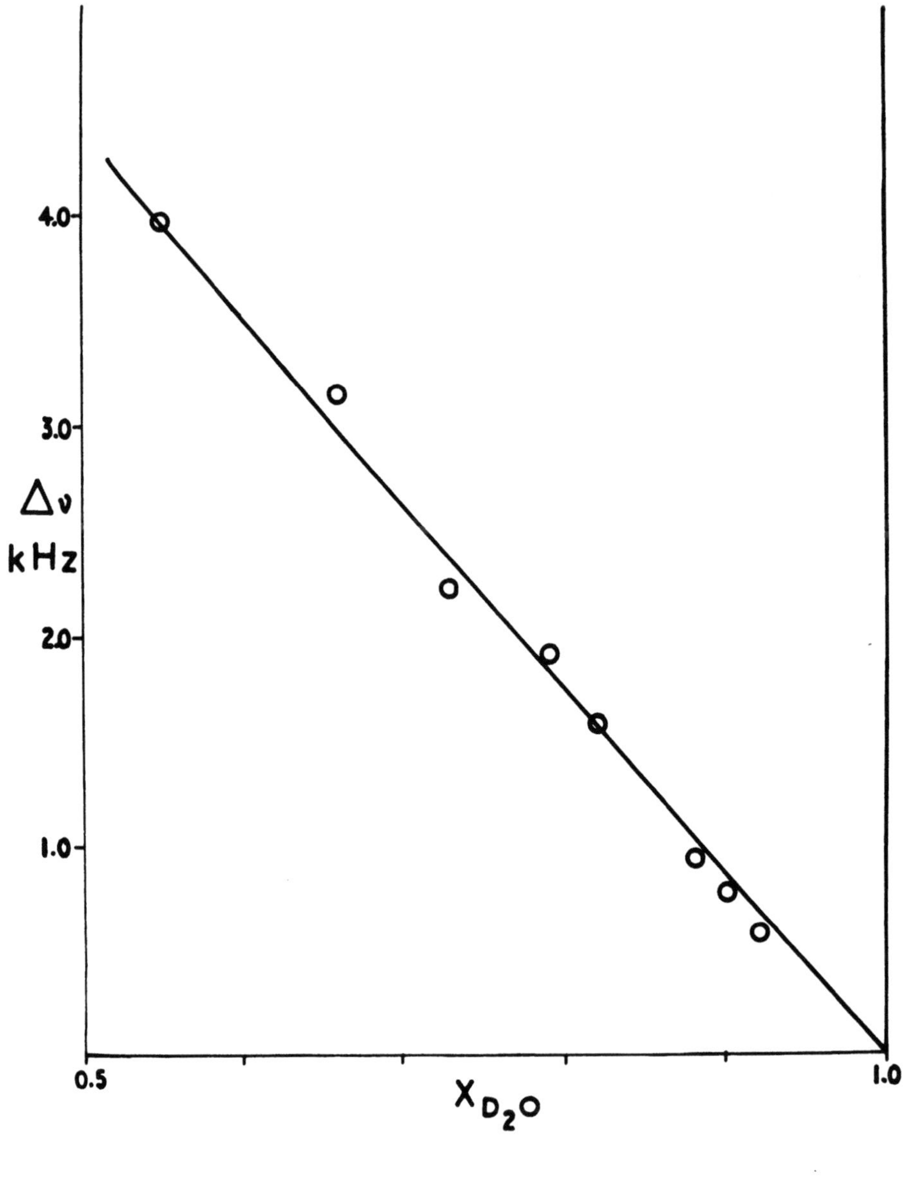

Fig. 5. Quadrupole splitting, Δν vs. mole fraction of D_2O, $X(D_2O)$, in the MG8 - D_2O system.

detailed study of this region. Preliminary observations of the temperatures above which two immiscible liquids are obtained do not agree with those of Larsson (1).

The values of $\Delta\nu$ at 20°C for the L_1 region and for the more intense doublet in the $L_1 + L_2$ region have been plotted against mole fraction of D_2O. Fig. 5 shows that a straight line can be drawn through the points which also cuts the mole fraction axis at the composition of pure D_2O if extrapolated through and beyond the region $L_1 + L_2$. There is no deviation from the line in the region of 0.2 MG8 + 0.8 D_2O as might be expected if a complex of that molecular ratio is formed.

The similar order of magnitude of the quadrupole coupling constants in this work to those observed by Lawson and Flautt in another l.c. system (7) suggests that here also there is rapid exchange between deuterons in two types of site. The lack of temperature dependence of the splitting means that within the l.c. phase the rate of deuteron exchange is always sufficiently rapid to lead to a time averaged quadrupole splitting and that the relatively small changes in temperature are insufficient to alter the distribution of deuterons in the two types of site.

In our case the two types may be the -OD sites on the monoglyceride molecule in which rotational freedom will be restricted, and those of liquid D_2O. Then, since there are two of each type of site on the respective molecules, we might assume under conditions of rapid exchange that the experimental value of $\Delta\nu$ is an average of $\Delta\nu(D_2O)$ and $\Delta\nu$ (MG8), the splittings at the two sites. Hence, in the same way as the observed chemical shift of water has been related to the shifts of hydrogen-bonded and non hydrogen-bonded protons in the liquid (20) :

$$\Delta\nu_{exp} = X_{MG8}\,\Delta\nu_{MG8} + X_{D_2O}\,\Delta\nu_{D_2O}$$

where X_{MG8}, X_{D_2O} are the mole fractions of MG8 and D_2O. Furthermore, if all the D_2O molecules are free to orient isotropically, $\Delta\nu(D_2O) = 0$, as found experimentally, see Fig. 5, and so :

$$\Delta\nu_{exp} = X_{MG8}\,\Delta\nu_{MG8}$$

Extrapolating in Fig. 5 to an MG8 mole fraction of unity gives $\Delta\nu$ (MG8) = 8.4 kHz. This value is much lower than the the value of ca. 100 kHz calculated for the case of simple reorientation about one axis (21).

The small values of the proton resonance line width and the deuteron quadrupole coupling constant both therefore confirm present views on the structure of the smectic l.c. state. Thus, although there is appreciable ordering of the molecular spaces in a bimolecular layer structure , the molecules themselves have many degrees of reorientational freedom within and between those spaces.

We wish to acknowledge the assistance of Mrs. L. Hemshall with the deuteron resonance measurements.

References

1. K. Larsson, Zeit. Phys. Chem. 1967, 56, 173
2. E.S.Lutton, J. Amer. Oil Chemists Soc. 1965, 42, 1068
3. A.W.Ralston, C.W.Hoerr, E.J.Hoffman, J. Amer. Chem. Soc. 1942, 64, 1516
4. A.S.C.Lawrence, M.P.McDonald, J.V.Stevens, Trans. Faraday Soc. in press
5. A.S.C.Lawrence, M.P.McDonald, Mol. Crystals, 1966, 1, 205
6. T.J.Flautt, K.D.Lawson, Adv. in Chem. 1967, 63, 26
7. K.D.Lawson, T.J.Flautt, J. Phys. Chem. 1968, 72, 2066
8. T.Malkin, M.R. El Shurbagy, J. Chem. Soc. 1936, 1628
9. D.A.Vassallo, J.C.Harden, Anal. Chem. 1962, 34, 132
10. B. Ellis, M.P.McDonald, J. Non-Cryst. Solids, 1969, 1, 186
11. J.D.Bernal, Z. Krist. 1932, 83, 153
12. A.Muller, Proc. Roy. Soc. 1932, 138A, 514
13. E.R.Andrew, J. Chem. Phys. 1950, 18, 607
14, J.D.Hoffmann, C.P.Smyth, J. Amer. Chem. Soc. 1949, 71, 431
15. A.S.C.Lawrence, M.A. Al-Mamum, M.P.McDonald, Trans. Faraday Soc. 1967, 63, 2789
16. A.S.C.Lawrence, A.Bingham, C.B.Capper, K.Hume, J. Phys. Chem. 1964, 68, 3470
17. D.Chapman, R.E.Richards, R.W.Yorke, J. Chem. Soc. 1960, 436
18. M.H.Cohen, F.Reif, Solid State Phys. 1957, 5, 321
19. C.Migchelsen, H.J.C.Berendsen, Colloque Ampere XIV, (North Holland Publishing Co. 1967) p.761
20. D.Eisenberg, W.Kauzmann, The Structure and Properties of Water (Oxford, 1969) p.196
21. D.D.Eley, M.J.Hey, H.F.Chew, W.Derbyshire, Chem. Comm. 1968, 23, 1474

Mesomorphism in Cholesterol-fatty Alcohol Systems

A. S. C. Lawrence

Emeritus Professor Chemistry

Sheffield University

During a study of mixing and unmixing of binary mixtures of polymorphic fatty alkanols in the solid state, the special case of cholesterol and these alcohols was examined, and a liquid crystalline phase was observed with C12,13,14,16 and 18 alkanols but not with the shorter decanol and octanol. Mlodziejowski[1] has reported liquid crystalline phases in the hexadecanol-cholesterol system but gives no figures for composition and temperatures. This has been done and fig. 1 shows the phase equilibria for this system. Extreme care needs to be taken in observing transitions owing to extreme supercooling of cholesterol in cooling runs and to its very slow rate of dissolution in heating runs; supercooling also occurs at the lc to solid transition but these temperatures were observed easily on the heating stage of a polarizing microscope. The points up to 25% of cholesterol were readily found by cooling curves; with 25% the arrest at the liquid to liquid crystal transition was only about one fifth of that at the freezing point. As usual, supercooling does not occur at the liquid to liquid crystal transition.

The constancy of the mesomorpic temperature range with increasing amounts of cholesterol suggests strongly that there exists only the 2:1 alkanol: cholesterol complex. Confirmation of this was provided by filtering a 45% solution at 61°, washing the cholesterol with warm methanol, drying and weighing; the weight was equal

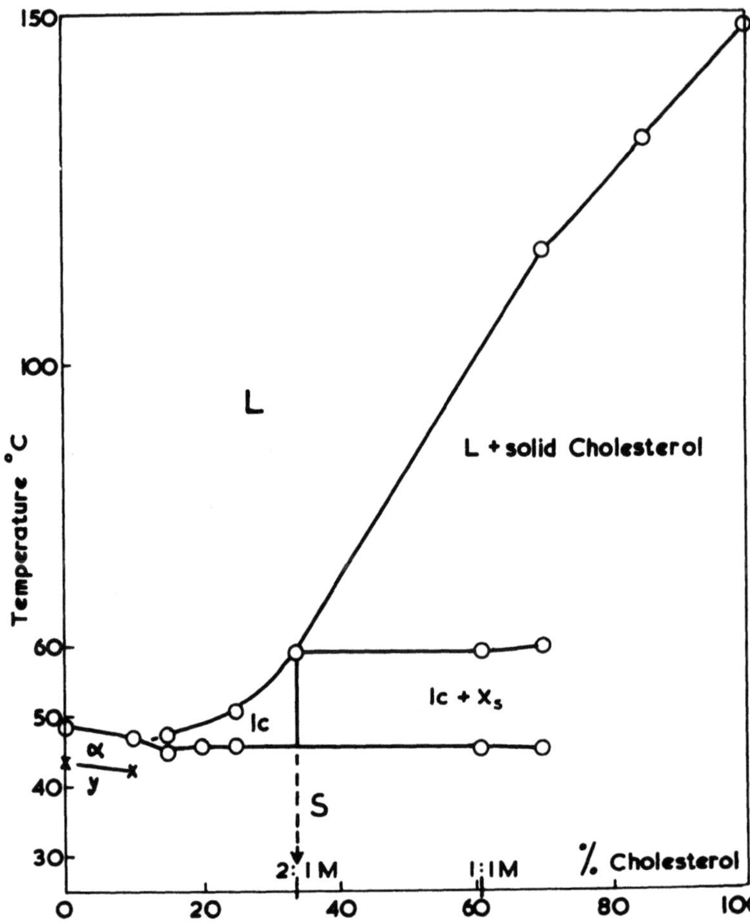

Fig. 1. Temperature-Composition equilibrium diagram for the system anhydrous hexadecanol and cholesterol.

to the additional amount above the 2:1 molar ratio and the filtrate gave the transition temperatures of the 2:1 composition. The binary alkanol-cholesterol systems dissolve water and form a ternary lc phase from which cholesterol separates between about 80° to 30° when cooled or heated sufficiently slowly; ca 5° per hour. On standing overnight at 25°, the semi-solid mass reverts to homogeneous liquid crystalline state. No liquid crystalline phase has been observed with fatty acids and cholesterol but it appears when water is added provided again that the chain length is 12 or more C atoms. Mono-decanoin behaves similarly but its full chain length is 13 C atoms; in this system the diol forms a lyotropic mesophase with water.

Conclusions

The formation of a liquid crystalline phase in a binary system of two components neither forming a mesophase is very unusual although Gaubert[2] has observed it again with cholesterol when succinimide and various organic acids are added. The hexadecanol case is clearly connected with the 2:1 complex whose existence requires the minimum alcohol chain length of 12 carbon atoms; i.e. a length roughly equal to the long dimension of the cholesterol molecule. The alkanols do not form a mesoform with either cholesteryl chloride or cholesterol acetate. It may also be noted that cholesterol has a large solubility in petroleum ether at room temperature when hexadecanol is present but addition of methanol precipitates it from the mixture. The easy conversion of cholesterol into a labile state whose freezing point is more than 100° below that of the pure compound is of special biological interest in view of the obvious need for some such mechanism.

References

1. Zeit. phys. chem (Leipzig) 135, 129, 1928.
2. Compt. rend. 156,149,1912.

LIQUID CRYSTALS IV.[1] ELECTRO-OPTIC EFFECTS IN p-ALKOXYBENZYLIDENE-p'-AMINOALKYLPHENONES AND RELATED COMPOUNDS

Joseph A. Castellano and Michael T. McCaffrey

RCA Laboratories, David Sarnoff Research Center

Princeton, New Jersey

A systematic study of the relationship between molecular structure and the electrical properties of liquid crystals has recently led us to the discovery of several new electro-optic effects in nematic[2,3] and cholesteric-nematic[4,5] materials. In the present work, which is a continuation of this study, a series of compounds with both nematic and smectic behavior were examined under externally applied electric fields. These new compounds were derived from p-alkoxybenzylidene-p'-aminoalkylphenones (I) and were prepared by conventional procedures.

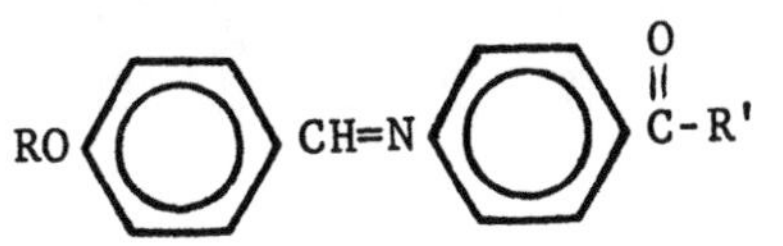

I

EXPERIMENTAL[6]

p-Alkoxybenzylidene-p'-aminoalkylphenones

The compounds of series I were prepared in 50-80% yields by the condensation of appropriate p-alkoxybenzaldehydes (purchased from Frinton Laboratories, South Vineland, N.J.) with p-aminoalkylphenones (Distillation Products, Eastman Organic Chemicals, Rochester, N.Y.) in refluxing benzene solution. A typical procedure is represented by the preparation of p-butoxybenzylidene-p'-aminopropiophenone: A mixture of 1.78 g (10 mmol) of p-butoxybenzaldehyde, 1.56 g (10 mmol) of p-aminopropiophenone and 0.1 g of benzenesulfonic acid was refluxed for four hours in 200 ml of benzene. The reaction

Table I

Substituted Benzylideneanilines Prepared for This Study

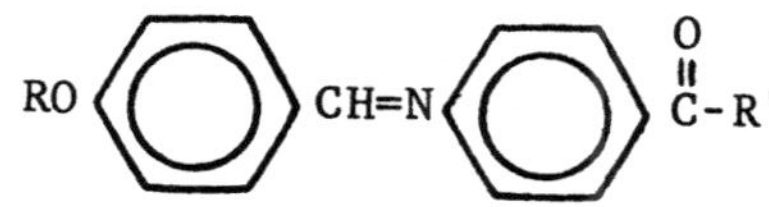

Compound	R	R'	Smectic Range, °C	Nematic Range, °C	Anal., % Calculated C	H	N	Found C	H	N
1	C_4H_9	CH_3	85-97	98-109[a]	---	--	--	---	--	--
2	C_6H_{13}	CH_3	74-108	109-111	77.98	7.79	4.33	77.84	7.85	4.25
3	C_8H_{17}	CH_3	71-113	114-116[a]	---	--	--	---	--	--
4	C_4H_9	C_2H_5	86-141	142-144	77.64	7.49	4.53	77.58	7.55	4.48
5	CH_3	CH_3	--	124.5[a,c]	---	--	--	---	--	--
6	CH_3	C_2H_5	--	115-132	76.38	6.41	5.24	76.49	6.35	5.15
7	CH_3	C_3H_7	--	101 (96)[b]	76.84	6.81	4.98	76.95	6.86	4.86
8	CH_3	C_4H_9	--	87-105	77.26	7.17	4.74	77.64	7.14	4.63

[a]Ref. 7; [b]Monotropic; [c]Not mesomorphic

was monitored by azeotropic removal of water. At the completion of the reaction the solvent was removed in vacuo and the crude product was recrystallized from hexane to yield 1.93 g (6.1 mmol, 61%) of the Schiff base. The other derivitives of I are listed in Table I.

Electrical Behavior of Materials

The experimental setup for examination of the materials under externally applied electric fields has been described previously.[2,3] The thickness of active area (~1 cm^2) in the present investigation was 12.5 microns. Nesa-coated quartz was again used for the electrodes.

DISCUSSION

It is well known that, when the mesomorphic transition temperatures for a homologous series of compounds, e.g., in a series of n-alkyl ethers or esters, are plotted against the number of carbon atoms in the alkyl chain, smooth curve relationships between even or odd members of the series are found to exist. This regular alternation of transition points has been explained by assuming that the alkyl chains adopt the "cog wheel" rather than the "zig-zag" conformation in the mesomorphic state.[8]

An example of this trend was provided[7] by the homologous series derived from I in which R' was fixed at CH_3. As the length of the alkyl chain in the ether portion of the molecule was increased, mesomorphic behavior appeared (Figure 1). An unusual feature of the phase transition plot for this series, however, was the increased nematic as well as smectic thermal stability as the chain length was increased from four to eight carbon atoms. This behavior has only been observed[8] with the series of 2- and 2'- substituted 4-p-n-alkoxybenzylidene-p'-aminobiphenyls in which the alkoxy chain had from seven to 18 carbon atoms. The trend of nematic-isotropic transition temperatures for these compounds, which contain lateral substituents, was postulated to occur as a result of very weak lateral cohesions brought about by the increases side spacing of the molecules. Since no lateral substituents are present in the series 1-8, one must assume that the acetyl group produces such strong terminal intermolecular attractive forces that the ratio of lateral to terminal cohesions is very low. The molecules may be firmly linked together by molecular attraction of their ends as previously postulated by Gray.[8] In the absence of adequate lateral interactions, however, chains of these molecules may be disrupted and will quickly break up into individual molecules as a result of thermal vibrations. As the lateral interactions are increased by lengthening the chain in the ether portion of the molecule from four to eight carbon atoms, disruption is prevented and nematic thermal stability rises. That

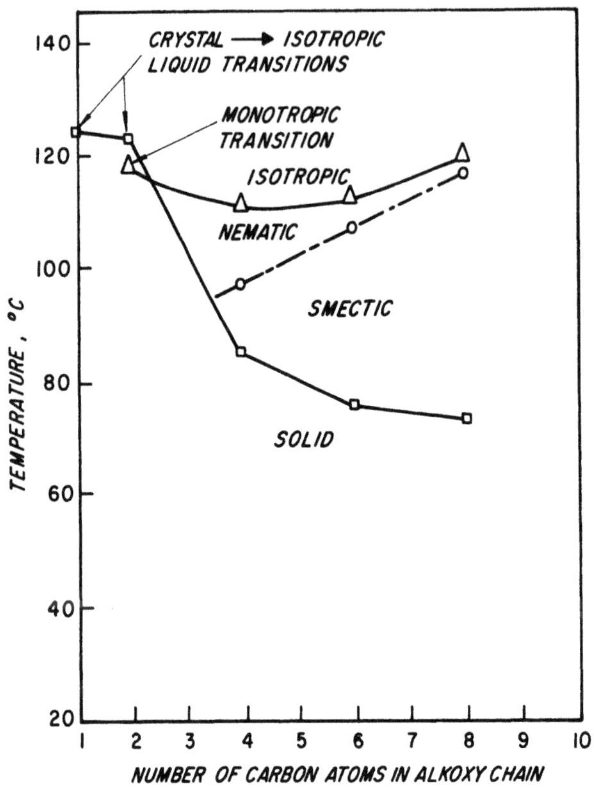

Fig. 1

Phase transition plot for the series: $C_nH_{2n+1}OC_6H_4CH{:}NC_6H_4\ COCH_3$.

these lateral interactions are indeed increased was illustrated by the higher smectic thermal stability.

Both the smectic and the nematic thermal stability can be further increased by extension of the chain in both the ether and the ketone portions of the molecule (as in 4). However, when the alkyl group in the ether portion remains fixed at one carbon atom and the chain in the ketone portion of the molecule was extended, no smectic behavior was observed (Figure 2). Hence it would appear that the chain attached to the carbonyl group exerts only a small influence on the lateral interaction and that terminal interaction between the ether oxygen and the carbonyl group remain high.

It is not surprising that the acetyl group produces such strong intermolecular attractive forces since it has a high permanent dipole moment. This fact enables the compounds of series I to exhibit interesting behavior under the influence of externally applied electric fields.

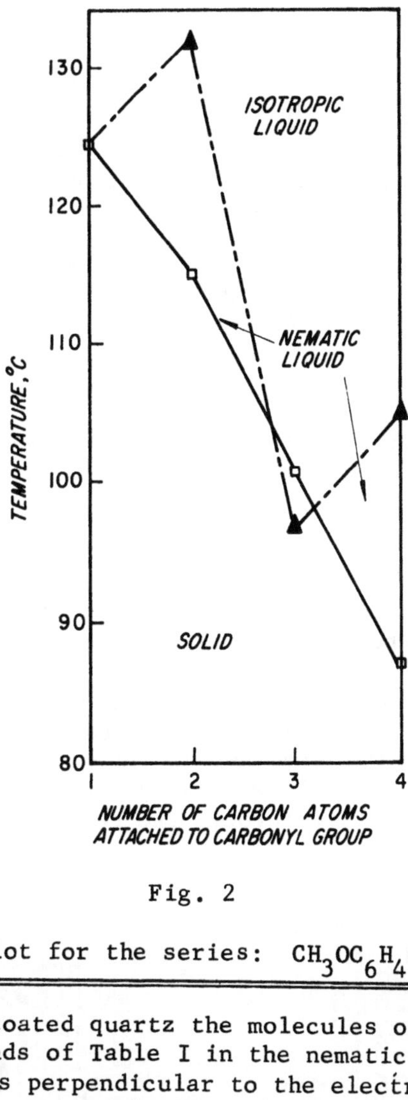

Fig. 2

Phase transition plot for the series: $CH_3OC_6H_4CH{:}NC_6H_4\ COC_nH_{2n+1}$

Between NESA coated quartz the molecules of any one of the mesomorphic compounds of Table I in the nematic state are oriented with their long axes perpendicular to the electrode surfaces (Figure 3a). Thus, between cross polarizers light was extinguished and the material appeared isotropic. This orientation presumably occurred as a result of the strong attraction between the carbonyl group and the tin oxide coating. Application of a low strength electric field (0.1-0.5 KV/cm) produced a reorientation of the molecules (Figure 3b) as they realigned with their dipole moments in the direction of the applied field. This process produced differently oriented regions of the birefringent swarms[9] and light scattering was observed. At higher field strengths (0.5-40 KV/cm) the reorientation process was complete, essentially all of the molecules were aligned in the direction of the field and the transmitted light was extinguished (Figure 3c). However, very weak dynamic scattering[2]

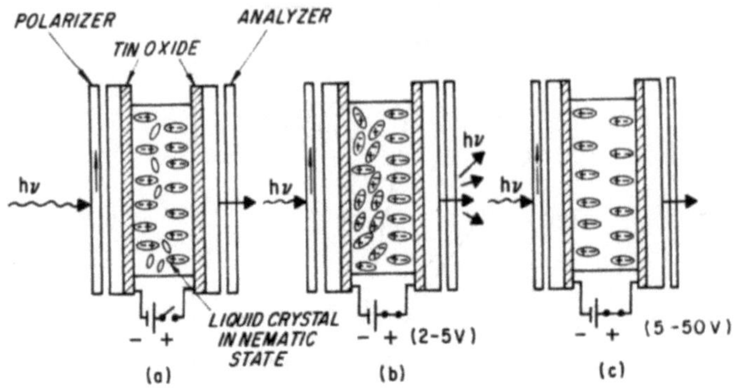

Fig. 3

Schematic representation of electro-optic cell with (a) no field applied, (b) a 2-5 volt DC field applied and (c) a 5-50 volt DC field applied.

was observed at the higher field strengths. This was presumably due to ions in transit through the nematic phase.

It was also possible to produce cooperative alignment of pleochroic dyes with these materials. As an example, a mixture of 1 and indophenol blue (1.0%) was examined in the normal cell configuration (Figure 4a). When these cells were subjected to fields of 0.1-0.5 KV/cm, the light transmitted through the cell was blue (Figure 4b).

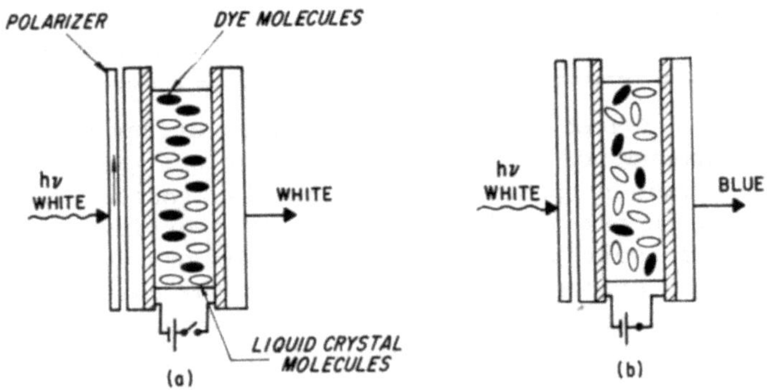

Fig. 4

Schematic representation of electro-optic cell containing liquid crystal: dye mixture with (a) no field applied and (b) a 2-5 volt DC field applied.

Thus, we were able to produce an electronic color switching effect which was the opposite of that previously observed in the nematic state of p-ethoxybenzylidene-p'-aminobenzonitrile.[3] All of the mesomorphic compounds in Table I exhibited this electro-optic effect when the material was kept in the nematic state. Since the dyes which were used previously[3] were soluble in these nematic hosts, it was possible to produce electro-optic cells with a wide variety of colors. The contrast ratios of the color changes, however, were not as high as those obtained in the previous work.

In the compounds which exhibited both smectic and nematic states (1-4), a second electro-optic effect was observed. Upon cooling these materials from their nematic states, smectic states formed at the transition temperature. This smectic state appeared as a milky white emulsion under cross polarizers (Figure 5a). This state bears a strong resemblance to the smectic schlieren texture (β smectic phase) described by Sackmann and Demus.[10] These workers also report the appearance of this phase upon quick cooling of the nematic melt. In contrast to this phase, the smectic state which appears on melting of the crystals, has the focal conic texture (α smectic phase)[10] (Figure 5b). Thus, the β smectic phase appears to be a monotropic liquid crystalline phase. When this state (between cross polarizers) was subjected to electric fields of the order of 40-160 KV/cm at 92° ± 1°C, the schlieren texture disappeared and the light was extinguished. The resultant state exhibited very weak dynamic scattering and appeared to be identical to the activated nematic state. The effect was strongly temperature dependent; as the temperature was lowered, the field strength required to activate the material increased. The maximum efficiency was obtained at a temperature which was near the smectic-nematic transition temperature. This result leads us to the tentative conclusion that the electric

Fig. 5
Microscopic appearance of the (a) focal conic texture and (b) smectic schlieren texture observed in the materials prepared for this study. Magnification 60X.

field produces a phase change from the smectic to the nematic state by virtue of breakdown of the lateral intermolecular Van der Waals forces. A simple calculation of the energies involved in this process supports this conclusion. The energy required to raise the temperature of the material from the smectic (92°) to the nematic state (98°) of compound 1 is determined from:

$$E = \text{Specific Heat } (T_2 - T_1)$$

Since most organic compounds have specific heats of 0.3-0.5 g-cal./g, we can assume that 1 has a specific heat of ~0.4 g-cal./g or 1.6 joules/g, which gives an energy of ~0.01 joules for a sample of 6 mg of material. The energy available at 200 volts DC is calculated to be 0.04 joules for material with a resistivity of 10^9 ohm-cm. Thus, enough energy is supplied to the smectic material to produce the phase change.

A comparable calculation for a change from the smectic (92°) to the isotropic (110°) state shows that 0.15 joules are required and indicates that higher voltages should be capable of affecting the latter phase change. This was established by experiments in which the material was subjected to field strengths of 200-400 KV/cm.

Studies of the electrical behavior of the smectic state has not received a great deal of attention in the literature. A recent report[11] describes the formation of domains in the smectic state of p-n-heptyloxybenzoic acid which has a smectic range of 92-98° and a nematic range of 98-146°. Their results were remarkably similar to those reported here and a similar mechanism may be operative with their material. A similar explanation may also be applicable to the results of Arora, et al.[12] These workers suggested the existence of a second nematic state partly on the basis of behavior of the material in an electric field. However, if this new state was actually the β smectic phase (or some other unusual smectic state) it might give the results which they observed.

Further experiments with these new materials are continuing.

CONCLUSIONS

A new series of liquid crystalline compounds which exhibited both smectic and nematic states which responded to externally applied electric fields was investigated. The material in the nematic state could be used to orient pleochroic dye molecules with an applied field. The unusual behavior of the smectic state under electrical excitation was attributed to a field induced phase change from the smectic to the nematic state.

REFERENCES

1. For the previous paper in this series, see J. E. Goldmacher and M. T. McCaffrey, Abstracts of Papers presented at 158th National ACS Meeting, New York, 1969.

2. G. H. Heilmeier, L. A. Zanoni, and L. A. Barton, Proc. IEEE, 56, 1162 (1968).

3. G. H. Heilmeier, J. A. Castellano and L. A. Zanoni, Mol. Crystals and Liquid Crystals, 5, in press. This paper was presented at the 2nd Int. Liquid Crystal Conference, Kent State Univ., Kent, Ohio, August 1968.

4. G. H. Heilmeier and J. E. Goldmacher, Appl. Phys. Letters, 13, 132 (1968).

5. G. H. Heilmeier, L. A. Zanoni, and J. E. Goldmacher, Abstracts of Papers, presented at 158th National ACS Meeting, New York, 1969.

6. All melting points are corrected. Elemental analyses were performed by B. Goydish and A. Murray of RCA Laboratories.

7. J. A. Castellano, J. E. Goldmacher, L. A. Barton, and J. S. Kane, J. Org. Chem., 33, 3501 (1968).

8. G. W. Gray, Molecular Structure and the Properties of Liquid Crystals, Academic Press, New York (1962).

9. This term is used merely to clarify the explanation of this phenomenon and not as an endorsement of the so-called "swarm theory".

10. H. Sackmann, D. Demus, Liquid Crystals, Gordon and Breach, New York (1967), p. 345. This work was presented at the 1st Int. Liquid Crystal Conference, Kent, Ohio, August 1965.

11. L. K. Vistin and A. P. Kapustin, Soviet Phy. Cryst., 13, 284 (1968).

12. S. Arora, J. Fergason, and A. Saupe, Abstracts of Papers presented at 2nd Int. Liquid Crystal Conference, Kent, Ohio, August 1968.

ACKNOWLEDGMENT

The research reported in this paper was jointly sponsored by the Electronics Research Center under NASA Contract NAS 12-638 and RCA Laboratories.

The authors would like to thank Dr. Wolfgang Helfrich and Edward F. Pasierb for many helpful discussions.

Infrared Spectroscopic Measurements on the Crystal-Nematic Transition

Bernard J. Bulkin and Dolores Grunbaum

Hunter College of the City University of New York

New York, New York 10021

INTRODUCTION

Although many studies have been carried out to measure the extent of order as a function of temperature in the liquid crystalline state of thermotropic liquid crystals, little work has been done to indicate the amount of order or disorder in solids as the liquid crystalline state is approached. Such studies as have been reported (1), mainly deductions based on the shape of thermal analysis curves in the pre-transition region, indicate that the crystal-nematic transition may be of greater than first order, or that there may be some pre-transition phenomena. The shapes of DTA and DSC curves are, however, very dependent on instrumental response, etc., often in ways which have yet to be fully explored.

In this paper we present results on the nature of the crystal-nematic transition derived from infrared spectroscopic measurements on the homologous series of 4,4' Bis(alkoxy)azoxybenzenes,I, (R=methyl, ethyl, butyl, pentyl, and hexyl.).

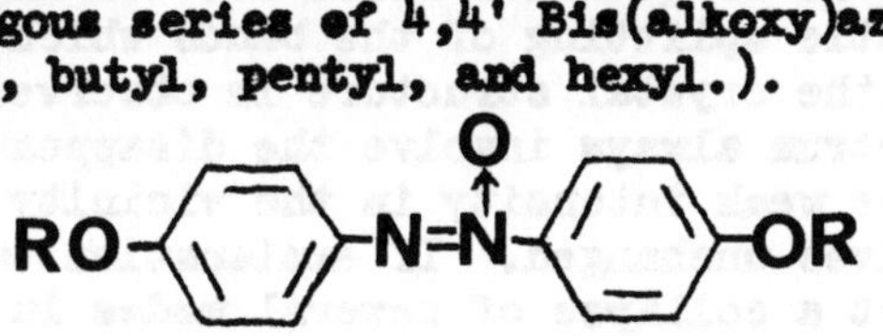

I

These results indicate that the crystal-nematic transition is indeed a gradual one, involving, in some cases, considerable changes in the solid as much as 25°C below the observed capillary transition temperature. Preliminary results on the crystal-smectic transition in 4,4' Bis(heptyloxy) azoxybenzene (I, R=heptyl) are also reported.

RESULTS AND DISCUSSION

The infrared spectra of I have been investigated in the 4000-600 cm^{-1} region by Maier and Englert (2). Despite the relatively low resolution used in those studies, changes were observed in the infrared spectra as the phase changed, these being especially marked at the crystal-nematic transition. Many more such changes in the spectra are found when a grating instrument is used and the spectral range is extended to 300 cm^{-1}.

The changes observed in the spectra may be summarized as follows: 1) The spectrum above 1600 cm^{-1} remains unaffected by the transitions from crystal to nematic (c-n) and nematic to isotropic liquid (n-l). 2) No new bands appear in the spectrum as the temperature is raised. 3) Several bands present in the crystal do not appear in the spectra of the nematic melts. 4) There are some changes in the relative intensities of bands in the liquid vs. the nematic melt. 5) The spectra of solutions of the solids I in CCl_4 and CS_2 are virtually the same as those of the isotropic liquid, once allowance is made for different half band widths due to the difference in temperatures.

The bands which disappear at the c-n transition are very interesting. Because the molecules are of low symmetry (point group C_s or C_1), there can be no degenerate or infrared inactive vibrations. The only other causes of the bands which disappear can be unambiguously associated with the lattice forces. They must be due either to 1) lattice vibrations; 2) a coupling between the molecules of the unit cell which leads to a systematic quadrupling of the bands as a consequence of the fact that there are four molecules per unit cell (3); or 3) combination modes between lattice vibrations and internal vibrations.

Explanation number three is probably the correct one. The bands in question occur at frequencies which are too high for lattice vibrations. No systematic splitting of the bands which might be expected on the basis of the crystal structure is observed. Further, the changes in the spectrum always involve the disappearance of sharp bands of medium or weak intensity in the vicinity of stronger bands which are themselves unchanged. If explanation two were correct, we would expect a collapse of several modes in a particular region of the crystal spectrum to one band in the isotropic liquid. Preliminary investigations of the lattice vibration Raman spectrum (Figure 1) of azoxydianisole indicates that all the bands which disappear can be assigned on the basis of explanation three.

Vibrational energy levels such as these provide a sensitive probe for the measurement of the coupling between individual molecule and the crystal lattice. This work reports an investigation of the temperature dependence of the absorbance for the bands which

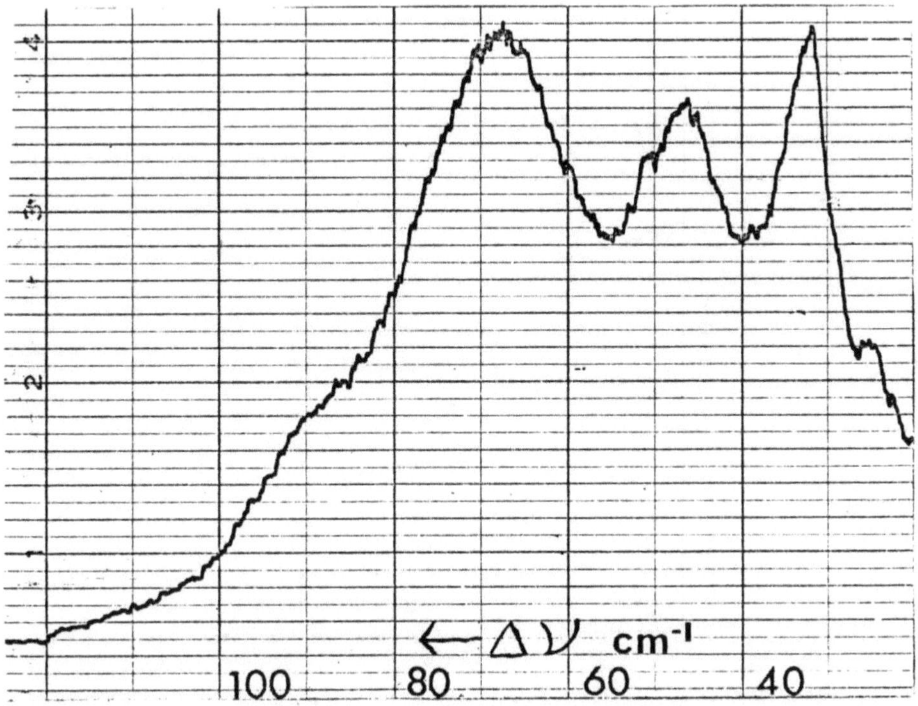

Figure 1. Raman spectrum of 4,4' Bis(methoxy)axoxybenzene in the lattice vibration region.

decrease markedly in intensity at the c-n transition. The temperature region of interest is the range of temperatures just below and above the transition temperature, because the behavior of crystal properties in this region can yield detailed information about the order of the transition(4). The results of this study are shown in Figure 2. For clarity the curves have been arbitrarily positioned on the absorbance scale, since only changes in absorbance are of interest. Data points were taken by first bringing the sample to a given temperature and then allowing ca. 15 minutes for temperature equilibration. We did not observe any of the strange hysteresis effects of mesomorphic samples which have been reported by other workers recently,(5,6), i.e., the absorbance reached a value characteristic of a particular temperature and did not fluctuate. Figure 2 shows that in each case there is a slope discontinuity in the curves at the observed transition temperatures. More interesting, however, is that the absorbance begins to decrease well below the observed transition temperatures, finally decreasing more and more rapidly until the first trace of liquid crystal can be observed visually. Note that the decrease in absorbance is not due to an increase in band width with increasing temperature, as demonstrated by the fact that the bands decrease relative to the other bands in the spectrum. Furthermore, it can be seen that some of the lower temperature transitions have the more gradually decreasing intensity.

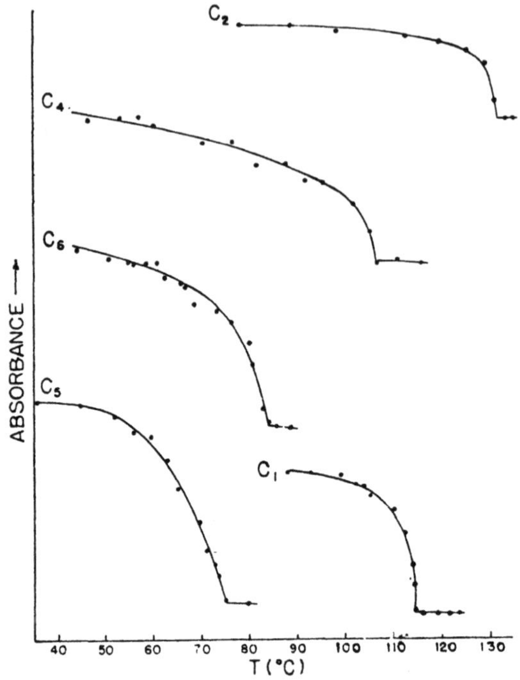

Figure 2. Temperature dependence of the absorbance of the "disappearing" infrared bands of 4,4' Bis(alkoxy)azoxybenzenes, labeled according to alkyl chain length.

The gradual decrease in absorption intensity illustrated in Figure 2 means that some of the molecules in the crystal must be vibrationally decoupled from the lattice forces well below the crystal-nematic transition temperature. How can this happen? The two possibilities seem to be 1) increased motional freedom for some molecules on the lattice site or 2) actual movement of a molecule from a lattice site to an interstitial site. The resolution of these alternatives is not yet unambiguously possible, but many aspects of the nematic phase favor the second explanation. In the nematic state we have taken the crystal structure and replaced it by a layered structure in which there is long range order only with respect to the orientation of long axes of the molecules. Such a structure can be reached from the crystalline state by taking molecules on lattice sites and moving them to interstitial sites, creating defects in the lattice.

From the point of view of theories of the solid state there are some interesting consequences of this result. It appears that the crystal-nematic transition is a sort of order-disorder transition of a one component system. Indeed, when the curves of Figure 2 are plotted on a reduced temperature scale (T/T_{c-n}) they resemble those predicted for a partially ordered system by Bragg and Williams (7). The situations are not completely analogous, of course, because that (7) treatment involved a two component system with two sets of sites, whereas the c-n transition can, at best, be thought of as a one component-two sets of sites system. It is particularly complicated, however, by the fact that we do not know how many interstitial sites there are; indeed, we do not even know how many interstitial sites there are; indeed, we do not even know if this is a fixed number. For this reason it is preferable to use a lattice defect model of the transition.

What is the energy barrier to creation of a lattice defect? From the curves of Figure 2 we can calculate an activation energy for the postulated process of moving a molecule from a lattice site to an interstitial site. This is done by plotting ln α, the fractional change in absorbance, vs. 1/T(°K). These plots are shown in Figure 3. The activation energies calculated from the slopes of the lines are summarized in Table 1, together with the c-n transitions temperatures.

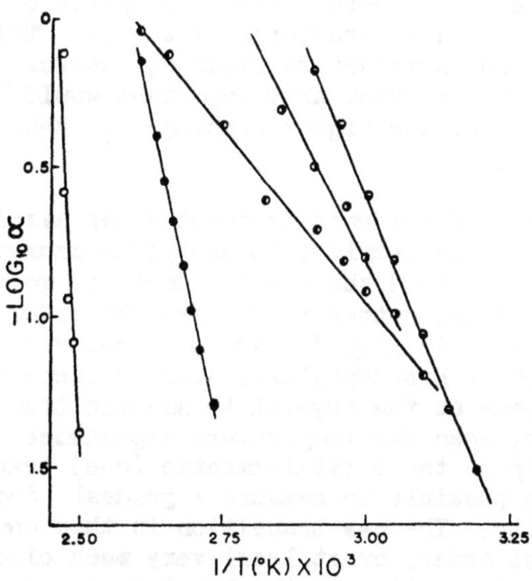

Figure 3. Arrhenius plots of log α vs. 1/T (°K) derived from Fig. 2. α is the fractional change in absorbance at a given temperature. The compounds (I) are ●=methyl; O=ethyl; O=butyl; O=pentyl; O=hexyl. The uncertainties in these values, which are shown in the table, were estimated from 1) reproducibility of the slopes between runs; 2) estimated temperature uncertainly of ±1°C 3) estimated absorbance uncertainty of ±.01. The activation energies follow the same alternating pattern characteristic of many properties of the homologous series of liquid crystals, which has been attributed by Gray(8) to an alternation in terminal group interactions.

TABLE I. ACTIVATION ENERGIES FOR CRYSTAL-NEMATIC TRANSITIONS IN 4,4' BIS(ALKOXY)AZOXYBENZENES

ALKYL GROUP	T_{c-n} (°C)	ACTIVATION ENERGY kCal	SPECTRAL REGION (cm^{-1})
METHYL	115	42 ± 6	1210-1120
ETHYL	133	128 ± 15	1000-860
BUTYL	108	10.6 ± 2	600-490
PENTYL	76	19.6 ± 3	1100-945
HEXYL	84	16.1 ± 2	600-495

In fact, our results indicate that the alternation in transition temperatures is possibly a more intimate property of the crystal, liquid crystal, and liquid structures as a unit. This point is discussed below in an introductory fashion, however, more work is needed to clarify it. Particularly important would be structural data on the crystals of the higher homologs (I, R=butyl, pentyl, hexyl, heptyl, etc.).

A certain amount of evidence is nonetheless available now. When one passes from the hexyloxy to heptyloxy azoxybenzene (and higher homologs thereafter) the crystal does not go over to the nematic phase initially, rather it follows the order crystal-smectic-nematic-isotropic liquid. Again we observe that the infrared spectrum of 4,4' - Bis(heptyloxy) azoxy benzene shows several bands which disappear at the crystal to smectic transition. In this case, however, when the temperature dependence of the absorption is examined up to the crystal-smectic (c-s) transition temperature, it is not possible to measure a gradual decrease in the absorption intensity. The c-s transition in this compound thus appears to be first order, or at least very much closer to first order than the c-n transition in the homologous compounds.

This behavior lends additional support to the lattice defect model of the c-n transition. The smectic phase involves maintenance of long range order within layers, beyond the order of the long axes of the molecules. Probably because of the long terminal chains, there is considerably less tendency to form the molecular lattice defect than to break an entire layer of crystal free. The lattice process, involving as it does a much greater mass, requires a much higher activation energy.

Finally, we wish to discuss the overall trend in activation energies which, despite the alternation, do decrease with increasing chain length from methyl to hexyl. When a molecule is moved from a lattice site to an interstitial site the activation energy which we have calculated herein is very likely a measure of both the barrier to moving the molecule into the interstitial site, as well as the weakening of the crystal lattice which results from the hole + interstitial molecule. If the defect is difficult to form, but once formed it greatly weakens the lattice, then this will hasten the collapse of the crystal to the nematic phase. The result is a high activation energy. In the situation which is observed here, the longer chain homologs are very likely more disordered than the methyl and ethyl compounds. (Experimentally we find that the hexyl and heptyl compounds are rather difficult to crystallize, for example). Creation of a lattice defect in a partially disordered structure does not perturb the overall lattice as much, so in these cases the collapse of crystal to liquid crystal is slower and a lower activation energy is observed.

CONCLUSIONS

This work confirms the earlier indications (1) that the c-n transition is not a first order transition. There are definitely pre-transition effects which appear when vibrational spectroscopy is used as a probe of the lattice forces. One of the points of this work is to demonstrate the power of infrared spectroscopy as a tool for the study of intermolecular interactions in condensed phases.

A lattice defect model offers a possible explanation of the observations concerning the c-n transition. It will be interesting to explore the behavior of other properties of mesomorphic materials just below the c-n transition temperature. The results reported recently by Labes (9) on conductivity changes at this transition are extremely interesting in light of the results discussed in this paper, and we are also pursuing measurements along these lines.

ACKNOWLEDGMENTS

This work was made possible by the generous support of the donors of the Petroleum Research Fund administered by the American Chemical Society (Grant 1198 - G2), the American Cancer Society (Grant P-504) and the City University of New York.

REFERENCES

1) a. H. Arnold, in Liquid Crystals, G. H. Brown, G. J. Dienes and M. M. Labes, eds., Gordon and Breach, New York, 1967, pp. 323-330 and refs. therein; b. A. V. Santoro and G. Spielholtz, Anal. Chim. Acta, 42, 537 (1968).

2) W. Maier and G. Englert, Z. Elecktrochem., 62, 1020 (1958).

3) J. D. Bernal and D. Crowfoot, Trans. Far. Soc., 29, 1032 (1933).

4) J. E. Mayer and S. F. Streeter, J. Chem. Phys., 7, 1019 (1939).

5) G. Durand and D.V.G.L. Narasimba Rao, Bull. Am. Phys. Soc., 17, 1054 (1967).

6) W. A. Nordland, J. Appl. Phys. 39, 5033 (1968).

7) W. L. Bragg and E. J. Williams, Proc. Roy Soc., A145, 699 (1934).

8) G. Gray, in ref. 1, pg. 142.

9) M. M. Labes and C. Kobayashi, paper presented at the Second International Symposium on Liquid Crystals, Kent, Ohio, August 1968.

The Alignment of Molecules in the Nematic Liquid Crystal State

John F. Dreyer
Polacoat, Inc.
Blue Ash, Ohio

The most distinctive characteristic of nematic liquid crystals is their optical property of anisotropy as evidenced by their birefringence while in a fluid state. They possess an organizational arrangement of their molecules which is intermediate between the disorder of an ordinary liquid and the three dimensional periodicity of crystals.

When the molecule is colored, this directionality is also evidenced by dichroism. Lack of discrete visible particles and homogenity between discontinuity lines are other visible characteristics. Orientation by contact with a rubbed surface is a unique property[1]. All of these characteristics are in part attributed to the arrangement of the molecules and the forces between them.

Not all liquid crystals are dichroic. Dichroism occurs when light vibrating in one direction is absorbed differently than that vibrating in a perpendicular direction. Both dichroism and birefringence provide a simple means for determining the direction of alignment of the molecules.

An organic molecule can absorb light vibrations only when the electric vector is parallel to the conjugated bond chain of the molecule. For most colored compounds, the long axis of the molecule contains the conjugated bond chain and is the principal color axis. Light vibrations perpendicular to the plane of flat molecules are not absorbed. The third axis may or may not absorb light depending on whether or not there are conjugated bonds on this axis[2].

When dichroic molecules are uniformly aligned, they form light polarizing filters. The property that nematic liquid crystals have of aligning themselves on an oriented surface provides a simple means of producing light polarizing filters[3]. These films have absorption for light vibrating along one direction and are transparent for light vibrating in a perpendicular direction.

The orientation pattern of these films is influenced by the character of the molecule and any impurities present. In addition to the characteristic Schlieren pattern and the discontinuity lines, serrations, varigations and domain patterns can occur (Figures I, II). These may be akin to the various arrangements found with soap[4] and called superphases by Zocher.

Color is evidenced by both isotropic and anisotropic materials. The absence of color on one axis is unique to anisotropic materials. Hence, for a film to be colorless for light vibrating on one axis, the molecules must be aligned with colorless axes parallel.

Various studies have been made on the spatial arrangement of flat colored molecules.

Morton[5], from studies of dichroism, suggests two arrangements for dye molecules, one with the long axes parallel and overlaping, the other with the long axes parallel and stacked.

Frank[6] suggests an arrangement of flat dye molecules as in a stack of coins.

Foerster[7] suggests that the molecules be with the flat sides together but crossed at angles.

Jelley[8] stated that an aromatic photosensitive dye formed threads which are analogous to the nematic form of liquid crystals. Further work by Sheppard[9] showed that the spacing between the molecules indicated that the molecules lay with flat sides together built up like a stack of cards with a spacing of 3.8 Angstroms. The increased absorption of light of longer wavelength is said to be due to this association of the molecules.

For liquid crystal materials containing a flat molecule, a line is not sufficient to show the complete alignment of the molecule. For molecules which are not symmetrical, the illus-

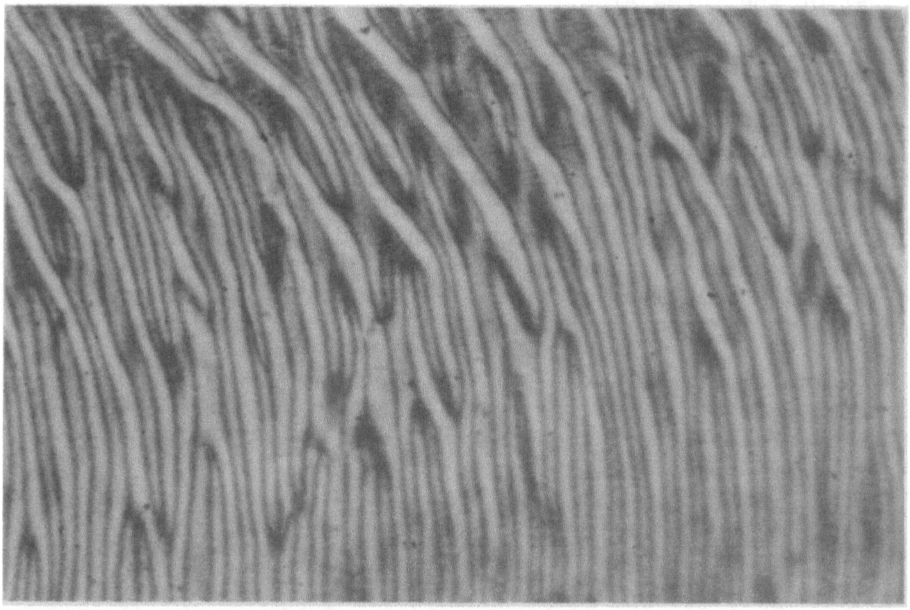

FIGURE I - DOMAIN PATTERN

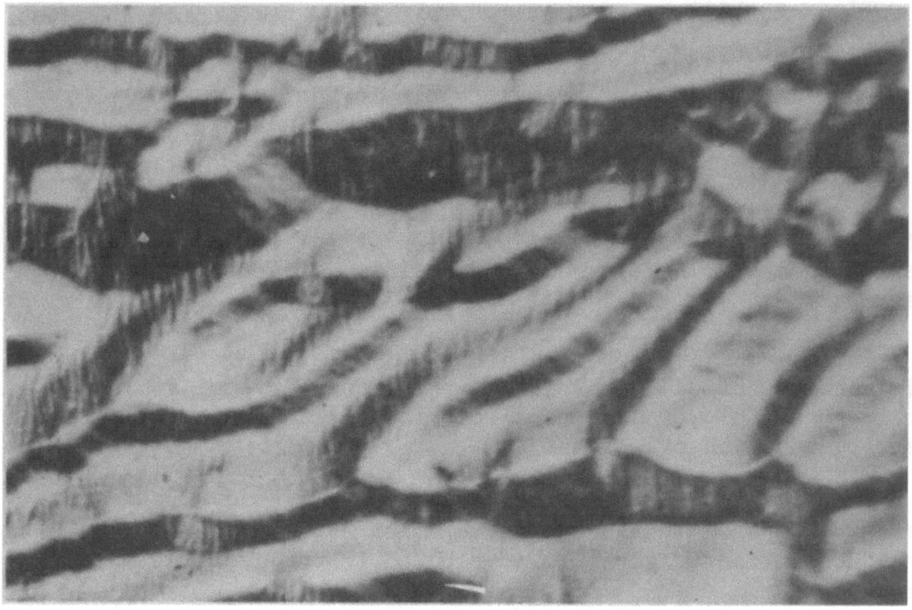

FIGURE II - DOMAIN AND SERRATIONS

tration should show the position of the head and tail as well as that of the three axes.

The dye molecule, Amaranth (Color Index No. 16185), has a lyotropic nematic liquid crystal phase with water as the solvent [10]. The molecule has the structure (Figure III) and can be illustrated by a flag shape with the axes marked "X," "Y" and "Z" in accordance with Calvin. The red color of this dye is due to the absorption of light vibrating along the principal "X" axis. There is no visible absorption along the perpendicular "Y" and "Z" axes. Hence, the formation of a dichroic film with one direction colorless does not tell us whether the "Y" or "Z" axes or both of them are parallel.

Dichroic materials are classified in two types, the positive and the negative[11]. For the negative type, the least absorbing axes are parallel. For the positive type, the strong absorbing axes are said to be parallel[12].

Many common thermotropic dichroic nematic liquid crystal materials such as para-azoxyanisole orient to give a positive dichroism. Salvarsan forms a positive lyotropic nematic liquid crystal[1]. That is, the light absorbing axis is in the direction parallel to the alignment of the support. Other liquid crystal materials give negative dichroism where the light absorbing axis is in a direction perpendicular to the orientation of the supporting surface.

Lyotropic dichroic nematic liquid crystals of the dyestuff type such as Amaranth or Methylene Blue orient on evaporation from solution on polished glass to produce a negative dichroism.

As the dye concentration increases during the evaporation of the solvent, a state is reached where the fluid becomes anisotropic. On continued evaporation of the solvent, there results a smearable film which maintains this anisotropic character.

The dried films produced are light polarizing films. They maintain the optical dichroism of the liquid crystal state and can be reversibly brought back into the fluid liquid crystal phase by addition of solvent.

Transmission curves of the dried polarizing dye film Amaranth compared to the dye in solution show increased wave (J band) absorption (Figure IV) attributed to the association of the molecules.

Stretching the dried film causes fracture similar to

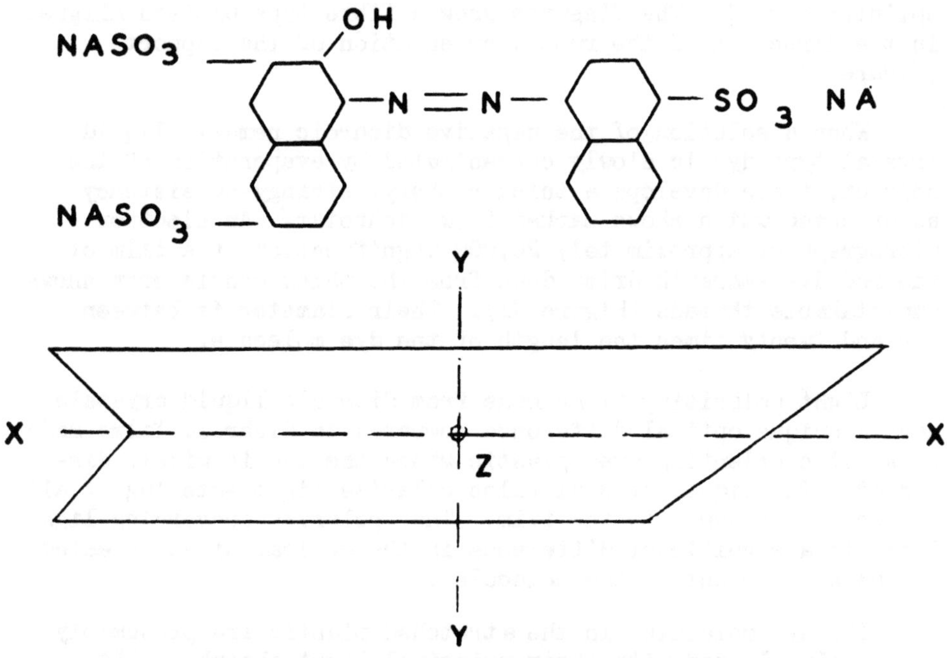

FIGURE III - DYE STRUCTURE AND SYMBOL

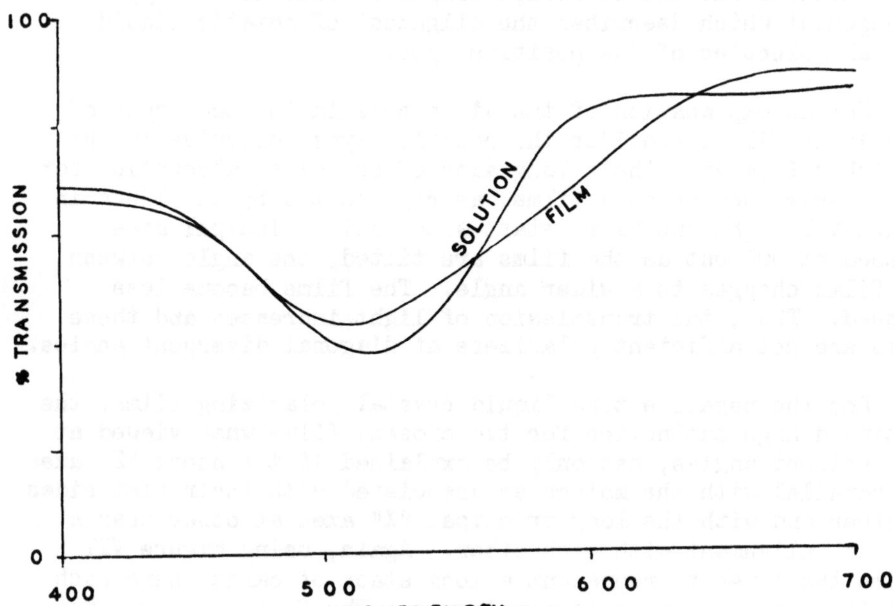

FIGURE IV - WAVELENGTH MM

splintered wood. The fissures show a fibre type pattern aligned in the direction of the rubbed orientation of the support (Figure V).

When a solution of the negative dichroic nematic liquid crystal type dye is slowly concentrated by evaporation of the solvent, there develops a thick rubbery, stringy-consistency super phase which shows marked flow dichroism. An electron micrograph of approximately 20,000 magnification of a film of the red dye Amaranth dried down from the thick consistency shows unmistakable threads (Figure VI). Their diameter is between ten and twenty times the length of the dye molecule.

Light polarizing films made from dichroic liquid crystals have a unique optical difference compared to dichroic films made by stretch orienting dyed plastic where the dye is widely dispersed. The liquid crystal films polarize light entering at all angles of incidence to the film. They polarize converging light. There is a significant difference in the optical effect created by the arrangement of the molecules.

The dye molecules in the stretched plastic are presumably separately aligned with their principal light absorbing "X" axis parallel but not otherwise aligned. This is the typical arrangement which describes the alignment of nematic liquid crystal molecules of the positive type.

For an explanation of the difference in the two types of polarizing films, consider the positive type molecules in the stretched film with their long axes of principal absorption for two superimposed crossed films, as represented by the lines in Figure VII. The two films start with their principal axes crossed at 90° but as the films are tilted, the angle between the films changes to a wider angle. The films become less crossed. Thus, the transmission of light increases and these films are not efficient polarizers at diagonal divergent angles.

For the negative type liquid crystal polarizing films, the continued high extinction for two crossed films when viewed at all incident angles, can only be explained if the short "Z" axes are parallel with the molecules associated with their flat sides together and with the long principal "X" axes at other than a parallel alignment with each other. Again, using Figure VII assume the lines to represent a long stack of cards where each individual card represents a molecule. The flat sides of the molecules are together and the short axes are parallel. The cards are stacked but not aligned as in a deck of cards. The principal colored axes of the molecules are then at various angles to each other (Figure VIII).

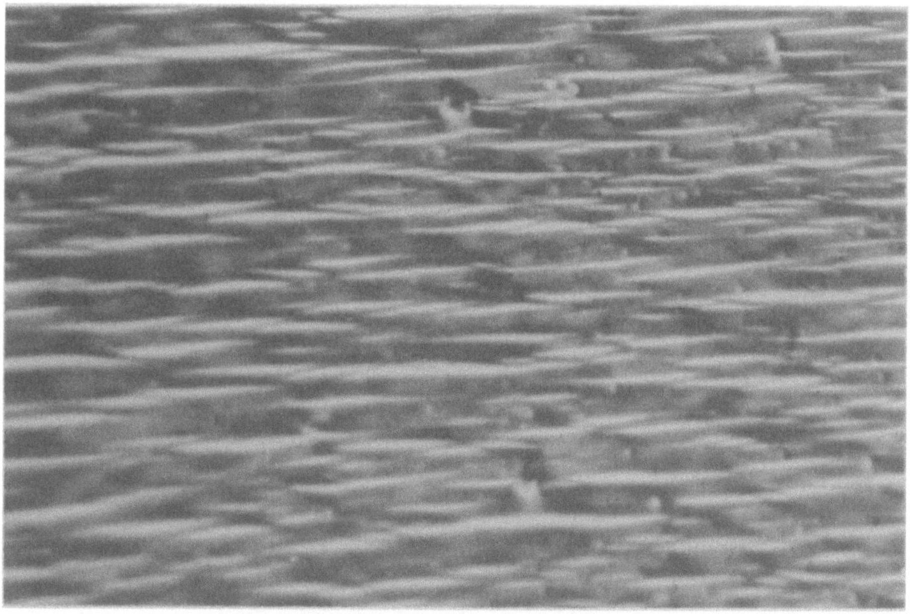

FIGURE V - SPLIT DICHROIC FILM

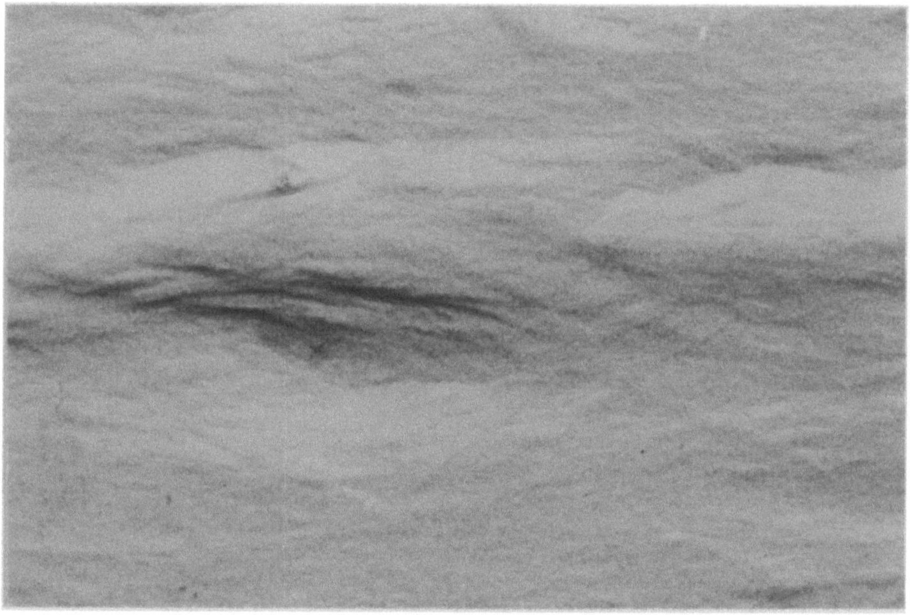

FIGURE VI - ELECTRON MICROGRAPH OF SUPERPHASE
(Courtesy of Procter and Gamble Co.)

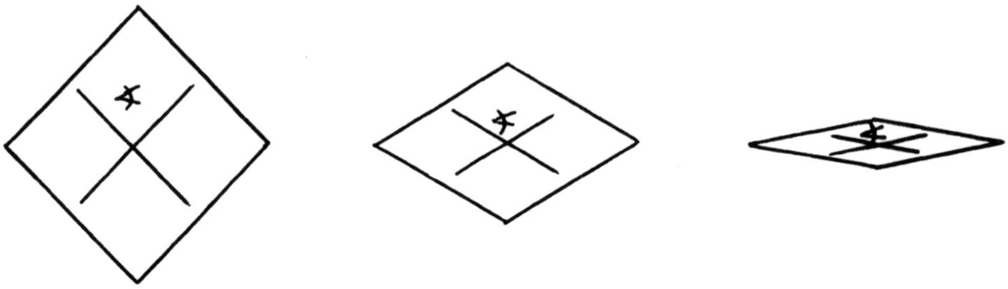

FIGURE VII - CROSSED FILMS TILTED

FIGURE VIII - STACKED MOLECULES

With this arrangement, the long axes of some of the molecules in one film will always be crossed at right angles with the long axes of molecules in the other film and there will be continuous extinction of the light at whatever angle the two crossed films are viewed. This is the only arrangement of the molecular axes that will give this condition. This arrangement conforms to that previously suggested for coplanor dye molecules but which had not previously been stated to be the liquid crystal negative type alignment.

Models of the molecule of the dye Amaranth can be easily stacked in loose association. Water associates with the sulfonic groups. It could lubricate the molecules and keep them further apart than in the crystal form. The stacks apparently associate in parallel array with hydrogen bonding the dominant force.

Cholesteric liquid crystals are said to be a variation on the nematic with the molecules built up in a stack with each adjacent molecule at an angle with the next to form a helix. They are then a special variation of the negative type nematic arrangement.

There appears then to be two distinctly different types of nematic molecules. The positive type which is said to have parallel arrangement of the principal axes only, as is most common with the thermotropic materials, and the negative type with parallel arrangement of the short axes only.

REFERENCES

1. H. Freundlich, R. Stern and H. Zocher, Biochem. Zeit 138, 307-317, 1923
2. Lewis & Calvin, Chem. Reviews 25, 273-328, 1939
3. J. Dreyer, U. S. Patent 2,400,877
4. F. B. Rosevear, J. Soc. Cosmetic Chemists 19, 581-594, 1968
5. T. H. Morton, J. Soc. Dyers Colourists 62, 272-80, 1946
6. H. P. Frank, J. Colloid Sci. 12, 480-495, 1957
7. T. Foerster, Naturwissenschaften 33, 166-195, 1946
8. E. E. Jelley, Nature 139, 631-632, 1937

9. S. E. Sheppard, Revs. Mod. Phys. 14, 303-340, 1942

10. J. Dreyer, Phys. & Colloid Chem. 52, 808-809, 1948

11. H. Zocher & F. C. Jacoby, Kolloid Chemische Beinhefte 24, 365-417, 1927

12. H. Zocher, private communication

POLYMORPHISM OF SMECTIC PHASES WITH SMECTIC A MORPHOLOGY

Sardari L. Arora, Ted R. Taylor, James L. Fergason

Liquid Crystal Institute, Kent State University

Kent, Ohio, 44240

Abstract

A number of 4-n- alkoxybenzylidene-4'-aminopropiophenones with different alkoxy groups have been synthesized. Many of these compounds show two smectic phases, both of which have a morphology identical to that of the classical smectic A of Sackmann and Demus. Both the phases show step drops, focal-conic structure and homeotropic form. No apparent change either in the mechanical or optical properties is observable at the transition temperature of these two phases. The heats of transition for smectic 1 - smectic 2 is much lower than what has been reported for smectic A - smectic B and higher than smectic A - smectic C. We propose a simple model for the phase transition based on the polar ordering of the molecules. Electric field effects are also presented.

We have synthesized a number of 4-n-alkoxybenzylidene-4'-aminopropiophenones which show ambiguities in the classification of phases. The chain length in the alkoxy substituent is C_1, C_3 - C_{10}, C_{12}, C_{14}, C_{16} and C_{18}. The compounds from methoxy to butoxy show a nematic phase. Compounds with chain length C_1, C_3, C_{16} and C_{18} show only one smectic phase, whereas those with chain length

C_4 - C_{10}, C_{12} and C_{14} exhibit two smectic phases. Both the smectic phases can be normally oriented and show the focal-conic texture typical of smectic A according to the classification scheme of Sackmann and Demus.[1]

The phase transition temperatures were determined both by differential thermal analysis (Du Pont DTA 900) and with a Leitz Panphot polarizing microscope using a Mettler FP-2 heating stage. Melting points (solid-liquid or solid-liquid crystal transition) have been regarded as the transitions with the highest transition energy. These are also always the transitions that can most easily be supercooled, whereas supercooling in the case of liquid crystal transitions is negligible. The assignments of the transition temperature were confirmed by the polarizing microscope except for the smectic 2 - smectic 1 transitions. The highest temperature smectic phase is always called smectic 1, the next lower one smectic 2, and so on. The transition temperatures for the various liquid crystal phases are listed in Table I. The error of the temperature measurements is estimated to be smaller than $\pm 2^{\circ}C$.

Smectic 1 is observed in all the compounds prepared in this series and occurs monotropically only in the case of 4-methoxybenzylidene-4'-aminopropiophenone. Smectic 2 is observed in compounds with chain lengths C_4 to C_{14} and exists as a monotropic phase in C_7, C_{11} and C_{14}. That smectic 2 disappears in C_{16} and C_{18} may be attributed to their high melting point; crystallization should occur at much higher temperatures than would be expected for the transition temperatures as extrapolated from a plot of the transition temperatures of the lower homologues. The nematic phase is observed only in compounds with chain lengths up to C_4. The DTA curve in Figure 1 shows the three liquid crystal phases observed for 4-butoxybenzylidene-4'-aminopropiophenone. In Figure 2, transition temperatures have been plotted and curves drawn on the basis of the differential thermal analysis data.

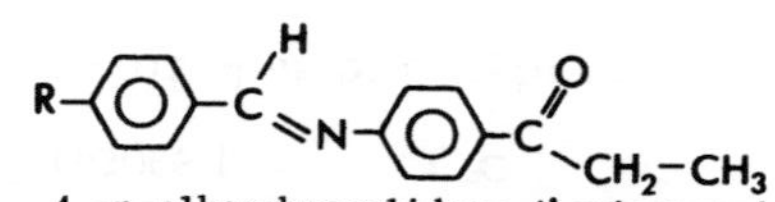

Table I. 4 -n-alkoxybenzylidene-4'-aminopropiophenones

Substituents	Transition temperatures, °C, from solid or preceding liquid crystal states to			
	Smectic 2	Smectic 1	Nematic	Isotropic
R = H_3CO	--	--	116 97[a]	133.5
H_7C_3O	--	103	134	139.5
H_9C_4O	79	87	143	146
$H_{11}C_5O$	82	83	--	145.5
$H_{13}C_6O$	80.5	82	--	148.5
$H_{15}C_7O$	--	98.5 83[b]	--	149.5
$H_{16}C_8O$	71	84	--	150.5
$H_{17}C_9O$	70	83	--	149.5
$H_{21}C_{10}O$	72	82	--	149
$H_{25}C_{12}O$	--	80 79.5[b]	--	146.5
$H_{28}C_{14}O$	--	84 77[b]	--	142.5
$H_{33}C_{16}O$	--	90	--	139
$H_{37}C_{18}O$	--	94	--	135

[a] Transition from monotropic smectic 1.
[b] Transition from monotropic smectic 2.

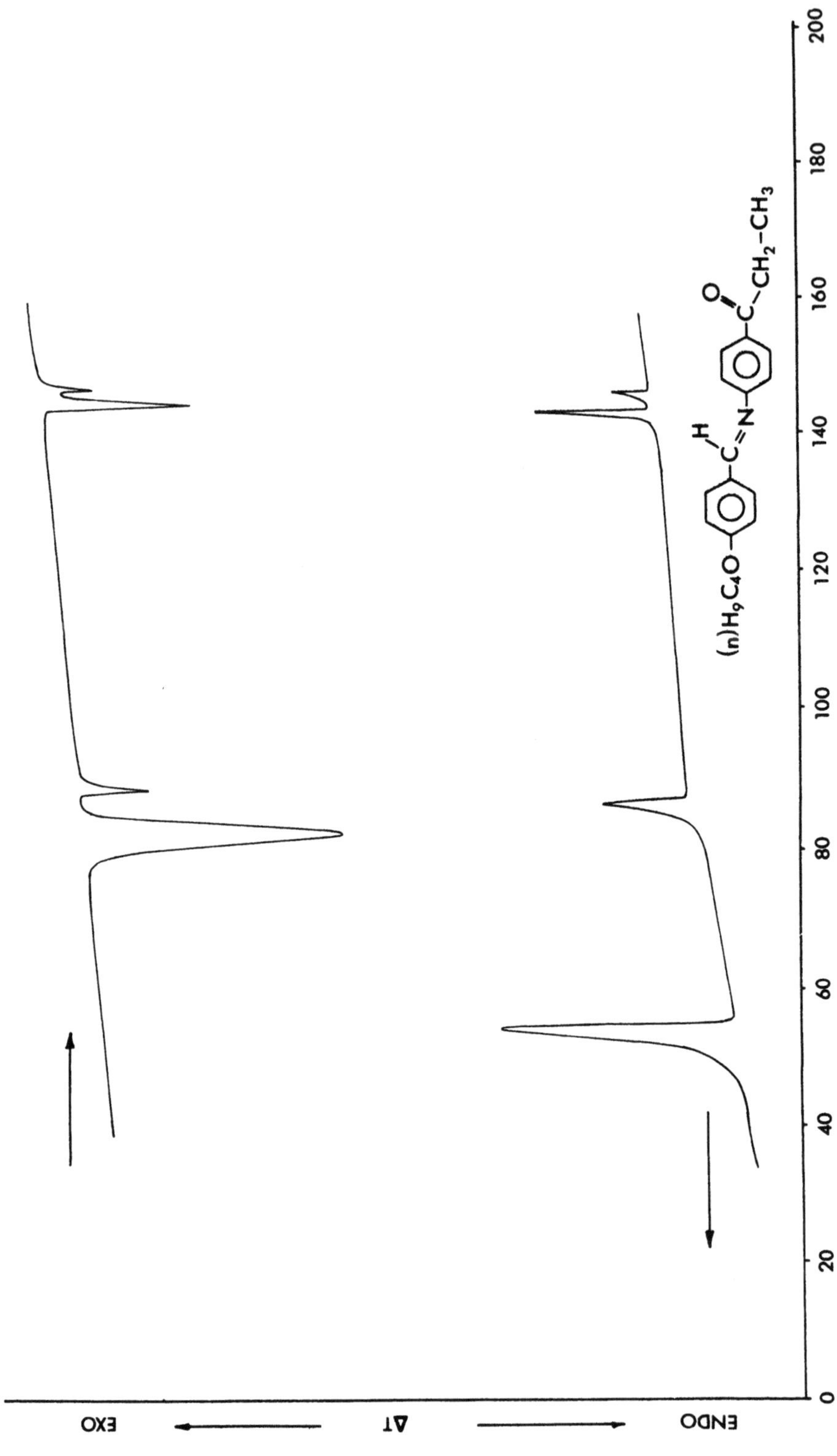

FIGURE 1.

Differential thermograms of 4-n-butoxybenzylidene-4'-aminopropiophenone

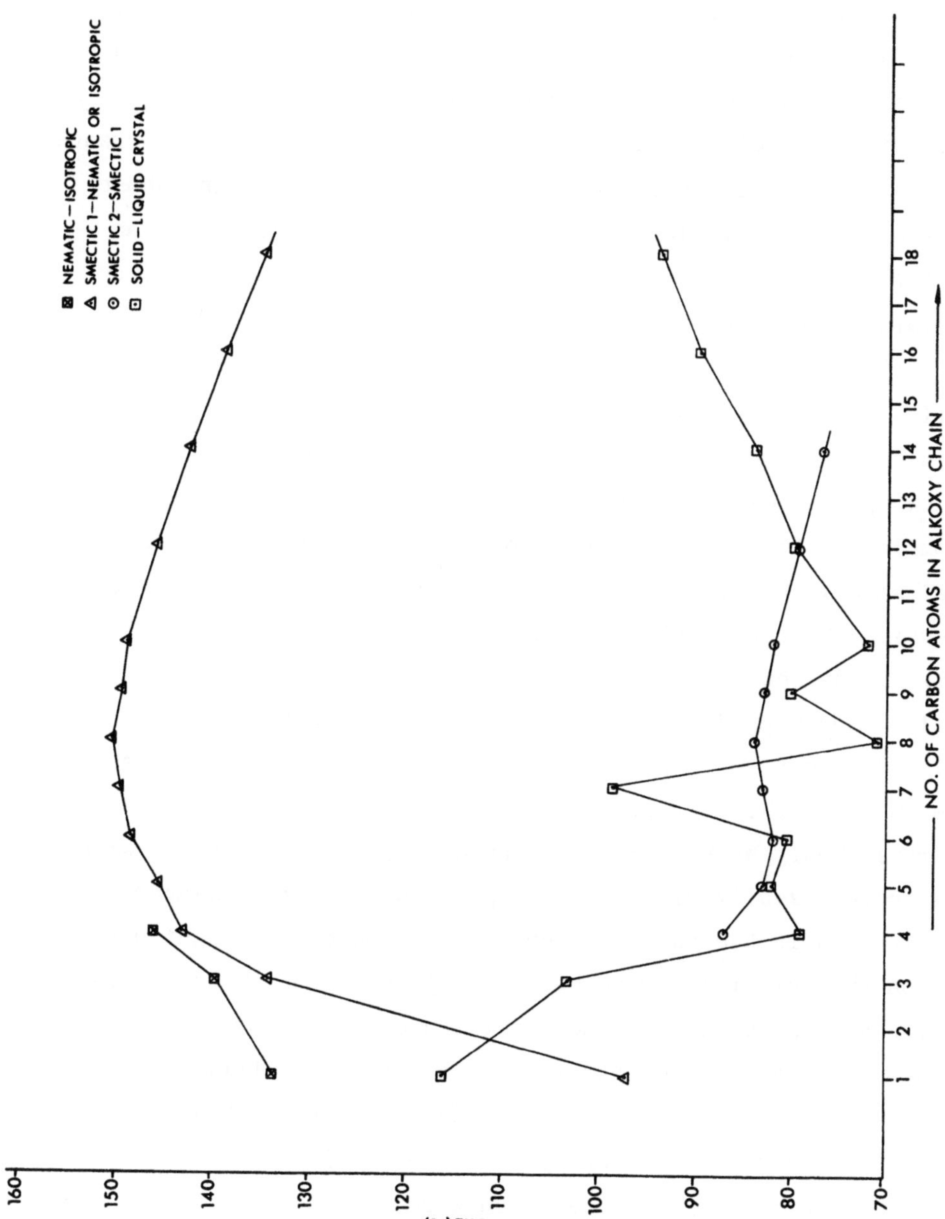

FIGURE 2.

Various transition temperatures of 4-alkoxybenzylidene-4'-aminopropiophenones plotted against the number of carbon atoms in the alkyl chain of the ether group.

The only systematic study of the smectic phase to date is due to Sackmann and Demus[1] who have classified a number of smectic phases on the basis of miscibility. If two liquid crystalline phases are identical at a given temperature, then they should mix and give an optically homogeneous preparation. The simplest case which Sackmann and Demus observed corresponds to the original smectic phase as delineated by Freidel,[2] that is, molecules arranged in layers whose position is undefined within the layer and whose long axis is perpendicular to the layer. This gives rise to an optically positive material which can be arranged in a step drop, has relatively high mobility in the plane texture and has the characteristic optical morphology known as focal-conic. This phase is called smectic A by Sackmann and Demus.

In the 4-n-alkoxybenzylidene-4'-aminopropiophenones, we have found two phases which fit into the morphology of the smectic A. The homologues from methoxy to butoxy show a nematic phase and a typical smectic A. In the butoxy, however, we encounter another phase which shows no optical discontinuity at the phase transition temperature as shown by the DTA. Both phases can be oriented normally, show step drops (Figure 3) without a microscope cover glass, and show a focal-conic optical texture. We have no apparent change in the mechanical properties nor the optical properties at this transition. We have not been able to observe the phase with a microscope visually or by using a photomultiplier to measure the depolarization in the normal orientation. Convergent light observation on the uniformly oriented normal preparation shows a uniaxial cross in both phases, and insertion of a quarter-wave or full wave plate gives the interference figure characteristic of a positive uniaxial material. The interference figure showed no change at the phase transition temperature. Thick samples of uniformly oriented normal preparation were observed at approximately a 20° angle to the surface. Observed at an angle, the sample is uniformly birefringent over the sample area. Even in

Figure 3. Step drops as observed in 4-n-octyloxybenzylidene-4'-aminopropiophenone (100x), without an analyzer

these thick samples no change in birefringence could be observed at the phase transition. The step drops show a remarkable mobility in both phases; but in our observations, we can see no discontinuity in the mobility at the transition temperature.

Recapitulating, the fact which we wish to emphasize is not that we see a lack of a significant difference between the phases; it is the fact that we see absolutely no discontinuous difference. This is the only case where we are aware of two smectic phases which are completely indistinguishable in both optical and mechanical properties within the limits of our measuring techniques.

The heats of transition for the smectic 1 - smectic 2 transition have been determined for the butoxy and octyloxy homologues. For octyloxy, the heat of transition is .04 kcal./M and .14 kcal./M for butoxy. These values are small in comparison to those listed by Saupe in his Review[3] for the typical smectic A - smectic C transition. The molecular structure of the 4-n-alkoxybenzylidene-4'-aminopropiophenones is unsymmetrical, with a relatively small dipole moment lying along the long axis. We, therefore, suggest that we are seeing a very simple phase transition involving the polar ordering of the molecules. That is, the transition corresponds to a change from a random arrangement of the alkoxy groups to a configuration where all the alkoxy groups are pointing in the same direction. This would be a phase transition from space group $D_{\infty h}$ to $C_{\infty v}$. The change in polar ordering at the phase transition is shown schematically in Figure 4. From elementary considerations, the change in entropy of such a transition would be R ln 2 or 1.38 cal./^{o}K. If in our case the supposition of a polar ordering is correct, we should have a phase transition energy less than 1.38 T. As we can easily see, this condition is met.

It is interesting to note that from our observations, the methods of Sackmann and Demus[1] Would fail to determine the phase transition we have observed. Since we would expect the phase transition to be much broader and therefore much more difficult to

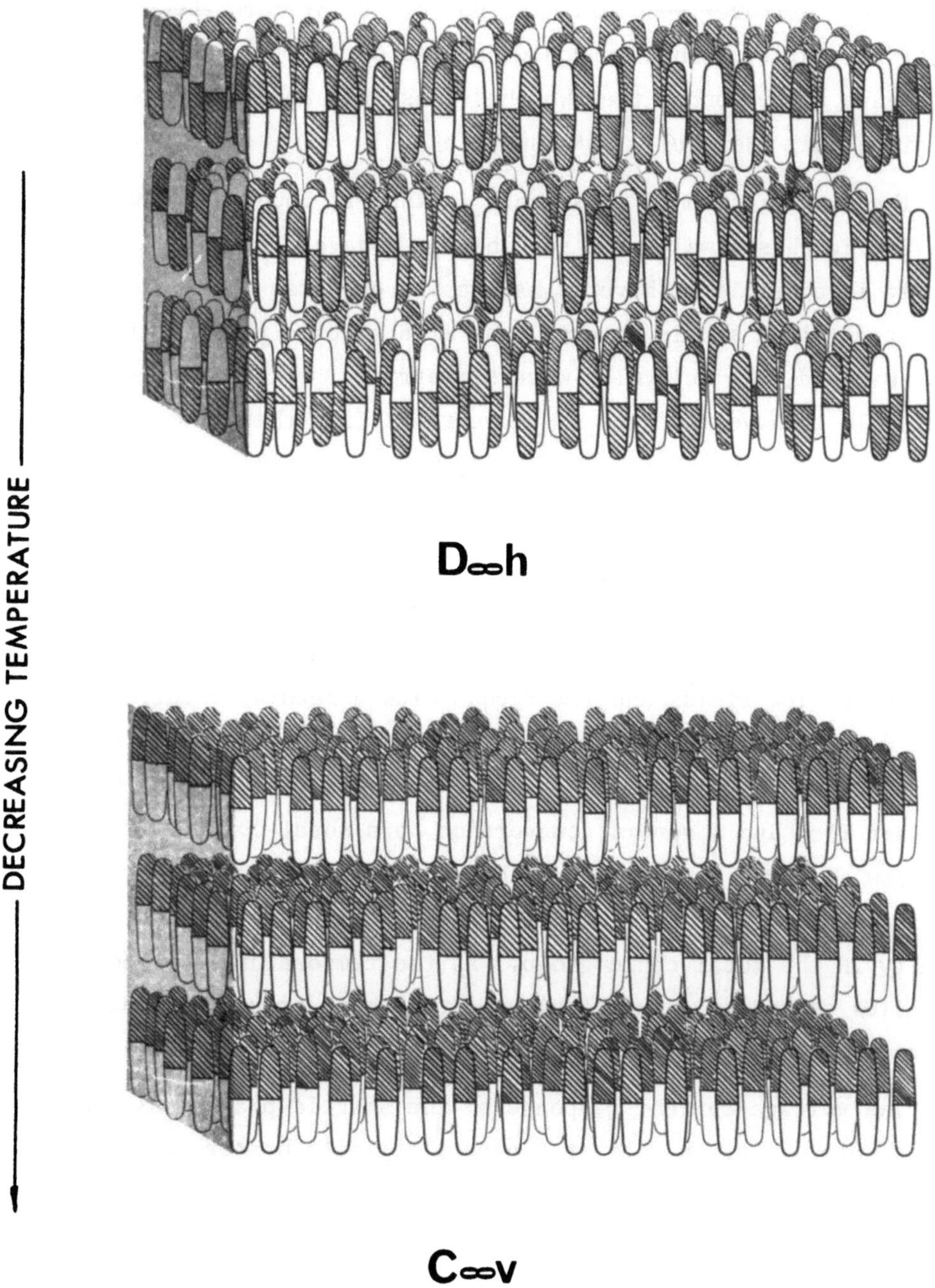

Figure 4. Schematic representation of the suggested change in the molecular structure at the smectic 1-smectic 2 transition temperature.

detect in a mixture, we would have no tool to discern the phase transition point. Since we could not use optical techniques, we would have to pronounce the two phases as indiscernible by Sackmann's technique and, therefore, both smectic A.

The significance of this phase is that it doubles the number of existing phases which are possible. We have previously found[4] that it is possible to have as many as seven first-order phase transitions in a single material, or five smectic phases. In the case we noted, the molecules were symmetrical. Thus, it was impossible to have the type of phase transition that we have proposed. We, therefore, feel that it would be possible to double the number of observed phase transitions.

In order to see if the polar ordering manifested itself in terms of a change in dielectric constant or response to electric field, we applied an AC triangular wave and measured the current. We found at low fields and low frequencies a saturable current which decreased as a function of temperature, but showed no discontinuity at the phase transition. At higher field strengths, we observed flow and a general conversion to a focal-conic texture. The flow occurred in a manner which indicated that it was occurring within layers; that is, at right angles to the hyperbolic discontinuity in the focal-conic group. As the temperature was decreased, this flow showed no discontinuity at the phase transition, again indicating no change in mechanical properties.

The voltage-current response of the smectic phases was in itself interesting in that it showed a definite hysteresis effect which was considerably different from anything previously observed for the nematic or isotropic liquid. (See Figure 5).

Preparation of materials

4-aminopropiophenone was recrystallized from commercially available material.

4-n-alkoxybenzaldehydes were prepared from p-hydroxybenzaldehyde

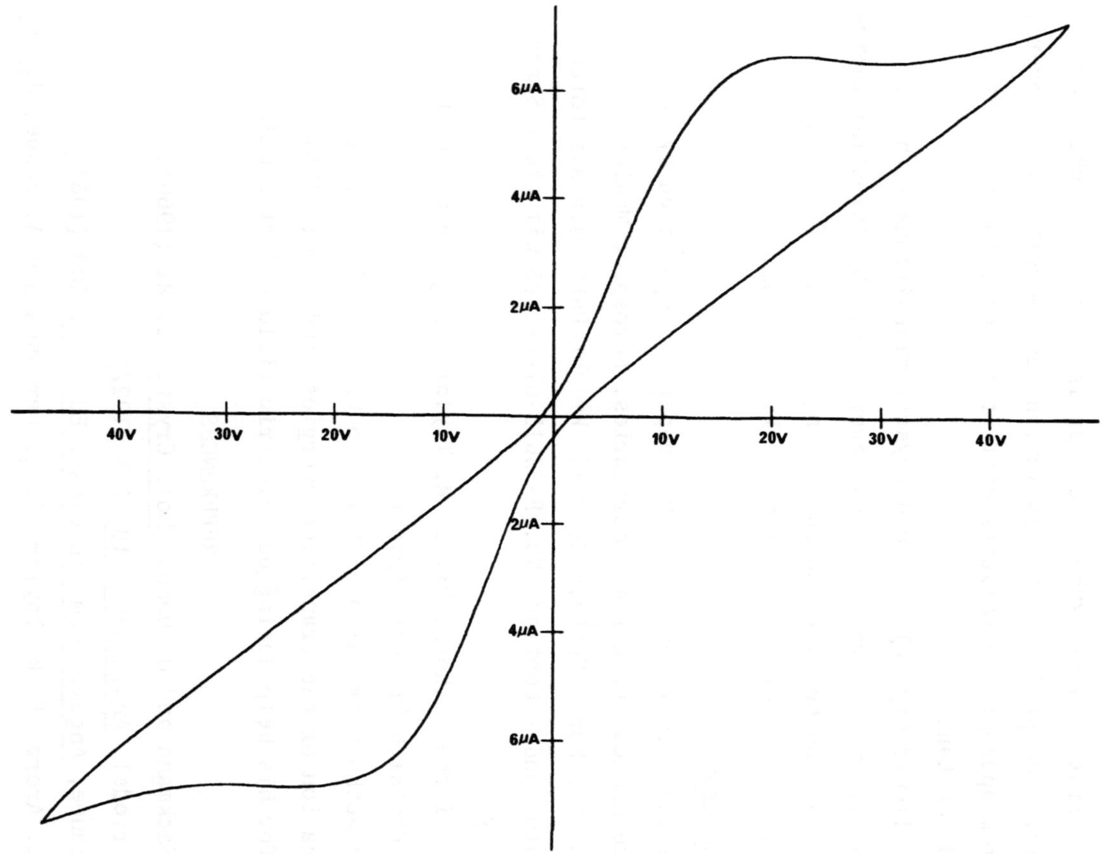

Figure 5. Current voltage relation for 4-n-octyloxybenzylidene-4'-aminopropiophenone at 95°C and a sweep rate of 2 volts/sec.

and various alkyl bromides, either according to the method of our earlier publication[5] or by that of Weygand and Gabler.[6]

Alkoxybenzylidene-aminopropiophenones were prepared by refluxing equimolecular quantities of the 4-aminopropiophenone and the appropriate 4-*n*-alkoxybenzaldehyde in absolute alcohol for 5-6 hours. The product after isolation was recrystallized several times from appropriate solvents until the transition temperatures remained constant.

The liquid crystal - liquid crystal transitions with the purified compounds were sharp and reversible. Differential thermal analysis gave on heating and on cooling, within a fraction of a degree, equal temperatures for these transitions.

Acknowledgment

The research reported in this paper was sponsored by the Aerospace Medical Research Laboratories, Aerospace Medical Division, Air Force Systems Command, Wright-Patterson Air Force Base, Ohio, under Contract No. F-33615-67-C-1496 with Kent State University.

One of the authors, Dr. T. R. Taylor, is grateful to the Advance Research Projects Agency.

The authors are also grateful to Dr. Paul Garn for the determination of the transition energies, and to Mr. Thomas Harsch for his help during our electric field effect studies.

REFERENCES

1. H. Sackmann and D. Demus, Mol. Cryst., 2, 81 (1966).
2. G. Friedel, Ann. Phys., 18, 273 (1922).
3. A. Saupe, Angew. Chem. internat. Edit. 7, 100 (1968).
4. S. L. Arora, T. R. Taylor, J. L. Fergason and A. Saupe, J. Am. Chem. Soc., 91, 3671 (1969).
5. S. L. Arora, J. L. Fergason and A. Saupe, Mol. Cryst. and Liq. Cryst. (In press).
6. C. Weygand and R. Gabler, J. Prakt. Chem., 155, 338 (1940).

Effect of Solvent Type on the Thermodynamic Properties of Normal Aliphatic Cholesteryl Esters*

Marcel J. Vogel,** Edward M. Barrall II,*** and Charles P. Mignosa**

The effects of purity on the phase transitions in cholesteryl esters have been observed by several workers (1, 2, 3). In addition, recent calorimetric studies have indicated that cholesteryl esters have more than one stable room temperature solid form (3). Barrall, Porter and Johnson found that for certain esters the transition temperature of the solid→mesophase conversion was very sensitive to the way in which the solid phase had been formed, from ethanolic solution or from the melt (3). Indeed, the propionate ester exhibited two overlapping endotherms for the solid→mesophase transition when the solid had been formed from the melt. The same solid formed from ethanolic solution exhibited only one endotherm. These and other reported effects indicate that the esters of cholesterol have an extensive solid-phase polymorphism.

Gas chromatographic studies using cholesteryl esters as the liquid phase have shown that certain materials, aromatic and halogenated hydrocarbons, have unusually high heats of solution in the esters (4,5). Therefore, it is possible that a significant portion of

---

*Part XX of a Series on Order and Flow of Liquid Crystals

**International Business Machines Corporation, Advanced Systems Development Division, Los Gatos, California
***International Business Machines Corporation, Research Division, San Jose, California

the recrystallization solvent may remain sorbed or bound in the solid phase. This solvent would act as an impurity in the solid crystal and significantly alter the thermal properties.

The purpose of this study is to determine the effect of recrystallization solvent type and impurity on the thermal properties of three cholesteryl esters, viz., propionate, palmitate and myristate.

EXPERIMENTAL

Starting Materials

The esters used in this study were synthesized from carefully purified starting materials. The cholesterol was obtained from van Schuppen Chemicals, Veenendaal, Holland. According to the analyses furnished by the manufacturer the isomeric purity was better than 99.5%. The acids were obtained from Eastman Distillation Products Industry, Rochester, New York. Prior to use the propionic acid was distilled, and the palmitic and myristic acids were recrystallized from ligroin.

Synthesis

The production of very pure cholesteryl esters requires more than ordinary attention to reaction conditions and times. The final product in each synthesis below is the "control" product discussed later.

Cholesteryl Propionate

To a 150 ml round bottom flask fitted with a Dean Stark tube and a condenser are added: 25.0 g (0.0646 mole) ligroin, recrystallized cholesterol, 8.0 g (0.108 mole) distilled propionic acid, 1 g p-toluenesulfonic acid and 150 ml ligroin (density 0.69-0.71). This mixutre was refluxed 60 minutes and the water monitored in the Dean Stark tube. The water yield is curve 1 of Fig. 1. The reaction mixture was washed with successive 10 ml portions of 20% ethanol-water until a neutral blue litmus reaction was obtained. The ligroin solution was dried over anhydrous sodium sulfate.

at room temperature overnight. The suspension was chilled in an ice bath and filtered under suction. The crystals of ester were washed with five 10 ml portions of 0°C ligroin. The suction-dried crystals were held in a vacuum desiccator at 10^{-3} Torr.

Cholesteryl Myristate

The following materials were placed in the apparatus described previously; 25 g (0.0646 mole) of ligroin recrystallized cholesterol, 16.5 g (0.0722 mole) ligroin recrystallized myristic acid, 1 g p-toluenesulfonic acid, and 150 ml ligroin. The mixture was refluxed 90 min. (see Fig. 1, Curve 3). The ester was treated in the same way as the palmitate.

Recrystallization

The samples used in the simple recrystallization studies were treated as follows. 1) 200 mg of the ester was weighed into a test tube. 2) With gentle heating the ester was dissolved in 50 cc of the desired solvent. 3) The solutions were allowed to evaporate to dryness. 4) Relatively perfect crystals were selected from the walls of the test tube under a stereo microscope. 5) These crystals were dried at 10^{-8} Torr for 24 hours. After the first few hours no loss in vacuum was noted. The control samples were washed with cold ligroin (no dissolution) and treated in vacuum as above.

Calibration of the Differential Scanning Calorimeter (DSC)

The temperature axis of the calorimeter was calibrated using the melting points of diphenyl amine, naphthalene, benzoic acid (NBS), indium (99.999%), tin and lead. Heating rates were 2.5 and 5°C/min. Purity analyses were carried out on DSC curves recorded at a heating rate of 2.5°C/minute and a sensitivity of 2 millical/sec. Mesophase data was taken at 5°C/minute and 1 millical/sec. Area was converted to calories using the heat of fusion of indium. All temperatures were extrapolated to the isothermal base line using the slope of the appropriate indium fusion endotherm (6).

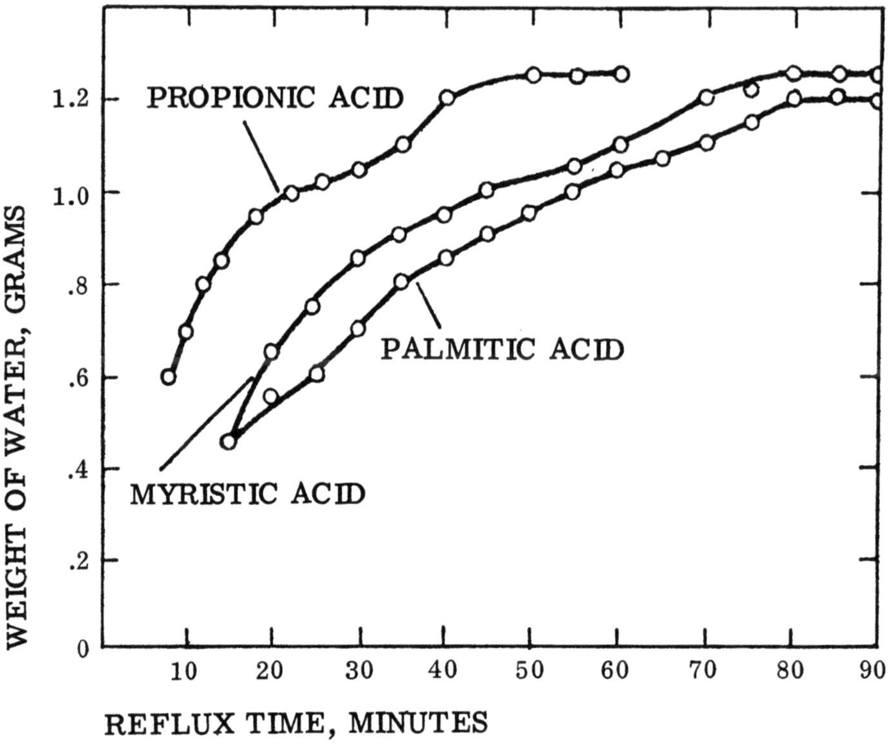
PROPIONIC ACID
PALMITIC ACID
MYRISTIC ACID
WEIGHT OF WATER, GRAMS
1.2
1.0
.8
.6
.4
.2
0
10
20
30
40
50
60
70
80
90
REFLUX TIME, MINUTES

Figure 1.

The heat of fusion for each ester was calculated from planimeter areas using a computer program developed at IBM. The purity calculations from the van't Hoff equation were carried out as described by Gray (7), Ennulat (2), and Plato and Glasgow (8). A computer program was developed for the linearization and calculation. Only the solid→mesophase transition (large thermal event) on the fourth heating was used.

RESULTS

Recrystallization Study

The results of the recrystallization study are shown in Table I. It is obvious that the choice of solvent has a great effect on the calculated purity of the individual esters. Since all of the solvents were applied to the control sample, the calculated purity of the individual samples should not be less than that of the control. This is irrespective of any judgment as to the applicability of the van't Hoff calculation or the absolute accuracy of the method. Three or more thermograms on samples demonstrated that the precision of the method was $\pm 4\%$ of the impurity. Table I indicates that solvent is irreversibly absorbed in the solid phase, and cannot be removed from the solid at 10^{-8} Torr. Benzene could not be removed from the liquid phase of the propionate at 10^{-4} Torr. The liquid phase propionate ester vacuum distills at 10^{-8} Torr to yield a relatively pure material. Of the solvents presented in Table I, acetone is the least retained and benzene and carbon tetrachloride the most retained. On standing for three weeks, the carbon tetrachloride and chloroform recrystallized samples deteriorated further. This may have been due to traces of hydrochloric acid formed by the decomposition of the solvents.

Table I. Effect of Solvent Recrystallization Type on Calculated Purity

Compound	Indicated Purity from Solid→Mesophase Transition					
	Control	Acetone	Hexane	Carbon Tetrachloride	Chloroform	Benzene
Cholesteryl Propionate	99.286	99.587	99.293	96.748	96.013	95.658
Cholesteryl Palmitate	97.659	98.266	97.234	97.184	96.369	94.031
Cholesteryl Myristate	96.783	97.804	97.610	95.212	96.400	97.242

Samples of the esters which had been recrystallized from ethanol two years ago, sealed under nitrogen, and stored in relative darkness, gave thermograms which were so broad and unusual that van't Hoff calculations on the melting endotherm were not possible. The myristate ester from these samples showed a very distinct endothermal peak near 154°C which could be due to free cholesterol. When these samples were originally recrystallized sharp thermograms characteristic of material of purity greater than 98% had been obtained (3).

In retrospect, strong absorption of aromatic and halogenated hydrocarbons could have been predicted from gas chromatographic studies (4, 5). The log elution time is a nonlinear function of the boiling point for aromatic materials (4) as it is for halogenated material (5). This is true of other non-cholesteryl liquid crystals (9, 10, 11).

Cholesteryl Propionate

In addition to gross effects of solvent on calculated purity, cholesteryl propionate exhibits a solvent dependent solid phase polymorphy and mesophase monotropism. Critical data are shown in Table II. Two forms of the solid can be seen in the polarizing microscope (Fig. 2A and B). The low-temperature solid (I), 95.2°c, is spherulitic in growth. The high-temperature form (II), 98.0°C, is dendritic. The low-temperature solid forms only in relatively solvent-free, pure materials. The high-temperature form is invariably present in materials of lower purity. The dendritic form also appears at lower temperatures when the impurity level is high enough to depress the melting point. From calorimetry and microscopy the following recrystallization path is suggested:

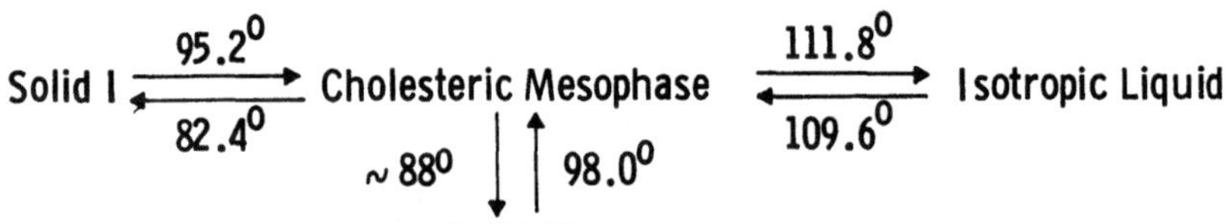

No Solid I⟶Solid II interaction could be found by microscopy on long standing at 95°C. Transition from

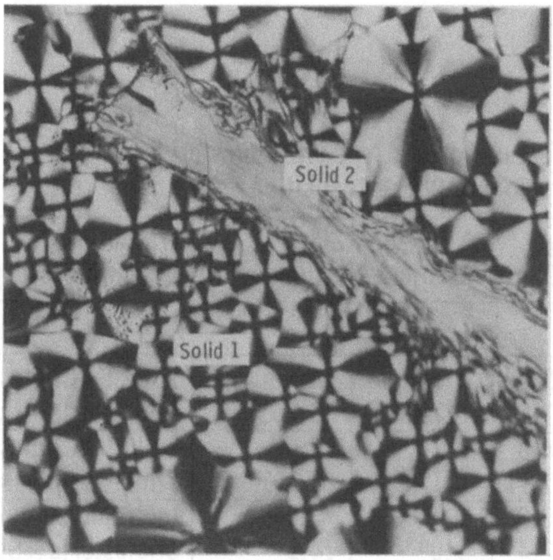

Fig. 2A. Cholesteryl Propionate, Solid I with an Inclusion of Solid II, Polarized Light 104X

Fig. 2B. Cholesteryl Propionate, Solid II, Polarized Light 104X

Table II. Effect of Solvent Type on Cholesteryl Propionate Phase Transitions

Mode	T_m (°C)	ΔH (kcal/mole)	ΔS (cal/mole/°K)	Transition (From Microscopy)
a. CONTROL				
1st Heat	95.2	5.34	14.5	Solid I→Cholesteric
	111.8	0.100	0.260	Cholesteric→Isotropic Liquid
2nd Heat	94.8	5.71	15.5	Solid I→Cholesteric
	111.71	0.150	0.391	Cholesteric→Isotropic Liquid
Cooling	110.0	-0.134	-0.349	Isotropic Liquid→Cholesteric
	82.4	-5.09	-14.3	Cholesteric→Solid I
b. ACETONE				
1st Heat	95.6	5.36	14.5	Solid I→Cholesteric
	113.12	0.180	0.466	Cholesteric→Isotropic Liquid
2nd Heat	95.5	5.61	15.2	Solid I→Cholesteric
	113.07	0.108	0.279	Cholesteric→Isotropic Liquid
Cooling	111.3	-0.144	-0.374	Isotropic Liquid→Cholesteric
	78.8	-4.91	-13.95	Cholesteric→Solid I
c. HEXANE				
1st Heat	94.5	5.37	14.6	Solid I→Cholesteric
	110.6	0.129	0.334	Cholesteric→Isotropic Liquid
2nd Heat	94.2	1.26	3.44	Solid I→Cholesteric
	97.4	3.71	10.0	Solid II→Cholesteric
	110.5	0.129	0.338	Cholesteric→Isotropic Liquid
	Both phases supercool			

Table II. Effect of Solvent Type on Cholesteryl Propionate Phase Transitions (continued)

Mode	T_m (°C)	ΔH (kcal/mole)	ΔS (cal/mole/°K)	Transition (From Microscopy)
d. CARBON TETRACHLORIDE				
1st Heat	88.8	3.21	8.89	Solid→Isotropic Liquid
	No Mesophase			
2nd Heat	85.9	3.36	9.36	Solid→Cholesteric
	92.6	0.191	0.522	Cholesteric→Isotropic Liquid
Cool	Freezing is depressed below instrument range.			

Mode	T_m (°C)	ΔH (kcal/mole)	ΔS (cal/mole/°K)	Transition (From Microscopy)
e. CHLOROFORM				
1st Heat	90.5	4.45	12.2	Solid→Cholesteric
	105.4	0.135	0.356	Cholesteric→Isotropic Liquid
2nd Heat	91.6	4.64	12.7	Solid→Cholesteric
	103.7	0.0843	0.224	Cholesteric→Isotropic Liquid
	Freezing is depressed below instrument range.			

Mode	T_m (°C)	ΔH (kcal/mole)	ΔS (cal/mole/°K)	Transition (From Microscopy)
f. BENZENE				
1st Heat	100.7	4.38	11.7	Solid→Isotropic Liquid
	No Mesophase			
2nd Heat	91.1	2.37	6.51	Solid I→Isotropic Liquid
	95.1	1.12	3.05	Solid II→Isotropic Liquid
	No Mesophase			
Cooling	76.9	4.43	12.7	Isotropic Liquid→Solid

Table III. Effect of Solvent Type on Cholesteryl Palmitate Transitions

Mode	T_m (°C)	ΔH (kcal/mole)	ΔS (cal/mole/°K)	Transition (From Microscopy)
a. CONTROL				
1st Heat	75.2	12.9	37.1	Solid I→Cholesteric
	77.6	0.332	0.948	Cholesteric→Isotropic Liquid
2nd Heat	75.1	12.8	36.7	Solid I→Cholesteric
	77.6	0.360	1.03	Cholesteric→Isotropic Liquid
Cooling	76.1	-0.305	-0.873	Isotropic Liquid→Cholesteric
	71.7	-0.388	-1.12	Cholesteric→Smectic
	48.1	-10.5	-32.6	Smectic→Solid I

Mode	T_m (°C)	ΔH (kcal/mole)	ΔS (cal/mole/°K)	Transition (From Microscopy)
b. ACETONE				
1st Heat	76.3	13.5	38.7	Solid I→Cholesteric
	79.8	0.374	1.06	Cholesteric→Isotropic Liquid
2nd Heat	76.2	13.0	37.2	Solid I→Cholesteric
	79.8	0.428	1.21	Cholesteric→Isotropic Liquid
Cooling	77.9	-0.374	-1.07	Isotropic Liquid→Cholesteric
	73.4	-0.401	-1.16	Cholesteric→Smectic
	56.8	-11.3	-34.3	Smectic→Solid I

Mode	T_m (°C)	ΔH (kcal/mole)	ΔS (cal/mole/°K)	Transition (From Microscopy)
c. HEXANE				
1st Heat	75.9	12.7	36.5	Solid I→Cholesteric
	78.7	0.393	1.12	Cholesteric→Isotropic Liquid
2nd Heat	75.7	12.4	35.6	Solid I→Cholesteric
	78.9	0.416	1.18	Cholesteric→Isotropic Liquid
Cooling	76.9	-0.208	-0.594*	Isotropic Liquid→Cholesteric
	72.5	-0.393	-1.14	Cholesteric→Smectic
	44.5	-11.7	-37.0	Smectic→Solid

*Poor definition

Table III. Effect of Solvent Type on Cholesteryl Palmitate Transitions (continued)

d. CARBON TETRACHLORIDE

Mode	T_m (°C)	ΔH (kcal/mole)	ΔS (cal/mole/°K)	Transition (From Microscopy)
1st Heat	76.0 No Mesophase	12.9	36.9	Solid II→Isotropic Liquid
2nd Heat	75.3 No Mesophase	12.8	36.6	Solid II→Isotropic Liquid
Cooling	No Mesophase Supercooling below instrument range			Isotropic Liquid→Solid II

e. CHLOROFORM

Mode	T_m (°C)	ΔH (kcal/mole)	ΔS (cal/mole/°K)	Transition (From Microscopy)
1st Heat	76.3 No Mesophase	13.5	38.5	Solid II→Isotropic Liquid
2nd Heat	75.6 77.9	12.9 0.363	37.1 1.04	Solid I→Cholesteric Cholesteric→Isotropic Liquid
Cooling	77.4 Cholesteric supercools	0.3113	0.888	Isotropic Liquid→Cholesteric

f. BENZENE

Mode	T_m (°C)	ΔH (kcal/mole)	ΔS (cal/mole/°K)	Transition (From Microscopy)
1st Heat	76.0 No Mesophase	12.5	35.7	Solid→Isotropic Liquid
2nd Heat	75.3 No Mesophase	12.6	36.2	Solid→Isotropic Liquid
Cooling	Broad peaks not suitable for calorimetry			

one solid form to another occurs only through the cholesteric mesophase. The control sample on fourth heating gave the following results:

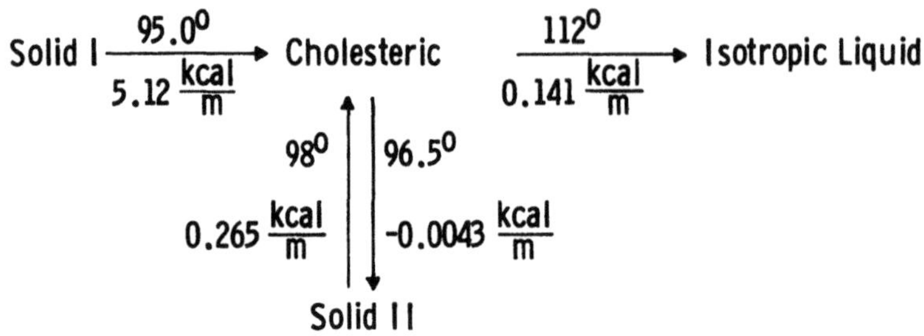

Thus, a small amount of recrystallization, -0.0043 kcal/mole, is indicated via the mesophase. No attempt was made to reach equilibrium between the two solids and the mesophase.

Solvents, which purity analysis indicated had remained in the solid, produced low melting Solid II and in some cases no mesophase formed on either heating or cooling. The benzene and aged ethanol recrystallized samples are excellent examples. Optical microscopy on the benzene sample failed to find a trace of the typical cholesteric texture on heating or cooling.

Cholesteryl Palmitate

Previous studies of this ester have shown the mesophases, smectic and cholesteric, to be monotropic (3) or partially monotropic (1). However, as Table III indicates, with sufficiently pure materials a lower melting solid forms, and the cholesteric isotropic liquid transition is no longer monotronic. Using acetone recrystallized material, the solid melts to form the cholesteric mesophase at 76.3°C. Earlier studies had indicated a higher melting point of 79.6°C, which is well above the mesophase range (1, 3). The lower melting form of the pure ester is distinctly spherulitic in the polarizing microscope. The higher melting form, obtained in an earlier study (3), is dendritic with only a few darker spherulites. The heat of fusion of the two forms is almost identical, 13.5 kcal/mole low melting to 14.2 kcal/mole high melting (3). The cholesteric isotropic liquid transition occurs at higher temperatures when there is a low solvent residue. The transition heats of the mesophases compare closely with previous literature values (see Table IV). On the basis of microscopy and calorimetry the following crystallization path is suggested:

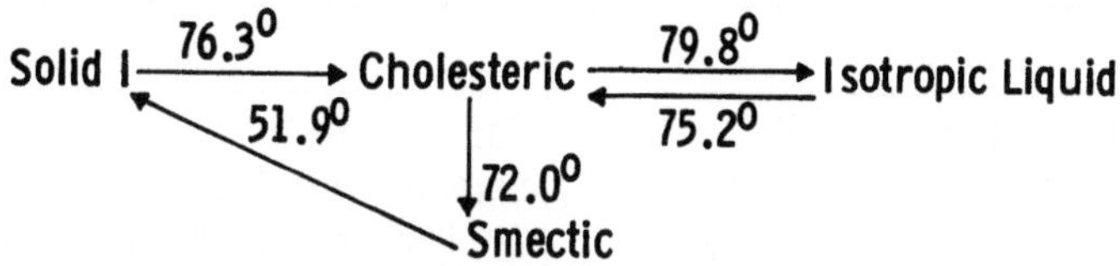

Cholesteryl Myristate

Due to the slow rate of reaction and insolubility of myristic acid, this ester was the least pure material studied. Vacuum sublimation produced material of high quality, but this is not the subject of the present study. The critical data on solvent recrystallized material is shown in Table V. None of these results should be considered as absolute since the highest purity reached was 97.8 mole%. It is interesting to note, considering the extreme sensitivity of other ester samples to impurity and solvent content, the insensitivity of the myristate ester. Indeed, this may account for the several good checks between laboratories using this ester (3, 12).

Table IV. Comparison of Acetone Recrystallized Cholesteryl Palmitate with Literature Values

	This Study	Literature	
	(Acetone)	Gray (1)	Barrall, Porter & Johnson (3)
T_1	[73.4]	[78.5]	[64.0]
T_2	76.3	79	79.6
T_3	79.8	83	[70.0]
ΔH_1	0.401	-	0.36
ΔH_2	13.5	-	14.2
ΔH_3	0.374	-	0.28

Key: T_1, T_2, T_3 Transition temperatures for the smectic⇄cholesteric, solid→cholesteric, cholesteric→isotropic liquid in ^{0}C. ΔH_1, ΔH_2, ΔH_3 Transition heats for the smectic⇄cholesteric, solid→cholesteric, and cholesteric→isotropic liquid in kcal/mole. [] indicate monotropism.

Table V. Effect of Solvent Type on Cholesteryl Myristate Transitions

Mode	T_m (^oC)	ΔH (kcal/mole)	ΔS (cal/mole/^oK)	Transition (From Microscopy)
a. CONTROL				
1st Heat	69.1	10.2	29.7	Solid→Smectic
	74.7	0.394	1.13	Smectic→Cholesteric
	80.3	0.351	0.992	Cholesteric→Isotropic Liquid
2nd Heat	69.0	9.97	29.1	Solid→Smectic
	74.8	0.329	0.945	Smectic→Cholesteric
	80.6	0.197	0.557	Cholesteric→Isotropic Liquid
Cooling	78.6	-0.307	-.872	Isotropic Liquid→Cholesteric
	72.6	-0.394	-1.14	Cholesteric→Smectic
	Supercools			

Mode	T_m (^oC)	ΔH (kcal/mole)	ΔS (cal/mole/^oK)	Transition (From Microscopy)
b. ACETONE				
1st Heat	69.1	10.5	30.8	Solid→Smectic
	76.0	0.446	1.28	Smectic→Cholesteric
	81.6	0.402	1.13	Cholesteric→Isotropic Liquid
2nd Heat	69.1	10.3	30.0	Solid→Smectic
	76.2	0.402	1.15	Smectic→Cholesteric
	82.8	0.335	0.941	Cholesteric→Isotropic Liquid
Cooling	79.8	-.290	-.822	Isotropic Liquid→Cholesteric
	74.0	-.357	-1.03	Cholesteric→Smectic

Mode	T_m (^oC)	ΔH (kcal/mole)	ΔS (cal/mole/^oK)	Transition (From Microscopy)
c. HEXANE				
1st Heat	68.9	10.3	30.2	Solid→Smectic
	75.2	0.615	1.76	Smectic→Cholesteric
	80.9	0.307	0.868	Cholesteric→Isotropic Liquid
2nd Heat	68.8	10.2	29.7	Solid→Smectic
	75.2	0.461	1.32	Smectic→Cholesteric
	81.0	0.373	1.05	Cholesteric→Isotropic Liquid
Cooling	79.1	-.329	-.935	Isotropic Liquid→Cholesteric
	73.2	-.329	-.951	Cholesteric→Smectic
	Supercools			Smectic→Solid

Table V. Effect of Solvent Type on Cholesteryl Myristate Transitions (continued)

Mode	T_m (°C)	ΔH (kcal/mole)	ΔS (cal/mole/°K)	Transition (From Microscopy)
d. CARBON TETRACHLORIDE				
1st Heat	67.4	9.49	27.9	Solid→Smectic
	74.8	0.681	1.96	Smectic→Cholesteric
	78.1	0.389	1.11	Cholesteric→Isotropic Liquid
2nd Heat	67.7	8.86	26.0	Solid→Smectic
	74.7	0.316	0.910	Smectic→Cholesteric
	79.1	0.292	0.830	Cholesteric→Isotropic Liquid
Cooling	67.9	-0.365	-1.07	Isotropic Liquid→Cholesteric
	63.6	-0.511	-1.52	Cholesteric→Smectic
	Supercools			Smectic→Solid

Mode	T_m (°C)	ΔH (kcal/mole)	ΔS (cal/mole/°K)	Transition (From Microscopy)
e. CHLOROFORM				
1st Heat	69.6	9.66	28.2	Solid→Smectic
	74.8	0.452	1.30	Smectic→Cholesteric
	78.6	0.226	0.642	Cholesteric→Isotropic Liquid
2nd Heat	68.0	9.33	27.3	Solid→Smectic
	72.5	0.497	1.44	Smectic→Cholesteric
	78.1	0.181	0.515	Cholesteric→Isotropic Liquid
Cooling	75.3	-0.339	-0.972	Isotropic Liquid→Cholesteric
	70.7	-0.361	-1.05	Cholesteric→Smectic
	Supercools			Smectic→Solid

Mode	T_m (°C)	ΔH (kcal/mole)	ΔS (cal/mole/°K)	Transition (From Microscopy)
f. BENZENE				
1st Heat	68.9	9.23	27.0	Solid→Isotropic Liquid
	No mesophase formed			
2nd Heat	67.4	9.15	26.9	Solid→Smectic
	72.0	0.457	1.33	Smectic→Cholesteric
	76.8	0.206	0.588	Cholesteric→Isotropic Liquid
Cooling	75.5	-0.343	-0.984	Isotropic Liquid→Cholesteric
	70.4	-0.396	-1.15	Cholesteric→Smectic
	Supercools			Smectic→Solid

Some inversion of smectic and cholesteric heats of transition is seen in Table V. Results on extremely pure samples indicate that the smectic ΔH should be less than the cholesteric ΔH (13). The mesophases are much more sensitive to impurity and solvent traces than the solid mesophase transition. This is predicted by the van't Hoff equation.

CONCLUSIONS

Traces of solvent sorbed on recrystallization by the solid phase have a great effect on the heats and temperatures of the mesophase transitions of the three esters studied. Some solvents, benzene and hydrocarbons, are sorbed to a much greater extent than oxygenated or saturated hydrocarbons. This study has indicated that great care must be used in removing the last traces of solvent and other impurities from mesophase-forming materials if meaningful thermal measurements are to be obtained. In addition, samples which showed reproducible transitions can decompose on standing under relatively inert conditions.

Cholesteryl esters appear to have a complex solid phase polymorphy which is highly dependent on sample purity. The lowest melting forms result when the sample is the most pure. Optical microscopy has indicated a difference in the crystal appearance between the low and high melting solid forms.

REFERENCES

1. G.W. Gray, J. Chem. Soc., 1956, 3733.
2. R.D. Ennulat, Analytical Calorimetry, R.S. Porter and J.F. Johnson eds. Plenum Press, New York, (1968) p. 219.
3. E.M. Barrall, R.S. Porter and J.F. Johnson, J. Phys. Chem., 71, 1224 (1967).
4. E.M. Barrall, R.S. Porter and J.F. Johnson, J. Chromatog., 21, 392 (1966).
5. D.E. Martier, P.A. Blasco, P.F. Carone, L.C. Chow and H. Vicini, J. Phys. Chem., 72, 3489 (1968).
6. E.M. Barrall and J.F. Johnson, Fractional Solidification Vol. II, M. Zief ed., Marcel Dekker Inc., New York (1969) p. 90.
7. A.P. Gray, Instrument News, 16, 9 (1966).

8. C. Plato and A.R. Glasgow, Jr., Anal. Chem., 41, 330 (1969).
9. H. Kelker, B. Scheurle, H. Winterscheidt, Anal. Chim. Acta, 38, 17 (1967).
10. H. Kelker, Z. Anal. Chem., 198, 254 (1963).
11. H. Kelker and H. Winterscheidt, Z. Anal. Chem., 220, 1 (1966).
12. J. Billard and J.P. Meunier, Compt. rend., 266, 937 (1968).
13. E.M. Barrall, R.S. Porter, and J.F. Johnson, Second Conference on Liquid Crystals, Kent State (1968) in press.

MOLECULAR STRUCTURE OF CYCLOBUTANE FROM ITS PROTON NMR IN A NEMATIC SOLVENT

Lawrence C. Snyder and Saul Meiboom

Bell Telephone Laboratories, Incorporated

Murray Hill, New Jersey

It is now well known that the nematic phase provides an anisotropic environment for solute molecules which permits high resolution NMR spectra.[1] The major structure of these solute NMR spectra is due to the intramolecular magnetic dipole-dipole interaction, which does not average to zero in an anisotropic environment. Because th. dipole-dipole interaction depends on the inverse cube of the distance between the nuclei, it provides information on molecular structure. The structure of several molecules has been determined from NMR in nematic solvents, including acetonitrile, methyl fluoride, benzene, cyclopropane, and bicyclobutane.[2]

Although the determination of the molecular structure of solute molecules is probably the most important result of the technique, it also provides information on the liquid crystal solvent. In the following, we report a structure determination for cyclobutane, and conclude with a short discussion of the implications of the observed spectra for the nematic solvent.

In analyzing the observed NMR spectra we have made use of the spin Hamiltonian which has been found adequate for high resolution NMR of molecules in nematic phases.[3] The spin Hamiltonian includes indirect spin-spin coupling of nuclear pairs i and j, given by the parameters J_{ij}. It also contains the intramolecular magnetic dipole-dipole interactions: the value for the pair of nuclei i and j is given by D_{ij}. The symmetry of the observed spectrum shown in Figure 2 implies that all protons in cyclobutane have the same diamagnetic shielding (chemical shift). This suggests the assumption that the spin Hamiltonian has symmetry D_{4h}. We made that assumption: it implies that there are five distinct J_{ij} and five distinct D_{ij} parameters in the spin Hamiltonian.

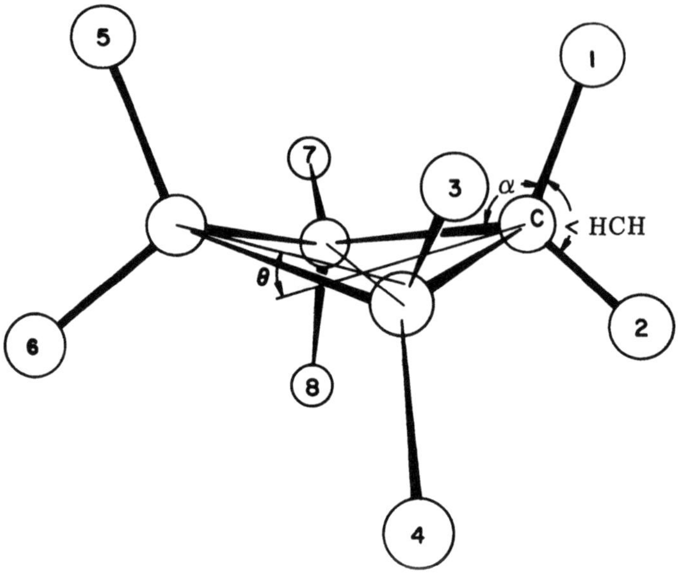

Figure 1: A bent conformer of cyclobutane and coordinates used.

Because of the high symmetry of cyclobutane, the dipolar interactions can be expressed in terms of a single motional constant,[3] $C_{3z^2-r^2}$, and the cartesian coordinates of the nuclei in the molecule fixed coordinate system of Figure 1.

$$D_{ij} = -2\times 5^{-\frac{1}{2}} K_{ij}\, C_{3z^2-r^2}\left[\left\langle\frac{(\Delta z_{ij})^2}{r_{ij}^{\,5}}\right\rangle - \frac{1}{2}\left\langle\frac{(\Delta x_{ij})^2}{r_{ij}^{\,5}}\right\rangle - \frac{1}{2}\left\langle\frac{(\Delta y_{ij})^2}{r_{ij}^{\,5}}\right\rangle\right]$$

Here r_{ij} is the distance between nuclei i and j. The constant K_{ij} is proportional to the product of the magnetic moments of the two nuclei.[3] The angular brackets indicate an average over the molecular vibrations.

The proton NMR spectra of cyclobutane were measured at 60 MHz in the nematic phase of p,p'-di-n-hexyloxyazoxybenzene. The spectrum was analyzed by an iterative trial and error method to give the spin Hamiltonian parameters in Table 1. These parameters give

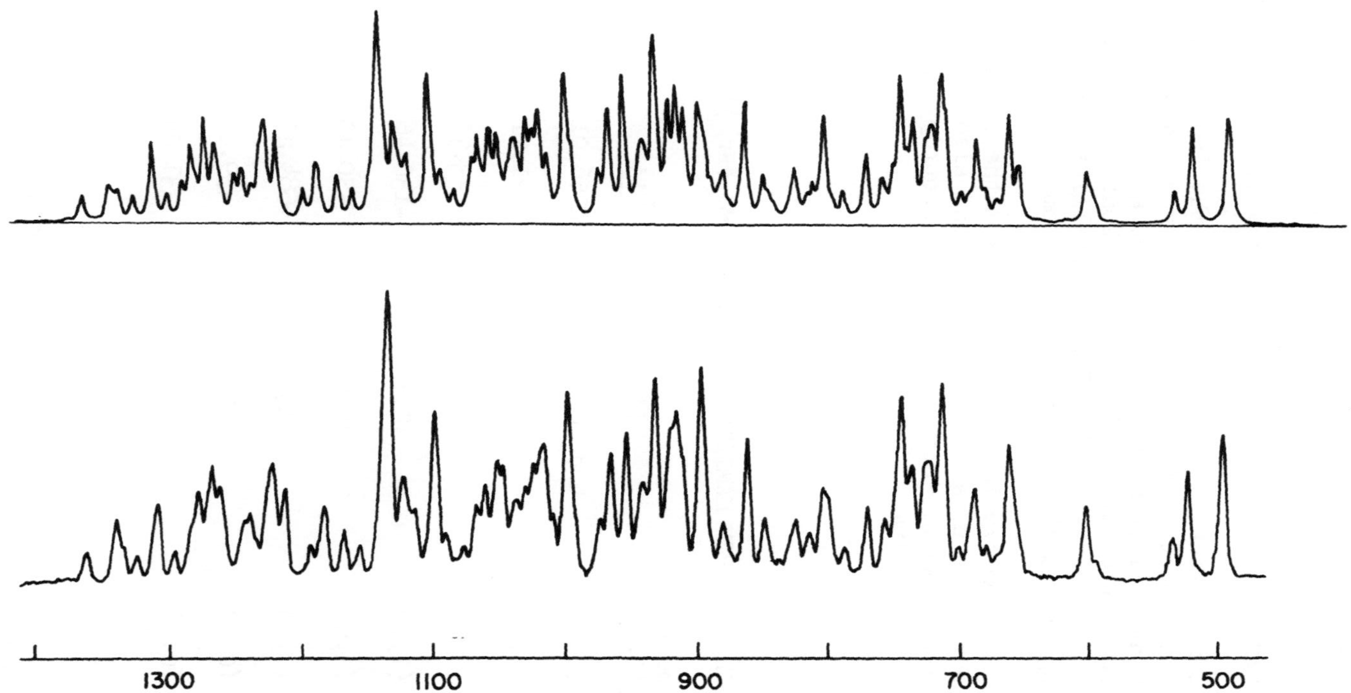

Figure 2: Low field part of the cyclobutane proton NMR spectrum. The complete spectrum is symmetrical. The frequency scale at the bottom of the figure has its origin at the center of symmetry. The lower spectrum is experimental, the upper is a theoretical simulation using the parameters given in Table I The frequency scale is in Hertz.

the theoretical spectrum shown in Figure 2. The search for dipolar interaction parameters D_{ij} was guided by the assumption that the molecule oscillates between two equivalent conformers having D_{2d} symmetry, as the one shown in Figure 1. Initially the proton positions were varied and finally the parameters D_{ij} themselves.

We also observed weak satellites[4] in both the inner and outer regions of the proton NMR due to molecules containing carbon-13 in the natural abundance. We adopted a value of the indirect coupling of the carbon-13 nucleus to bonded protons which was obtained by Burke and Lauterbur.[5] The satellite spectra were fitted by varying the position of the carbon atom.

In deducing a geometry for cyclobutane from the constants D_{ij}, we have ignored the effect of vibrations and also the possibility of pseudo-dipolar interactions due to anisotropy of the indirect spin-spin coupling.[3]

TABLE I. SPIN HAMILTONIAN PARAMETERS (Hz.)[a]

Atom Pair (i,j)	J_{ij}	D_{ij}
(1,2)	Undetermined	+1256.01
(1,3)	+10.4	-221.21
(1,4)	+4.95	+3.73
(1,5)	+2.5	-130.46
(1,6)	+0.5	-30.10
(C,1)	+136[b]	+684.23
(C,3)	0[b]	-39.20
(C,5)	0[b]	-31.24

[a] All parameter values are believed accurate to ± .7 Hz.

[b] Assumed values.

One should note that four coordinates are required to specify the location of the protons of cyclobutane. The ratios of the indirect couplings to one another are determined by the ratios of those coordinates to one another. We may form three ratios of coordinates to the fourth. There are five ways that one may choose four D_{ij}'s in order to form ratios of three of them to the fourth. As a consequence, we have varied the ratios of coordinates in five ways to fit the ratios of D_{ij}'s in their five subsets. By computing the change in each ratio of coordinates caused by an error (change) of 1 Hz in each of the D_{ij}'s, we have computed a value of minimum variance for each of the three coordinate ratios. From these we have computed our final geometry of cyclobutane.

One unfortunate aspect of our study is that it has not been possible to observe carbon-13 proton NMR lines which underlie the main spectrum. However, only lines in this region depend on the dipolar interaction of carbon-13 with protons on other carbons. This circumstance makes it impossible for us to determine one of the two coordinates required to give the position of a carbon atom. We are thus forced to make an assumption to fix this coordinate. We have done this in two ways. The first is to assume that the two C-H bond lengths of protons to a carbon are equal. The second is to assume the tetrahedral value for the ∠HCH. The corresponding geometries of cyclobutane are given in Table II along with that determined by electron diffraction. Because we have no way to evaluate the motional constant, we are only able to determine the geometry of a molecule up to a scale factor. In Table II we have fixed the scale to give the C-C bond length its electron diffraction value of 1.548 Å.[6]

We believe that our study shows conclusively that cyclobutane is bent, and that it oscillates rapidly between the two equivalent bend conformers. Theoretical spectra computed with a bent geometry differing only slightly from those given in Table II are shown in Figure 3 to support this view. The planar form is obtained with $\theta = 0°$.

The dihedral angle of 22.9° to 27.0° which we give is somewhat smaller than the value of 35° given in our preliminary report.[6] It is also smaller than the value of 35° obtained from electron diffraction[7] and the value of $33.3 \pm \frac{1}{2}°$ deduced in the analysis of a difference band in the near infrared spectra of cyclobutane.[7] More accurate determinations of molecular geometry from NMR in nematic solvents must await a more complete accounting of the effect of vibrations.

We now discuss some other conclusions that can be drawn from the observed NMR spectra. One striking fact, evident from Figure 2, is the high resolution obtainable. The observed linewidth is

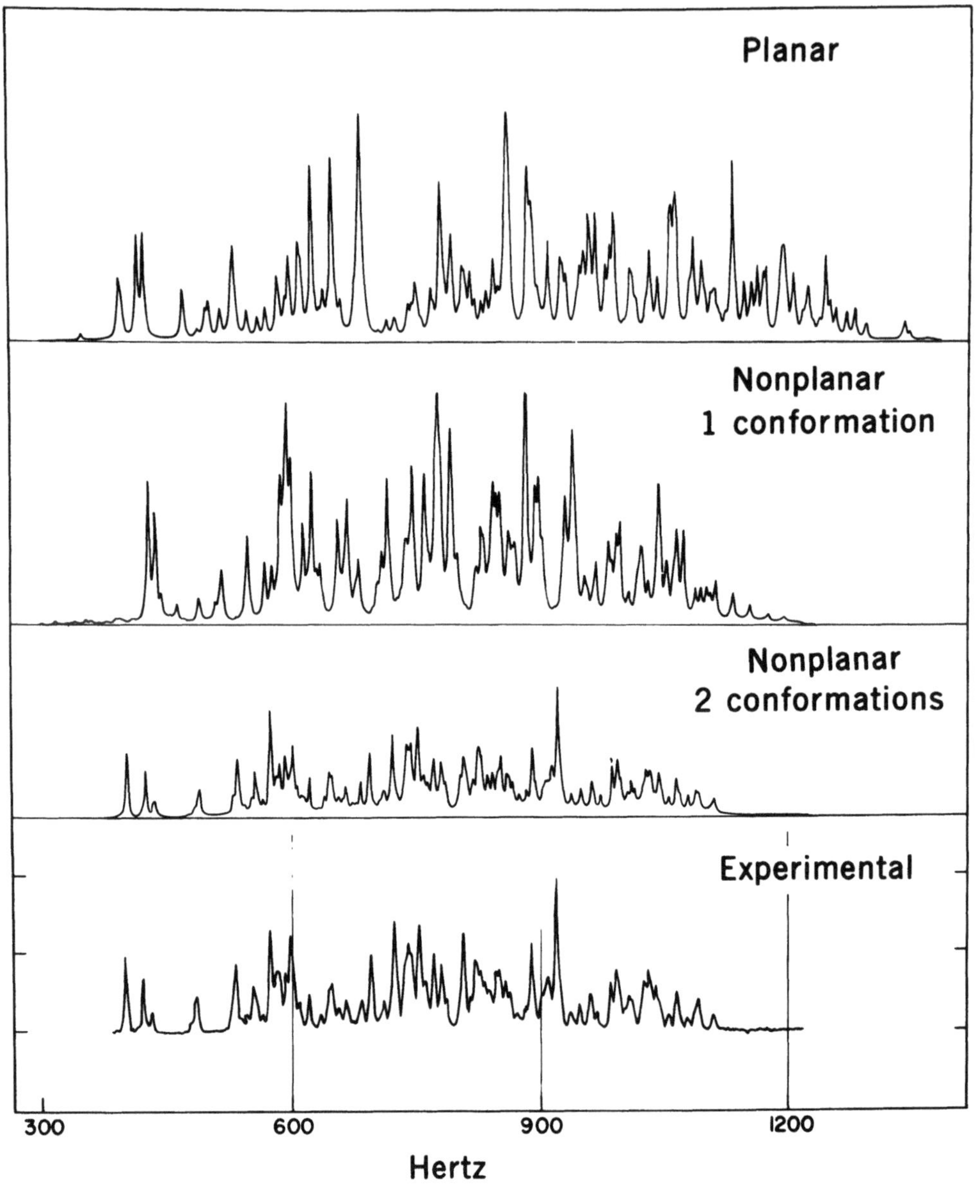

Figure 3: Cyclobutane spectra. The top three are calculated for the molecular structures indicated.

a few Hertz, and the main contribution to this figure is the magnetic field inhomogeneity (which is relatively large, because the sample cannot be spun). With very careful field adjustments, one can obtain linewidths of about 1 Hertz. The overall width of the spectrum is about 2000 Hz. As the spectrum width is directly proportional to the degree of orientation of the solute molecules, any variation in orientation results in an inhomogeneous broadening of the lines. One can therefore conclude that the degree of orientation is uniform over the sample to better than one part in a thousand, surely a high degree of uniformity.

The above conclusion requires some qualification. The orientation referred to above is a time average, that is, the liquid crystal could have a non-uniform structure, provided there are rapid fluctuations producing a uniform average. The excess linewidth, Δ , caused by such a process can be estimated from the equation[9]

TABLE II. CYCLOBUTANE GEOMETRY

Coordinate	Assumption $R_{CH1}=R_{CH2}$	Assumption $\angle HCH=109.47$	El. Diff.
R_{CH1}	1.133 Å	1.123 Å	1.092 Å
R_{CH2}	1.133	1.085	1.092
$R_{C-C'}$	(1.548)[a]	(1.548)[a]	1.548
$\angle$ HCH	108.1°	(109.47°)[a]	110°
$\angle\alpha$	122.1	118.7	(125°)[b]
$\angle\theta$	27.0	22.9	35°

[a] Assumed.

[b] The methylene group was apparently assumed to be bisected by the plane of its carbon atom and the two bonded to it.

$$\Delta = \frac{\pi}{2} f^2 \tau$$

where f is the amplitude of the fluctuations in NMR frequency and τ the correlation time of the fluctuations. Taking $\Delta < 1$ Hz, f = 1000 Hz, one obtains that $\tau < 10^{-6}$ sec. If one assumes a diffusion coefficient of 10^{-5} cm^2 $sec.^{-1}$, this time gives a maximum diffusion length of 3×10^{-6} cm. For the required averaging to occur, the domains in a "swarm" theory should be smaller than this figure, or about 10,000 molecules. This seems to rule out the much larger domains that have been proposed in the past to explain light scattering and magnetic and dielectric properties.

Finally, a study of the magnitude of the motional constants can give information on the degree of orientation of the solvent. We have studied the variation of motional constants with temperature of bicyclobutane dissolved in p,p'-di-n-hexyloxyazoxybenzene. The nematic range of the solution is between 70°C and 87.5°C. Excellent solute spectra can be obtained over this entire range. The overall width of the spectra, which is a measure for the degree of orientation, decreases from 1750 Hz just above the melting point (70°C) to 1100 Hz half a degree below the isotropic point (87°C). Extrapolation to the isotropic point gives a width of 1050 Hz. The nematic phase thus preserves a high degree of orientation and uniformity up to the isotropic transition; the orientation disappears discontinuously at the transition.

REFERENCES

1. A. Saupe and G. Englert, Phys. Rev. Letters 11, 462 (1963).

2. For reviews on the subject see: A. Saupe, Angew. Chem. 80, 99 (1968); G. R. Luckhurst, Quarterly Rev. London 22, 179 (1968); S. Meiboom and L. C. Snyder, Science 162, 1337 (1968).

3. L. C. Snyder, J. Chem. Phys. 43, 4041 (1965).

4. L. C. Snyder and S. Meiboom, J. Chem. Phys. 47, 1480 (1967).

5. J. J. Burke and P. C. Lauterbur, J. Am. Chem. Soc. 86, 1870 (1964).

6. S. Meiboom and L. C. Snyder, J. Am. Chem. Soc. 89, 1038 (1967).

7. A. Almenningen, O. Bastiansen and P. N. Skancke, Acta. Chem. Scand. 15, 711 (1961); also P. N. Skancke, Thesis 1960,

Institute for Theoretical Chemistry, Norges Teckniske Hogskule, Trondheim, Norway.

8. T. Ueda and T. Shimanouchi, J. Chem. Phys. 49, 470 (1968).

9. Pople, Schneider and Bernstein, High-resolution Nuclear Magnetic Resonance, McGraw-Hill, New York 1959 - Chapter 10.

MAGNETIC ALIGNMENT OF NEMATIC LIQUID CRYSTALS

J. O. Kessler†

Department of Physics, The University of Arizona, Tucson, Arizona 85721

It has generally been taken for granted that the magnetic alignment of nematic liquid crystals is due to their anisotropic diamagnetic susceptibility. On a per-molecule basis the orientation-dependent magnetic interaction energy is much smaller than kT. It is therefore necessary to invoke some mechanism which reduces the effect of thermal agitation, if the latter is not to destroy the magnetic alignment. The spontaneous ordering of the molecular axes in the nematic state provides just such a mechanism. Thus, the usual descriptions of magnetic alignment depend on the interaction of the field with self-ordered macroscopic regions of the liquid. These regions are thought of as essentially rigid in the swarm theory, and deformable, but coherent, in the continuum theory.[1,2]

One may consider magnetic alignment in terms of the interaction of the field with individual molecules, i.e., as re-formation of the nematic phase with its preferred axis along the field. Under these circumstances, the possibility of a new type of magnetic interaction suggests itself. The hypothesis is made that the primary interaction of the magnetic field with the mesophase is not the usual one, via the susceptibility, but is an interaction which modifies the phase-organizing forces. The Saupe-Maier theory,[3] which has been very successful on the whole, explains the nematic phase in terms of the anisotropic dispersion force. The effect considered here is a modification of that electrostatic force by the applied magnetic field. This modification would take the form of a spatial anisotropy. The situation envisaged is somewhat analogous to ferromagnetic anisotropy due to a crystal lattice.

What justification or need is there for the introduction of this hypothesis? It must be admitted that so far there exist few published data which may be interpreted in this fashion, to the exclusion of the usual point of view. Three possible cases are cited below.

1. The magnetic susceptibility anisotropy alignment theory predicts that the molecular axis of minimum diamagnetic $|\chi|$ should be aligned along the applied magnetic field. In the case of para-azoxyanisole (PAA), for instance, this means that the long molecular axis should point along $\vec{H}$, as is indeed the case. If the applied magnetic field causes a spatial anisotropy in the phase-organizing cooperative dispersion force, then one might expect spatial orientation effects other than just minimum $|\chi|$ alignment. Such an effect is implied by x-ray measurements which show that the magnetic field intensity produces a change in molecular packing of PAA through rotational ordering of molecules around their long axis.[4] The standard magnetic alignment theory provides no explanation for such an effect.

2. It has been shown[5,6,7,8,9] that magnetic fields can convert a cholesteric mesophase to a nematic one. Here again, the long molecular (presumably minimum $|\chi|$) axes tend to align along the magnetic field. Thus the phase transition is generally seen as an increase of helix pitch to infinity, which is the nematic state. There is one well known[9] exception. A cholesteryl chloride-cholesteryl myristate mixture orients itself with the helix axis along the magnetic field! Thus, either the $|\chi|$ is not a minimum in the long molecular axis direction--or the alignment is not due to the usually accepted mechanism.

3. The normal magnetocaloric effect ("adiabatic demagnetization") is proportional to $(\partial\chi/\partial T)_H$. This quantity is very small for PAA for instance,[10] except possibly at the nematic isotropic transition temperature. On the other hand, a measurable magnetocaloric effect has been reported.[11,12] If the magnetic field interacts not only with the induced moment, but also modifies the phase-organizing forces, one would expect an additional term in the magnetocaloric effect coefficient. Magnetocaloric measurements are in progress in our laboratory. Very preliminary results indicate that an

effect may be present--but a definite statement will have to await an improved apparatus. It should also be noted that a magnetocaloric effect larger than the standard prediction may have an alternative explanation in terms of magnetoelastic coupling.

†Work supported by a NASA Institutional Grant.

REFERENCES

1. *Molecular Structure and the Properties of Liquid Crystals*, G. W. Gray, Academic Press, New York 1962, Ch. III.
2. A. Saupe, Angew. Chem. 80, 99 (1968) is one of the most complete recent review papers.
3. W. Maier and A. Saupe, Z. Naturforsch. 13a, 564 (1958); 14a, 882 (1959); 15a, 287 (1960).
4. I. G. Chistyakov and V. M. Chaikovsky, Sov. Phys.--Crystallog. 12, 770 (1968).
5. P. G. DeGennes, Solid State Commun. 6, 163 (1968).
6. R. B. Meyer, Appl. Phys. Letters 12, 281 (1968); 14, 208 (1969).
7. Durand, Leger, Rondelez, and Veyssie, Phys. Rev. Letters 22, 227 (1969).
8. Sackmann, Meiboom, and Snyder, J. Am. Chem. Soc. 89, 598 (1967).
9. Sackmann, Meiboom, Snyder, Meixner, and Dietz, J. Am. Chem. Soc. 90, 3567 (1968).
10. G. Foex, Trans. Farad. Soc. 29, 958 (1933).
11. W. Moll and L. Ornstein, Proc. Acad. Amst. 21, 259 (1919).
12. M. Miesowicz and M. Jezewski, Physik. Z. 36, 107 (1935).

THE AGGREGATION OF POLY-γ-BENZYL-L-GLUTAMATE IN MIXED SOLVENT SYSTEMS

JOHN C. POWERS, JR.

HUNTER COLLEGE OF THE CITY UNIVERSITY OF NEW YORK

(WORK DONE AT IBM RESEARCH LAB., SAN JOSE, CAL.)

Electro-optic methods have been widely employed in investigating the structures and properties of biopolymers. From decay measurements[1] it is possible to estimate molecular dimensions. The saturation of birefringence[2] or dichroism[3] allows the estimation of dipole moments and the anisotropy of polarizability for rigid axially symmetrical molecules.

Much of the initial work has been performed on synthetic polymers, such as poly-γ-benzyl-L-glutamate (PBLG), which dissolve in organic solvents, and which can assume the conformation of a rigid rod. The Kerr constant of PBLG was first measured in ethylene dichloride (EDC) by Tinoco[4] in 1956. In this solvent it was presumed to exist as an unaggregated rigid rod at low concentrations.

Problems due to the aggregation of PBLG have been recognized since its first reported synthesis. Doty, Bradbury and Holtzer[5] mentioned the extremely high viscosity of solutions of the polymer in benzene and dioxane and Tinoco[4] also stated that solutions of PBLG in EDC gel after a period of weeks. Wada[6] examined the dielectric dispersion of PBLG in dioxane and dioxane-DMF mixtures and formulated a theory to explain his results involving both head-to-tail and side-by-side (anti-parallel) associations. Recently Watanabe[7] has reported some measurements of the Kerr constant of PBLG in mixed solvent systems and found evidence for several types of aggregation. At about the same time we reported our studies of the aggregation of PBLG in various pure solvents.[8] We also found evidence of antiparallel association in both benzene and dioxane solutions and for head-to-tail association in concentrated solutions in EDC.

The results reported here are a study of the aggregation of PBLG in benzene and dioxane and mixtures of these solvents with DMF and EDC. These experiments were performed on a low molecular weight polymer in contrast to previous work in order to maximize any effects due to the number of polymer ends in solution. The work of Wada and Watanabe was conducted on high molecular weight polymer.

EXPERIMENTAL

A. Apparatus

The apparatus for measuring the Kerr constants is of conventional design and uses a photoelectric method to detect the induced phase change δ. It has been described previously.[8,9] All determinations of birefringence were made at 35°C as were the corresponding measurements of solution viscosity. The latter were measured in Cannon viscometers and the intrinsic viscosity obtained in the usual way from a plot of η_{sp}/c versus c.

B. Materials

Poly-γ-benzyl-L-glutamate was prepared by the method of Blout and Karlson[10] by polymerization of the benzyl-L-glutamate-N-carboxyanhydride with sodium methoxide in dimethylformamide (DMF). The molecular weight of the polymer was 37,000 as estimated from the measured intrinsic viscosity in dichloroacetic acid at 25°C.[11]

The solvents (benzene, dioxane, ethylene dichloride and dimethylformamide) were all purified by standard methods and redistilled before use.

C. Procedures

The Kerr constants (B) were obtained from Kerr's equation (1) by plotting values of versus F^2

$$B = \frac{\delta}{2\pi \ell F^2} \qquad (1)$$

(the cell length ℓ is in cm. and the applied field F is statvolts/cm.) Rectangular DC voltage pulses from one to three milliseconds in duration were applied at one second intervals; these pulses ranged in size from 80 to 1000 V. Since the electrode spacing is 0.4 cm., the maximum applied voltage was 2500 V/cm.

The plots were linear up to the highest values of applied field indicating that little or no saturation is occurring. Generally 8 to 10 points were taken per determination. The amount of polymer was varied from 0.0005 g/ml. to at least 0.005 g/ml; in the two solvent systems containing benzene the amount of polymer was increased to 0.01 g/ml. The mixed solvent systems comprised a non-polar solvent (benzene, dioxane) and a polar solvent (dimethylformamide, ethylene dichloride), the compositions being 0.5, 1, 2, 5 and 50% by volume of the polar component.

In order to establish a standard state of reference for the individual solvent systems, the values of the specific Kerr constant (B/c) at each polymer concentration were plotted against concentration and extrapolated to zero concentration. This produces a value of an intrinsic specific Kerr constant ([B/c]) for each solvent system. These plots were reasonably linear except for those in the most non-polar solvent systems. In those systems, which contained large amounts of either benzene or dioxane, the plots tended to slope upwards slightly; a smooth extrapolation could be made in every case however. Values of the intrinsic specific Kerr constant and intrinsic viscosity of PBLG in the various solvent systems are given in Table I. Plots of these data will be found in the discussion section.

DISCUSSION

The specific Kerr constant for a system,which has a polarizability ellipsoid having an axis of revolution and a permanent dipole oriented parallel to this symmetry axis, can be written in terms of molecular parameters as[2]

$$B/c = \frac{2\pi}{15n\rho\lambda}(g_1 - g_2)\left(\frac{\mu^2}{k^2T^2} + \frac{\alpha_1 - \alpha_2}{kT}\right) \quad (2)$$

where n and ρ are the index of refraction and density of the solvent and solute respectively, λ is the wave length of the light, μ the permanent dipole moment of the polymer, $(\alpha_1 - \alpha_2)$ is the anisotropy of electrical polarizability and the factor $(g_1 - g_2)$ is defined as the anisotropy of optical polarizability or the optical factor.

Since O'Konski, Yoshioka and Orttung[2] have shown that the second term in brackets can be neglected for the case of a polypeptide such as PBLG, the magnitude of the specific Kerr constant is dependent on the permanent dipole moment of the polymer and the optical factor. Estimates of the optical factor can be obtained from saturation measurements and are available for three of the four solvents used. These values at 5000 A are listed in Table II along with estimates of the dipole moment obtained from Equation (2). The value for benzene was estimated previously by us to be 3×10^{-2}, a figure which is probably high; we shall assume for discussion here a value of about 10^{-2}.

TABLE I

INTRINSIC KERR CONSTANTS AND INTRINSIC VISCOSITIES FOR PBLG IN MIXED SOLVENT SYSTEMS

A - Benzene/DMF

Vol. % DMF	[B/c] cm^4/v^2g	[η] cm^3/g
0	0.020	650
0.5	0.011	245
1	0.010	310
2	0.0029	84
5	0.0021	28
50	0.0104	28
100	0.0193	31

B - Benzene/EDC

Vol. % EDC	[B/c] cm^4/v^2g	[η] cm^3/g
0	0.020	650
0.5	0.012	450
1	0.020	800
2	0.032	650
5	0.020	260
50	0.028	78
100	0.0155	54

C - Dioxane/DMF

Vol. % DMF	[B/c] cm^4/v^2g	[η] cm^3/g
0	0.0071	92
0.5	0.0051	90
1	0.0073	174
2	0.0088	110
5	0.0111	41
50	0.0124	19
100	0.0193	31

D - Dioxane/EDC

Vol. % EDC	[B/c] cm^4/v^2g	[η] cm^3/g
0	0.0071	92
0.5	0.012	94
1	0.0095	132
2	0.0078	140
5	0.0134	212
50	0.016	33
100	0.0155	54

TABLE II

OPTICAL AND PHYSICAL CONSTANTS FOR PBLG IN VARIOUS PURE SOLVENTS

Solvent	(g_1-g_2)	Ref.	B/c ($cm.^4/v^2g.$)	(Debye)
EDC	4.1×10^{-3}	2,4	0.0155	1080
DMF	2.8×10^{-3}	12	0.0193	1450
Benzene	10^{-2}	8	0.020	805
Dioxane	2.8×10^{-3}	12	0.0071	875

As can be seen from Table II, the effective dipole moment of the polymer varies widely from solvent to solvent. The value in DMF probably represents the best estimate for the free polymeric rod. Some form of association obviously exists in the other solvents which is reducing the effective dipole moment.

These observations and the general behavior of the polymer in mixed solvents can be described in terms of several model types of aggregation similar to the approach used by us[8] and other workers.[6,7] The predicted deviations of both [B/c] and [η] from the ideal case of a free peptide molecule for the major modes of aggreagation can be listed as follows:

	Aggregation Type	Change in [B/c]	Change in [η]
(1)	Linear, head-to-tail	Increase	Increase
(2)	Linear, head-to-head	Decrease	Increase
(3)	Lateral and parallel	No change	Small decrease
(4)	Lateral and anti-parallel	Decrease	Small decrease

These types of association would not be expected to be of equal strength. Head-to-tail association would be more stable than head-to-head aggregation due to the possibility of hydrogen bond formation between the two complementary chains. Head-to-head aggregation would involve not only an intermolecular dipole-dipole repulsion, but also a geometric problem due to the reversal of order of residues in one helix relative to the other. Similarly anti-parallel lateral association is favored over parallel lateral association because of the dipole-dipole interactions.

Discussion of the experiments in terms of the above concepts fails to take account of the most striking features of the plots in the accompanying figures, namely, the occurrence of sharp maxima in both [B/c] and [η] as a function of solvent. Since PBLG is known to form stable liquid crystals[13] (presumably cholesteric in very concentrated solutions), it seems reasonable to assume that these sharp changes in the experimental quantities represent the occurrence of actual phase changes brought about by modifying the polar character of the solvent. Similar maxima have been observed in the viscosity-melt curves of all three types of liquid crystal, i.e. nematic, smectic and cholesteric.[14] The observations reported here would then constitute the lyotropic analog of the thermally induced phase changes. Since the electric field induces an anisotropy in the systems, the changes shown by the variation of [B/c] with concentration will not be expected to occur at the same solvent composition as the changes in [η]. The individual systems will now be discussed separately.

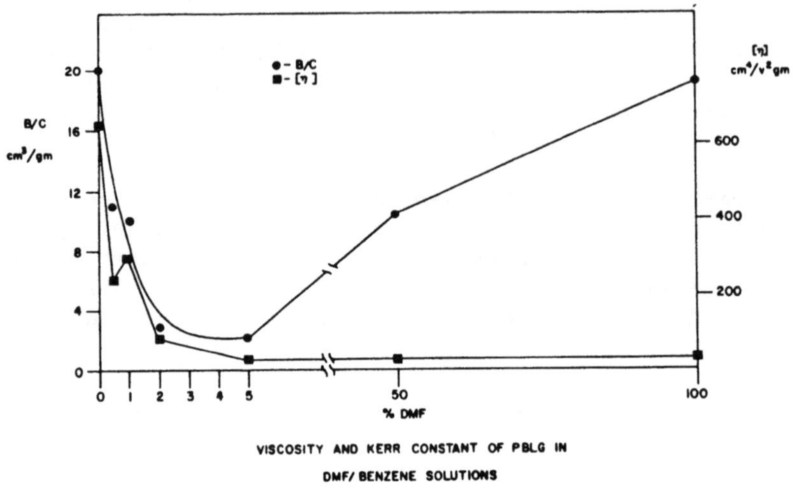

Figure 1. Viscosity and Kerr constant of PBLG in DMF/Benzene solutions.

A. DMF/Benzene

Both viscosity and Kerr constant are extremely high in pure benzene; the addition of even a small amount of DMF causes a substantial decrease in both quantities. In benzene the conformation of the polymeric aggregate much be such as to generate a large birefringence but yet a small average dipole moment. In view of the large viscosity observed, the conformation is probably that of a smectic liquid crystal. This phase then undergoes a change to a more isotropic system in which the major form of association seems to be anti-parallel and to involve only a few molecules (both [B/c] and [η] are small). On further addition of DMF this breaks up into individual solute molecules and [B/c] rises. Thus the changes in this system are smectic, anti-parallel, free polymer molecule. DMF is a (relatively) very polar solvent and only a small amount is necessary to effect large changes.

B. EDC/Benzene

Ethylene dichloride is a much less polar solvent than DMF and thus it might be expected that lyotropic changes in this system would be more gradual than those observed above. Such indeed is the case. Clearly defined maxima are observed whereas in DMF/benzene these maxima were not resolved. The observations can again be accounted for by assuming a smectic phase is present initially in pure benzene and that this undergoes a transition to an anti-parallel type of association. The magnitude of [B/c] in pure

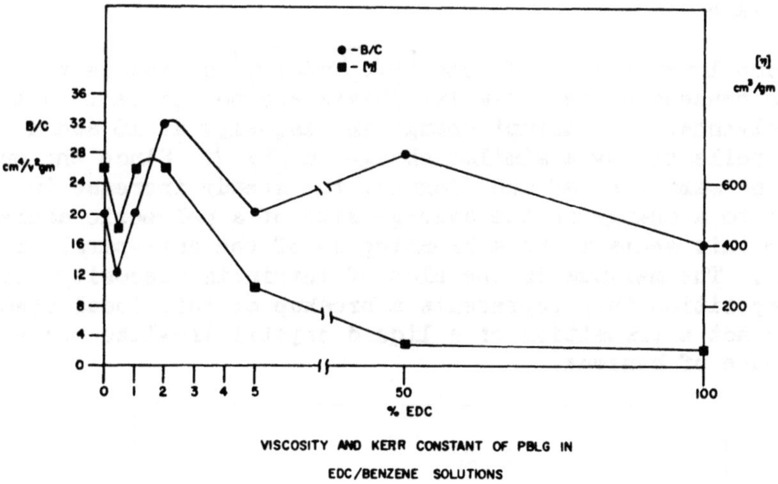

Figure 2. Viscosity and Kerr constant of PBLG in EDC/Benzene solutions.

EDC suggests that this form of aggregation persists throughout the entire range of solvent composition for this system. Unfortunately these changes in [B/c] are also dependent on the optical factor difference in the two solvents (a large difference); this would introduce complications of an indeterminate character.

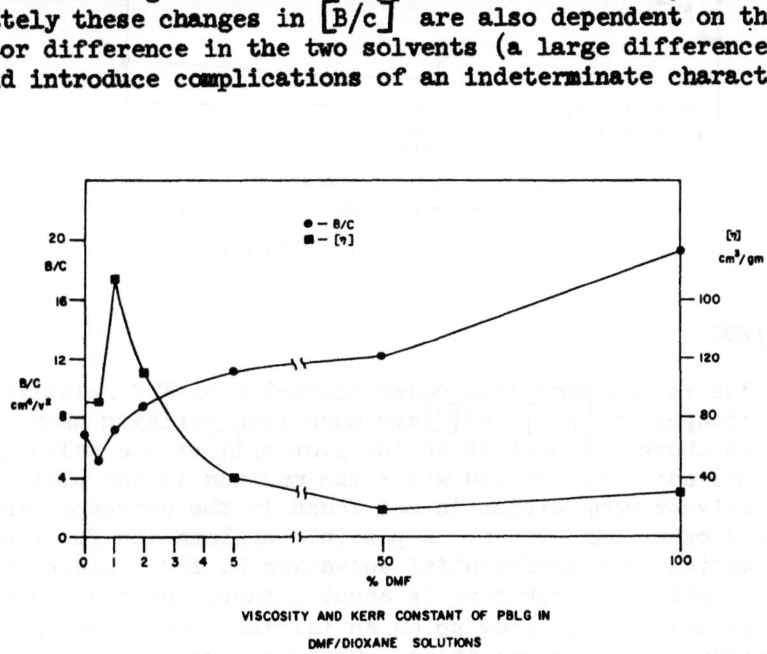

Figure 3. Viscosity and Kerr constant of PBLG in DMF/Dioxane solutions.

C. DMF/Dioxane

The much lower values of both [B/c] and [η] in dioxane relative to those in benzene suggest similar phases are not present in the two pure solvents. The abrupt change in viscosity at 1% added DMF is not reflected by a similar change in [B/c]. Since the optical factor is the same for DMF and dioxane, the steady increase in [B/c] is due to a change in the average size of a polymeric aggregate. Here this seems to be a breaking up of the anti-parallel association. The maximum in the plot of intrinsic viscosity versus solvent composition then represents a breakup of this local association and not a transition of a liquid crystalline-like phase as in the case of benzene.

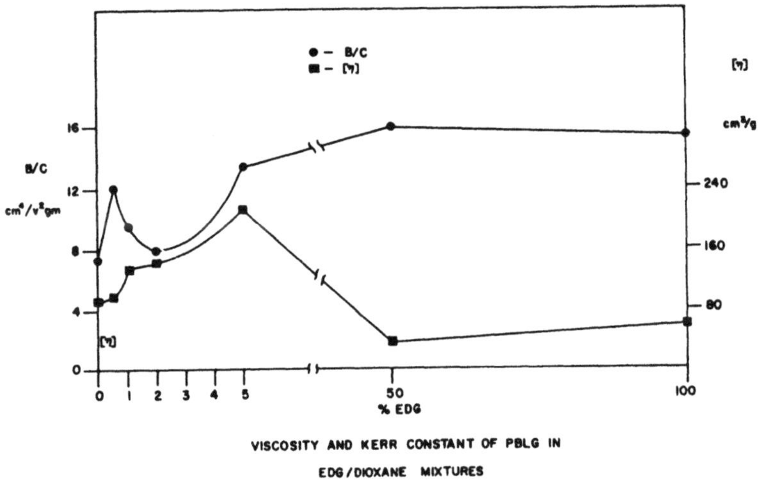

Figure 4. Viscosity and Kerr constant of PBLG in EDG/Dioxane mixtures.

D. Dioxane/EDC

Again, due to the much less polar character of EDC relative to DMF, the changes in [B/c] and [η] are much less striking than those observed above. A maximum in the plot of [η] versus solvent composition has not been reached while the maximum in the plot of [B/c] versus solvent composition is not found in the corresponding plot for the dioxane/DMF system. A possible explanation for the latter observation is a preferential solvation by EDC. Since the difference in optical factor here is about a factor of two, solvation of an aggregate could show up as an initial jump in the plot of [B/c] versus solvent composition. Clearly some additional work is needed to resolve this problem; a shortage of polymer prevented any extension of this work beyond the reported values.

CONCLUSIONS AND SUMMARY

The behavior of this low molecular weight PBLG, enen in pure solvents, is now seen to be rather complex. In EDC there are evidences of association at low concentrations, but in DMF the polymer seems to be substantially free and unaggregated. There is evidence for a birefringent and highly viscous phase in benzene (probably smectic in nature), but in dioxane the aggregation appears to be only short range and predominantly anti-parallel. The sharp maxima in the plots of both $[B/c]$ and $[\eta]$ versus solvent composition imply strongly that lyotropic phase changes are occurring. Since DMF is a more polar solvent than EDC, these changes are more dramatic when DMF is added. More than one transition seems to occur in benzene/polar solvent mixtures. In the benzene/DMF system the birefringent phase changes to one similar to that found in dioxane before complete dissociation occurs.

It is rather remarkable that aggregation of this low molecular weight polymer is so pronounced that it persists to very low dilutions. A further study of high molecular weight PBLG under similar conditions would help to clarify the nature and occurrence of these lyotropic transitions. A wider range of solvent compositions needs to be studied in this regard since the transitions can be exceedingly sharp as our experiments have shown.

REFERENCES

1. C. T. O'Konski and A. J. Haltner, J. Am. Chem. Soc., 78, 3604 (1956).
2. C. T. O'Konski, K. Yoshioka and W. H. Orttung, J. Phys. Chem., 63, 1558 (1959).
3. C. Houssier and E. Fredericq, Biochem. Biophys. Acta, 88, 450 (1964).
4. I. Tinoco, Jr., J. Am. Chem. Soc., 79, 4336 (1957).
5. P. Doty, J. H. Bradbury and A. M. Holtzer, J. Am. Chem. Soc., 78, 947 (1956).
6. A. Wada, J. Polymer Sci., 45, 145 (1960).
7. H. Watanabe, J. Chem. Soc. Japan, 86, 179 (1965).
8. J. C. Powers, Jr. and W. L. Peticolas in "Ordered Fluids and Liquid Crystals," Advances in Chemistry #63, American Chemical Society, Washington, D.C., 1967, p. 217.
9. J. C. Powers, Jr., J. Am. Chem. Soc., 89, 1780(1967).
10. E. R. Blout and R. H. Karlson, J. Am. Chem. Soc., 78, 941 (1956).
11. J. C. Mitchell, A. E. Woodward and P. Doty, J. Am. Chem. Soc., 79, 3948 (1957).
12. K. Yamaoka, Thesis, University of California (Berkeley), 1962.
13. C. Robinson, Trans. Faraday Soc., 52, 571 (1955).
14. G. W. Gray, "Molecular Structure and the Properties of Liquid Crystals," Academic Press, New York, 1962, pp. 97-105.

LIQUID CRYSTALS III.[1] NEMATIC MESOMORPHISM IN BENZYLIDENE ANILS CONTAINING A TERMINAL ALCOHOL GROUP

Joel E. Goldmacher and Michael T. McCaffrey

RCA Laboratories, David Sarnoff Research Center

Princeton, New Jersey 08540

Most of the past research in nematic liquid crystals has concerned itself with the synthesis of compounds possessing a high nematic thermal stability. However, we have continued and directed our investigation toward the preparation of low molecular weight Schiff's bases in order to depress the nematic-isotropic transition point. This work has led to the discovery of the alcohol function as a terminal group in nematic liquid crystals, and in particular, to materials possessing low crystal nematic transitions.

EXPERIMENTAL SECTION

General

Transition temperatures were measured in open capillary tubes with an Arthur H. Thomas Model No. 6406-K melting point apparatus and are all uncorrected. *p*-Alkoxy benzaldehydes were prepared from *p*-hydroxybenzaldehyde and various alkyl iodides in methanolic KOH by the method of Weygand and Gabler[2] to give 50-60% yields of the known compounds. The *p*-aminobenzyl and phenethyl alcohols were purchased from Sapon Laboratories and used without further purification. Combustion analyses were performed by B. L. Goydish of these laboratories.

Substituted Anils

In the general procedure a mixture of 0.010 mol each of the appropriate *para*-substituted benzaldehyde and *para*-substituted aniline in 100 ml of benzene containing 0.1 g of benzenesulfonic acid was refluxed for 2-4 hours. The water was removed azeotropically and was collected and measured in a Dean-Stark trap. After the calculated amount of water was collected, the solvent was removed in vacuo and the residue recrystallized from isopropanol (Table I). Yields of product obtained after one recrystallization ranged from 50 to 80%. All of the nonmesomorphic compounds were recrystallized to constant melting point while the nematic materials were recrystallized until the nematic-isotropic transition temperatures were constant and reversible.

RESULTS

The preparation of 16 new anils was carried out by the acid-catalyzed condensation of p-alkoxybenzaldehydes with p-aminobenzylalcohol and p-aminophenethylalcohol, respectively. These series exhibit the general mesomorphic properties found in other similar systems, namely the lower homologs are nonmesomorphic while the middle homologs are purely nematic. In addition these compounds represent the first example of liquid crystalline behavior occurring in a molecule containing a terminally substituted alcohol group. The properties of these compounds are presented in Table I along with the analytical data.

DISCUSSION

In general one does not expect mesomorphic behavior in any system which will give rise to intermolecular hydrogen bonding. This type of association not only increases the melting point drastically, but also has a tendency to encourage a nonlinear arrangement of the molecules. For example it was found[3] that phenolic compounds are never mesomorphic, however, elimination of the hydrogen bonding by replacement of the phenolic hydrogen by an alkyl group may give a mesomorphic ether. The following examples illustrate this point.

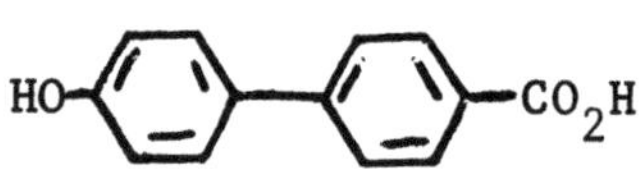

(dimer) Solid $\xrightarrow{294.5}$ isotropic[3]

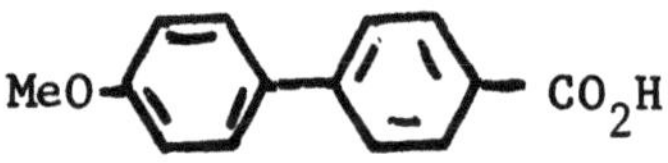

(dimer) Solid $\xrightarrow{258}$ nematic $\xrightarrow{300°}$ isotropic[3]

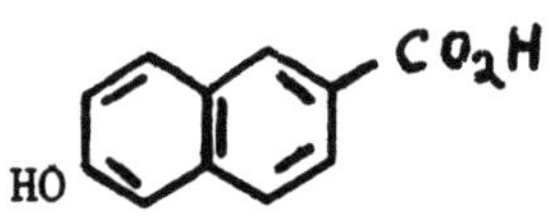

(dimer) Solid $\xrightarrow{250}$ isotropic[4]

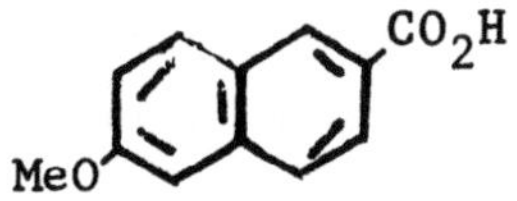

(dimer) Solid $\xrightarrow{206}$ nematic $\xrightarrow{219}$ isotropic[4]

The first pair of compounds shows that the effect of the hydrogen bonding is not solely because of the high melting point which it places upon the hydroxy compound.

A well recognized example of intermolecular hydrogen bonding has the reverse effect, increasing the tendency to mesophase formation by greatly lengthening the molecule. This, of course, refers to the dimerization[6] of carboxylic acids by intermolecular hydrogen bonding of the carboxyl groups. This type of association preserves the linearity of the molecule and increases molecular length, maintaining the intermolecular attractions at a relatively high level. The fact that this type of association is necessary in order that the p-n-alkoxybenzoic acids may exhibit mesophases is clearly demonstrated by the fact that the corresponding esters are nonmesomorphic. Only when we are dealing with the analogous biphenyl compounds[3] are the lengths and polarizabilities of the monomeric esters great enough for them to be mesomorphic, although the thermal stabilities of their mesophases are much lower than those of the corresponding biphenyl acids.

The results of our investigation lend further support to the above mentioned results on hydrogen bonding in liquid crystals. We have found that p-alkoxybenzylidene-p'-aminophenols (Table II) are nonmesomorphic due to strong intermolecular hydrogen bonding which destroys the linearity of the molecule. However, insertion of one or two methylene groups between the hydroxy group and the benzene ring eliminates intermolecular association, thereby restoring the anisotropic nature of the molecule. Infrared studies on these molecules clearly show the presence of free OH groups. The infrared spectrum of compound 6 was recorded neat at 65°C and showed a strong band at 3660 cm^{-1}, consistent with data reported by Bellamy.[5] The general mesomorphic properties found in other similar systems, namely that when the mesomorphic transition temperatures for a homologous series of compounds are plotted against the number of carbon atoms in the alkyl chain, smooth curve relationships between even or odd members of the series are found to exist. Figures 1-3 clearly show this trend.

Table I
Substituted Alcohols Prepared for This Study

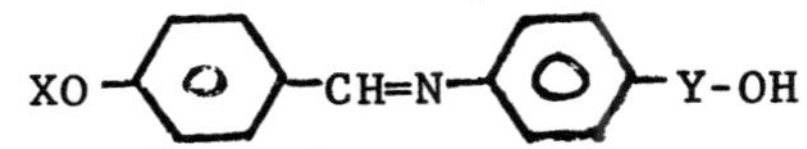

Compound	X	Y	M.P.°C	Nematic Range °C	Anal. % Calculated C	Calculated H	Calculated N	Anal. % Found C	Found H	Found N
1	CH_3	CH_2	91-2		74.66	6.27	5.81	74.38	6.31	5.70
2	C_2H_5	CH_2	93-4		75.27	6.71	5.49	75.01	6.60	5.40
3	C_3H_7	CH_2	66-7		75.81	7.11	5.20	75.90	7.26	5.35
4	C_4H_9	CH_2		63-70	76.29	7.47	4.94	76.45	7.50	4.90
5	C_5H_{11}	CH_2		61.5-63	76.73	7.80	4.71	76.61	7.59	4.81
6	C_6H_{13}	CH_2		58-73	77.13	8.09	4.50	77.10	8.11	4.59
7	C_7H_{15}	CH_2		70-71.5	77.50	8.36	4.30	77.75	8.49	4.41
8	C_8H_{17}	CH_2		70-78	77.84	8.61	4.13	77.97	8.73	4.21
9	CH_3	C_2H_4		(81-2)[a]73	75.27	6.71	5.49	75.61	6.82	5.20
10	C_2H_5	C_2H_4		91-93	75.81	7.11	5.20	75.88	7.30	5.16
11	C_3H_7	C_2H_4	95-6		76.29	7.47	4.94	76.28	7.74	4.88
12	C_4H_9	C_2H_4		(91-2)[a]88	76.73	7.80	4.71	76.68	8.02	4.57
13	C_5H_{11}	C_2H_4		(85-6)[a]80	77.13	8.09	4.50	77.10	8.11	4.81
14	C_6H_{13}	C_2H_4		(89-90)[a]84	77.50	8.36	4.30	77.80	8.38	4.25
15	C_7H_{15}	C_2H_4	91-2		77.84	8.61	4.13	77.48	8.82	4.13
16	C_8H_{17}	C_2H_4		(92-3)[a]88	78.14	8.84	3.96	78.02	9.00	3.79

[a]Monotropic

Table II

p-Alkoxybenzylidene-p'-aminophenols

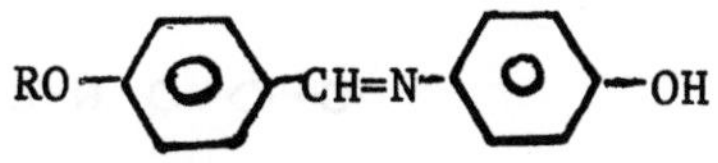

Compound	R	M.P.°C
17	CH_3	193-5
18	C_2H_5	187-8
19	C_3H_7	177-9
20	C_4H_9	150-1
21	C_5H_{11}	140-1
22	C_6H_{13}	148
23	C_7H_{15}	130-1
24	C_8H_{17}	125-6

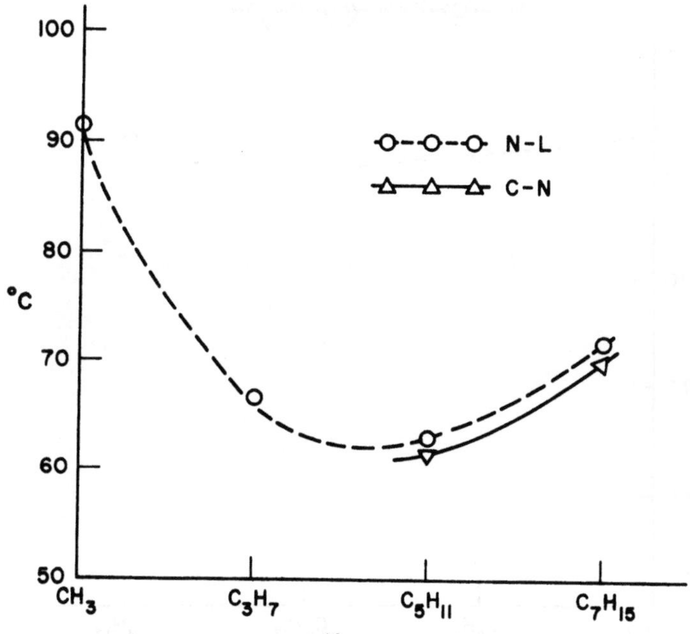

FIG. 1 PHASE TRANSITION TEMPERATURES FOR:

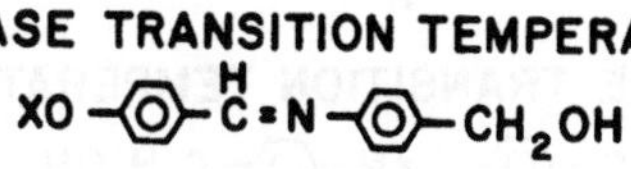

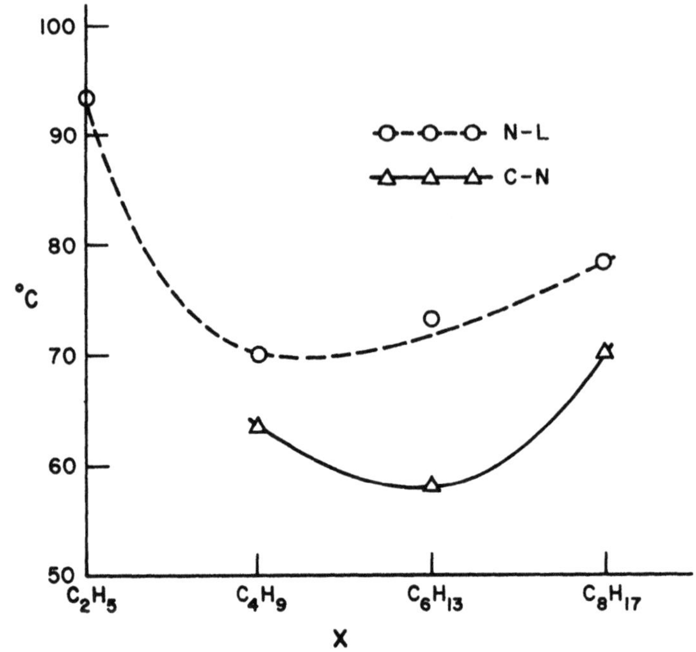

FIG. 2 PHASE TRANSITION TEMPERATURES FOR:

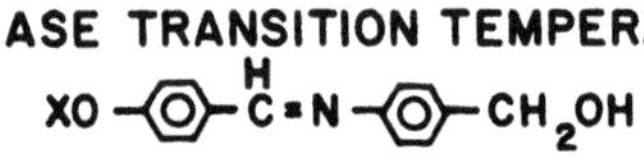

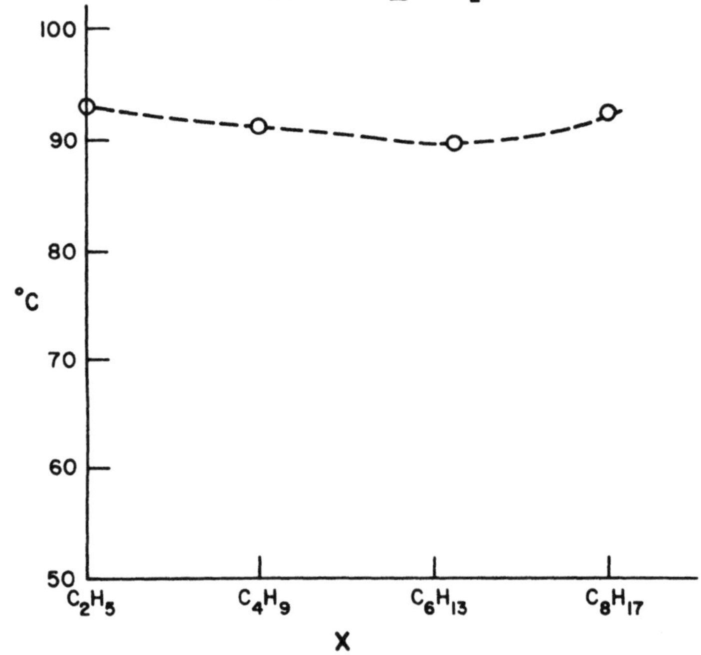

FIG. 3 PHASE TRANSITION TEMPERATURES FOR:

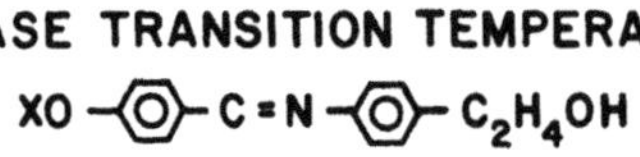

REFERENCES

1. For the previous papers in this series, see J. Org. Chem., 32, 476 (1967) and Ibid., 33, 3501 (1968).
2. C. Weygand and R. Gabler, J. Prakt. Chem., 155, 338 (1940).
3. G. W. Gray, et al., J. Chem. Soc., 1412 (1955); G. W. Gray, Ibid., 393 (1957).
4. G. W. Gray and B. Jones, Ibid., 236 (1955).
5. L. J. Bellamy, The Infrared Spectra of Complex Molecules, Methuen & Co., Ltd. London, England; Chapter VI.
6. G. W. Gray and B. Jones, J. Chem. Soc., 4179 (1953).

MESOMORPHIC PROPERTIES OF THE HETEROCYCLIC ANALOGS OF BENZYLIDENE-4-AMINO-4'-METHOXYBIPHENYL

William R. Young, Ivan Haller, and Larry Williams

IBM Watson Research Center, Yorktown Heights, N. Y.

In order to ascertain the effect of heterocyclic rings on liquid crystal stability, a series of Schiff bases derived from 4'-methoxy-4-biphenylamine and aromatic heterocyclic aldehydes has been prepared. The ability of each compound to form a liquid crystal phase, either by melting the crystal or cooling the isotropic liquid, has been determined, and the appropriate heat of transition and corresponding entropy change have been measured by differential scanning calorimetry. The results will be discussed in terms of geometrical considerations, substituent effects, and intermolecular interactions.

INTRODUCTION

Intrinsic factors which determine the physical states of pure chemical substances are numerous and difficult to apply in the case of polyatomic molecules. Although several approximate rules of thumb are available to the chemist for predicting properties such as melting point or boiling point, these guidelines are extremely general and must be employed with considerable caution.

In this regard, much interest has been generated in recent years concerning mesophase formation by organic compounds.[1] Specifically, a number of investigators have attempted to provide data with which to predict whether a particular material will display mesomorphism and to what extent. The more successful studies have used the technique of making small and systematic changes in molecular structure and interpreting the corresponding variation in physical properties in terms of this "simple" structural modification. Unfortunately, small and systematic changes are frequently not very "simple," since many of the parameters which affect liquid crystal behavior, e.g., length, breadth, permanent dipoles, polarizability, and molecular planarity, are varied simultaneously. In order to obviate this difficulty to a large extent, and at the same time elucidate the roles of permanent dipoles and polarizabilities in the absence of significant geometrical modifications, we have begun an investigation of the nematic properties of compounds in which benzene rings have been replaced by aromatic heterocyclic analogs.[2]

We have selected benzylidene-4-amino-4'-methoxybiphenyl, I,

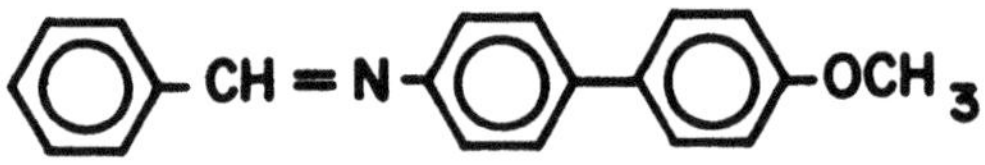

I

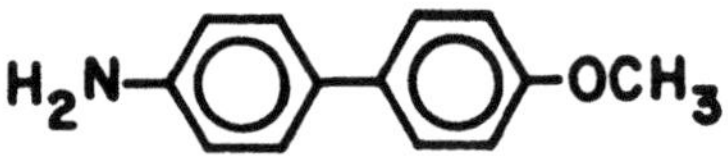

II

as our principal comparison compound. This Schiff base and its precursor 4'-methoxy-4-biphenylamine (II) are well known in the literature;[3] in addition, I and its para-substituted derivatives have been the subject of an elegant and detailed investigation of substituent effects on nematic stabilities.[4] Compound I has a reported nematic range of 174 to 176°,[4] and therefore it was postulated that condensation of II with such readily available aldehydes as furfural or 3-pyridinealdehyde might yield mesomorphic Schiff bases. Accordingly, amine II was prepared in large quantities,[5] by the method shown in Fig. 1, and condensed with several aromatic heterocyclic aldehydes. The resulting anils were analyzed for liquid crystal behavior.

EXPERIMENTAL

Melting points are corrected. Infrared spectra were obtained on a Perkin-Elmer 137B Infracord. Nuclear magnetic resonance spectra were obtained on a Varian HA-60-IL Spectrometer. All new compounds had satisfactory spectral properties and elemental analyses. Mesomorphic transition temperatures, accurate to within 0.5°, and phase identifications were performed on a Leitz Ortholux-POL Polarizing Microscope equipped with a Koeffler Hot Stage. Calorimetric measurements were made on a calibrated Perkin-Elmer 1B Differential Scanning Calorimeter, and are believed accurate to 5%.

4-Biphenylbenzoate (IV): To a vigorously stirred mixture of p-phenylphenol (Eastman White Label, 168g, 1 mole), 10% sodium hydroxide (240 ml, 1.1 mole) and water (500 ml) was added dropwise benzoyl chloride (155 g, 1.1 mole) at such a rate that the reaction

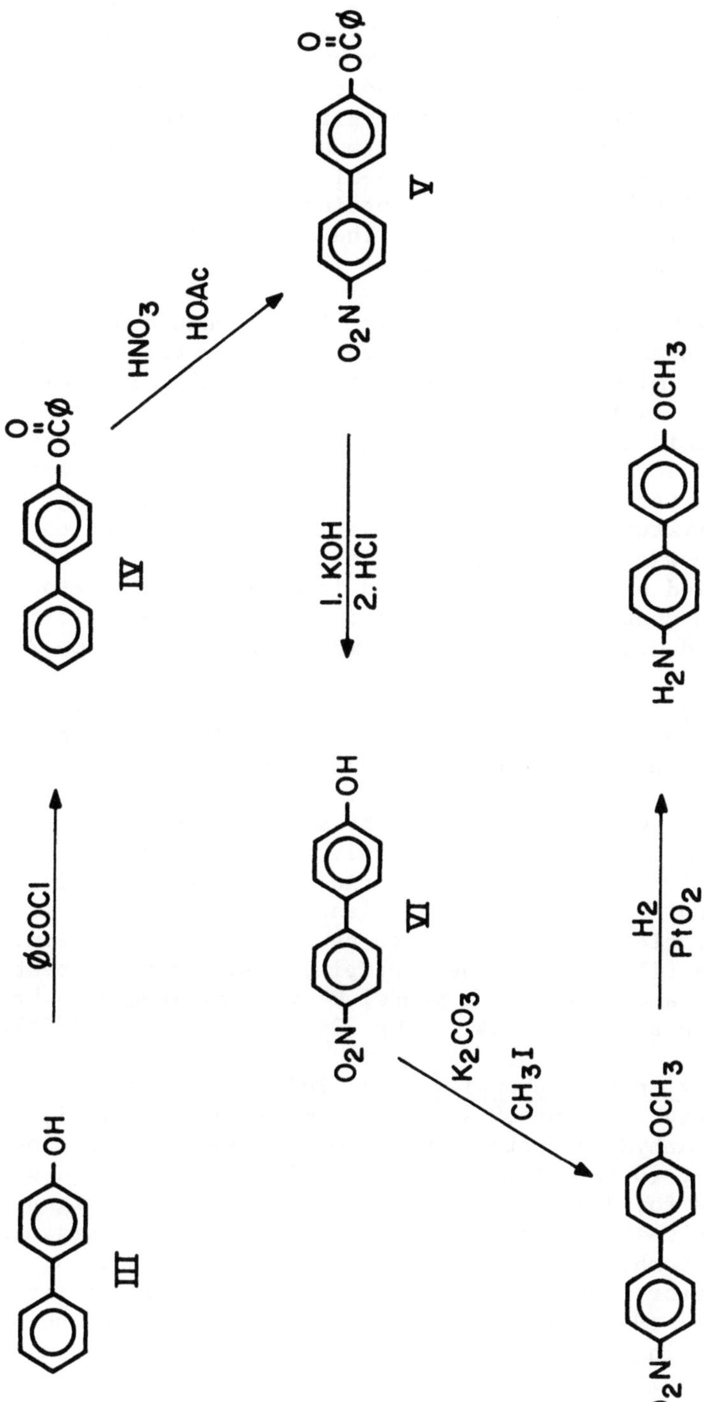

Fig. 1. Synthetic scheme for the preparation of 4'-methoxy-4-biphenylamine.

temperature was maintained at 30-35°. After the addition, the mixture was kept at 35° for 1 hr and at 20° for 1.5 hrs. The mixture was made basic with 50% sodium hydroxide solution. The resulting precipitate was filtered and thoroughly washed with warm water until the filtrate gave no precipitate when tested with 5% hydrochloric acid. The crude product was obtained as an off-white powder (199 - 214 g, 73-80%). Pure 4-biphenylbenzoate (IV) (124 g, 46%, mp 149°) (lit.[6] 150-151°) was obtained by recrystallization from 900 ml of glacial acetic acid. The mother liquors afforded an additional 49 g of the product.

4'-Nitro-4-biphenyl Benzoate (V): The procedure of Jones and Chapman[7] was modified to prepare compound V. Ester IV (124 g, 0.45 mole), 1100 ml of glacial acetic acid, and 260 ml of water was heated to 70° with stirring. To this was added 285 ml of 90% fuming nitric acid at a rate which maintained the 70% temperature. When 80% of the acid had been added, the mixture turned orange and soon thereafter a yellow precipitate formed. Following the addition, the mixture was stirred 2 hrs at 70° and cooled to 20°. The precipitate was collected and washed with acetic acid and water. The dried product was digested in a minimum of acetone for 20 min. and the suspension was filtered to afford 100 g (75%) of the desired product, a yellow powder, mp 209-210° (lit.[7] mp 214°).

4-Hydroxy-4'-nitrobiphenyl (VI): Following a modified procedure of Jones and Chapman,[7] 99 g of ester V (0.31 mole) and 495 ml of absolute alcohol were allowed to reflux. To the stirring mixture was added 10% alcoholic potassium hydroxide (690 ml). The deep red solution was refluxed for 1 hr after the addition. Cooling afforded the blue potassium salt which was collected, washed thoroughly with alcohol, and dried. The salt was dissolved in 2000 ml of hot water, and a sufficient quantity (ca. 70 ml) of 38% hydrochloric acid was added to neutralize the solution. The hot mixture was filtered and the precipitate was washed repeatedly with hot water and dried. The phenol VI was obtained as a bright yellow powder in 82% yield (44 g), mp 204-205° (lit.[7] mp 205°).

4-Methoxy-4'-nitrobiphenyl (VII): Following a general procedure of Gray and Jones,[8] 44 g (0.25 mole) of phenol VI, 135 g (0.72 mole) of anhydrous potassium carbonate, 220 ml of cyclohexane, and 61 g (0.42 mole) of methyl iodide were refluxed with rapid agitation for 3 hrs, cooled, and filtered. The inorganic precipitate was washed several times with ether and the organic filtrates were combined. After stripping the solvents on a rotary evaporator, the yellow residue was recrystallized from absolute alcohol (950 ml), following treatment of the hot solution with charcoal. The desired methyl ether VII was obtained in 89% yield (42 g), mp 109-110° (lit.[9] mp 109-110°).

4'-Methoxy-4-biphenylamine (II): To a Parr bottle were added 35 g (0.15 mole) of the nitro derivative VII, 0.71 g of platinum oxide, and 400 ml of absolute alcohol. The mixture was hydrogenated under 1 atm of hydrogen at 70°. After the theoretical amount of hydrogen had been absorbed (ca. 12 hrs), crude product crystallized upon cooling as a beige powder (30 g, mp 134-135°). The product was redissolved in absolute alcohol and an excess of 6 M sulfuric acid was added to precipitate the amine sulfate. After filtering, the salt was washed with alcohol, dried, and redissolved in 200 ml of water. The solution was made basic with 10N sodium hydroxide, the aqueous layer was saturated with sodium bicarbonate, and the organic layer was extracted twice with benzene. The combined organic layers were washed with saturated sodium bicarbonate and water, and dried over magnesium sulfate. The benzene was removed on the rotary evaporator and the residue was recrystallized from absolute alcohol to give 24 g (81%) of 4'-methoxy-4-biphenylamine (II), mp 143-144° (lit.[10] mp 146.5-147°).

Schiff Bases (General Procedures): (A) An equimolar mixture of 4'-methoxy-4-biphenylamine (II) and the freshly purified aromatic aldehyde was allowed to (a) reflux two hours, or (b) stir overnight in a minimum amount of absolute alcohol. In most cases, the product would crystallize at 20°; if not, cooling to -15° was sufficient to induce crystallization. The Schiff bases were recrystallized to constant melting point from alcohol (charcoal treatment), benzene and/or methylcyclohexane. (B) Alternatively, some Schiff bases were prepared by heating a mixture of amine II and a 20% molar excess of the aromatic aldehyde to 125° in an open vessel for 10 to 20 minutes. The cooled, crystalline product was washed with alcohol and petroleum ether, followed by recrystallizations as in procedure A.

The choice of procedure A or B was purely arbitrary. In every case, thin layer chromatography showed the presence of only one component. The Schiff bases and their physical properties are listed in Table I.[11]

RESULTS AND DISCUSSION

The syntheses performed were, for the most part, routine. Purification steps were eliminated to increase yields when it was ascertained that the purity of the subsequent products was not decreased. Several methods of mineral acid/metal reduction of the nitro compound VII were attempted prior to platinum oxide-catalyzed reduction. The latter method was far superior in both yield and product quality.

Several of the Schiff bases listed in Table I show nematic mesophases. Employing Gray's definition[1] that nematic stability

TABLE I. Physical Properties of Anils of 4'-methoxy-4-biphenylamine.

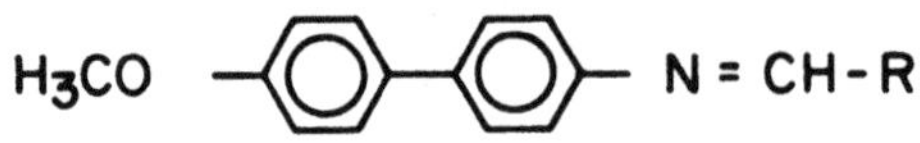

R	No.	Transition	T (°C)	ΔH (kcal/mole)	ΔS e.u.
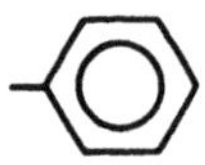	I	C → N N → I	170.0 175.5	8.40 0.094	18.77 0.209
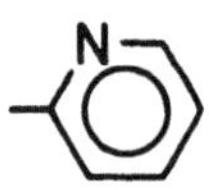	VIII	C → N N → I	116.8 118.8	4.89 0.047	12.56 0.120
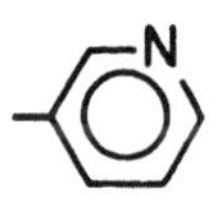	IX	C → N N → I	175.7 195.2	3.70 0.095	8.25 0.203
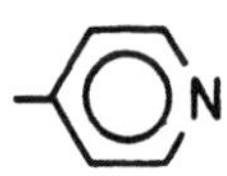	X	C → I N → I	193.8 181.3	8.65 0.084	18.53 0.184
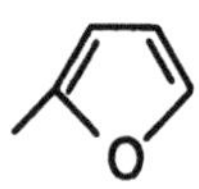	XI	C → I	138.6		
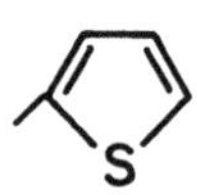	XII	C → I	160.2		
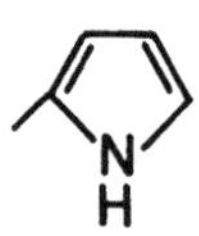	XIII	C → N N → I	186.0 187.5(dec)		
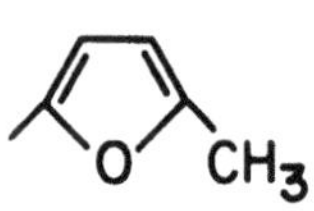	XIV	C → I	158.5		
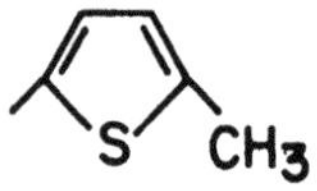	XV	C → N N → I	160.1 211.0	7.51 0.153	17.34 0.317

increases with the temperature of the nematic-isotropic transition, replacement of the terminal benzene ring in compound I by an unsubstituted heterocyclic ring containing a permanent dipole moment results in no appreciable increase in this stability. In fact, the nematic-isotropic transition temperature is lowered significantly in some instances. For example, anil XI undergoes a crystal-to-isotropic transition at 138.6°; the liquid supercools until crystals are reformed at 131.5°. Therefore, a nematic mesophase could not exist above the latter temperature. This should be compared with 175.5° for the benzene analog. The extra permanent dipole associated with the heterocyclic ring, therefore, does not necessarily enhance mesomorphism, assuming negligible geometric modification.[12]

The enthalpies and entropies for the nematic-to-isotropic transitions were measured calorimetrically for those compounds having stable nematic phases. The data, given in Table I, vary regularly: the compounds with lower nematic stabilities absorb less energy and display less change in gross order upon undergoing the transition. Some smaller and less polarizable nematic compounds display significantly higher transition enthalpies and entropies.[13] For example, anisilidine-p-acetoxyaniline turns isotropic from the nematic phase at 109° with the absorption of 0.22 kcal/mole and an increase of 0.58 e.u. These values are typical for p,p'-disubstituted derivatives of benzylidene-aniline and contrast sharply with the results for Schiff base VIII which has a similar transition temperature. Apparently, an extra aromatic ring (compound VIII) is less important energetically than polar substituents at the molecular termini.

It is rather difficult to explain the relative nematic stabilities of the three pyridine derivatives VIII-X. These results do, however, suggest the following conjecture. If one makes the tantalizing but tenuous assumption that the two nitrogens in anil VIII are part of a 1,4-diaza-1,3-butadiene moiety (in the presumably favored s-trans conformation), it can be postulated that the two C=N group dipoles will partially cancel each other and hence reduce the net intermolecular dipole interaction. The greater separation of the nitrogens in compounds IX and X implies that this cancellation may not be important in the latter two materials. Then for the 3-pyridinealdehyde Schiff base, IX, there are three dipoles acting across the molecule (including one at each terminus) which help to enhance nematic properties. For anil X, one of the lateral dipoles is replaced by a dipole along the molecular axis. The experimental result here is a slight reduction in nematic stability (versus IX), although a rationalization for this is not easily forthcoming.

A tentative speculation must again be invoked in interpreting the experimental results for compounds containing the five-membered heterocyclic rings. In the case of the pyrrole derivative XIII, we have a material which is capable of intramolecular hydrogen bonding, as in the structure XIIIa. Pyrrole itself undergoes

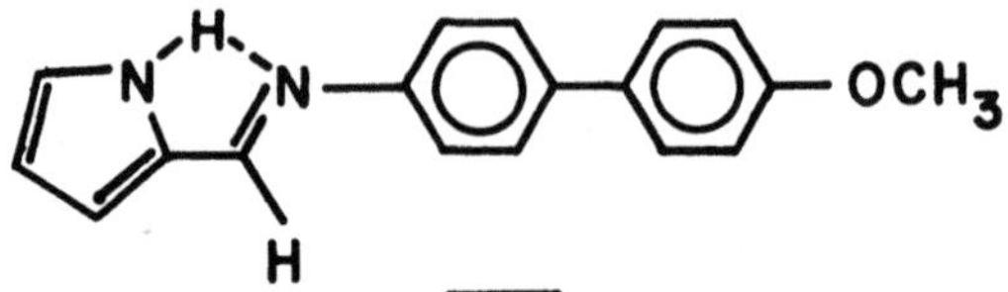

XIIIa

intermolecular hydrogen bonding, as evidenced by its acidic properties and its boiling point.[14] If XIII underwent intermolecular hydrogen bonding, one would expect a much higher melting point and a possible extinction of mesomorphic properties.[15] Therefore, structure XIIIa appears to be reasonable, and the nematic properties for this compound are plausible. Corroborating evidence has been obtained from the N-H stretching region of the infrared absorption spectrum.

Because anils XI and XII are not capable of hydrogen bonding in a manner reflected in structure XIIIa, a comparison between the three compounds is most difficult at this time. Assuming negligible differences in geometries,[16] the major difference in mesomorphic properties for furan and thiophene derivatives can be considered as arising from electronic variations. Although oxygen is more electronegative than sulfur, the dipole moments for furan and thiophene are similar, 0.67 and 0.53 D respectively (the heteroatom being at the positive end of the dipole in both cases).[16] One major difference between the two parent compounds, however, is polarizability. Refractive index measurements[17] indicate that thiophene is about 30% more polarizable than furan. Also, thiophene is known to be more aromatic in its chemistry than furan, a reflection of higher electronic delocalization.[18] Therefore, it would be expected[19] that thiophene compounds would have higher nematic stabilities than comparable furan derivatives.

Unfortunately, neither compound XI nor XII shows a nematic mesophase. This is somewhat surprising since these two Schiff bases melt at a lower temperature than compound I. Although the overall length:breadth ratio and molecular shape are approximately the same for the three materials, the polarizability and/or very slight geometric differences between five- and six-membered rings may be responsible for this result.

However, it is possible to test the hypothesis that thiophene compounds should have higher nematic stabilities than furan derivatives. Of the corresponding Schiff bases derived from 5-methyl-2-furfural and 5-methyl-2-thiophenealdehyde, XIV and XV, only the latter shows a nematic mesophase, although both melt at about 160°.

Albeit by no means conclusive, this result is consistent with the hypothesis.

The enhancement of mesophase stability by a methyl group placed at the terminus of a benzene ring on a long molecule is well known. This same enhancement, most likely due to increased polarizability, is apparently operative in the thiophene compound under discussion.

Additional series of anils based on furan, thiophene, pyrrole, and the pyridines are currently in preparation. These materials will subsequently test the speculative proposals and hypotheses presented herein.

ACKNOWLEDGMENTS

The authors take pleasure in acknowledging the able assistance of A. Aviram for the preparation of compound XI, and H. A. Huggins for performing the microscopy and calorimetry.

FOOTNOTES

1. G. W. Gray, Molecular Structure and the Properties of Liquid Crystals (Academic Press, New York, 1962).
2. Several heterocyclic mesomorphic materials have been reported. See, for example, Reference 1, p. 157.
3. L. F. Trefilova and I. Y. Postovskii, Doklady Akad. Nauk S.S.S.R., 114, 116 (1957).
4. Reference 1, pp. 131-133.
5. We are indebted to Prof. G. W. Gray for providing his synthetic scheme.
6. F. F. Blicke and O. J. Weinkauff, J. Am. Chem. Soc., 54, 330 (1932).
7. B. Jones and J. Chapman, J. Chem. Soc., 1952, 1829.
8. G. W. Gray and B. Jones, ibid., 1954, 1467.
9. G. W. K. Cavill and D. H. Solomon, ibid., 1955, 1404.
10. C. Ivanov and I. Panaiotov, Doklady Akad. Nauk S.S.S.R., 93, 1041 (1953).
11. Compound I has a reported nematic range of 174-176° (Reference 1, p. 133). Compound X is a monotropic nematic liquid crystal. Compound XI has a reported melting point of 126-127° (Reference 3). The authors are unaware of reports concerning the remainder of the Schiff bases listed in Table I. C = crystal, N = nematic phase, I = isotropic liquid.
12. Molecular shapes for compounds I and VIII-XIII are very similar. No broadening is detected on Dreiding models.
13. a. E. M. Barrall, R. S. Porter, and J. F. Johnson, Molec. Crystals, 3, 299 (1968).

b. E. M. Barrall, R. S. Porter, and J. F. Johnson, J. Phys. Chem., 68, 2810 (1964).

14. R. M. Acheson, An Introduction to the Chemistry of Heterocyclic Compounds, Second Edition (Interscience Publishers, London, 1967), pp. 63-64, 93, 120.

15. Reference 1, pp. 161-163.

16. M. H. Palmer, The Structure and Reactions of Heterocyclic Compounds (Edward Arnold Ltd., London, 1967), p. 255.

17. N. A. Lange, Editor, Handbook of Chemistry, Ninth Edition (Handbook Publishers, Inc., Sandusky, Ohio, 1956), pp. 1323, 1358.

18. This aromaticity difference is attributed to sulfur 3-d orbital participation. See Reference 16, p. 257.

19. Reference 1, pp. 148-155.

EFFECT OF END-CHAIN POLARITY ON THE MESOPHASE STABILITY OF SOME SUBSTITUTED SCHIFF-BASES

Ivan Haller, IBM Watson Research Center, Yorktown Heights, New York

Robert J. Cox, IBM Research Laboratory, San Jose, California

The mesophase transition temperatures and heats of transitions of two series of novel substituted benzylideneaminoacetophenones were measured. The geometry of corresponding members of the two series was kept closely similar while changing their polar character by selecting n-alkoxy groups as substituents for series A and by replacing a methylene group by an oxygen atom in series B. Oxygen substitution reduces nematic stability relative to the smectic phase, and lowers the transition temperatures of the mesophases throughout the series. The heats of transitions and the associated entropy changes are generally higher in series B. The thermal properties are insensitive to the location of the oxygen atom in the end-chain. The data indicate that dispersion forces dominate over permanent dipole interactions in determining mesophase stability.

INTRODUCTION

A fundamental problem in the understanding of mesomorphic behavior is the identification of the molecular forces responsible for liquid crystalline order. Repulsive forces, dispersion forces, and interactions between permanent dipoles have been postulated separately as the dominant forces required for mesomorphic behavior. Various theoretical treatments[1-3] of the nematic phase assume a different one of these three interactions for their models; yet they correctly predict the existence of the order-disorder transition.

Extensive experimental studies by Gray and coworkers established correlations[4] between molecular structure and the stability of liquid crystalline phases. It is clear from Gray's work that the molecular shape, i.e. the repulsive forces, greatly influences the mesophase transition temperatures, obscuring, unless special precautions are taken, the more subtle effect of a simultaneous change in the molecular polarity. The prime purpose of the present and of the preceding[5] paper was to synthesize series of related liquid crystalline compounds differing in their dipole moments and polarizabilities but closely similar in molecular geometry.

In this paper we report the synthesis and the measurement of the mesomorphic transition temperatures and heats of transitions of two homologous series of benzylideneaminoacetophenones (I). In series A the substituent, R, was chosen to be an alkyl group of

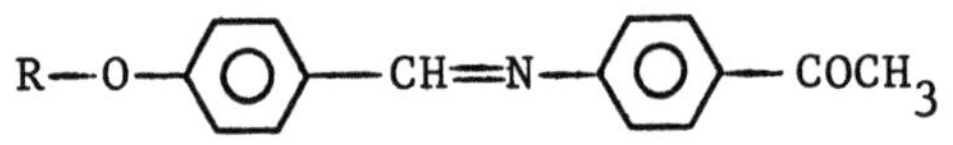

(I)

varying length. Corresponding members in series B contain an ether oxygen atom in place of one of the methylene groups of the alkyl substituent. As the $C \diagup O \diagdown C$ and $C \diagup CH_2 \diagdown C$ bond angles and bond lengths in most known compounds are nearly identical,[6] no change in molecular geometry, except possibly in conformation, is expected between corresponding members of series A and B. At the same time, replacement of a methylene with an oxygen introduces a local dipole moment of about 1.2 D;[7] this, of course, adds vectorially to the dipole moments of the other parts of the molecule, and the resultant will depend on the location of methylene replaced and on the conformation. Finally, the introduction of an oxygen atom, independent of its position and of the chain conformation, results in a substantial reduction in the polarizability of the molecule.

EXPERIMENTAL

Synthesis of Substituted Benzaldehydes. A solution of the sodium salt of p-hydroxybenzaldehyde in ethanol was prepared. An equivalent amount of the appropriately substituted alkyl iodide in ethanol was added to this solution, and the reaction mixture refluxed for several hours. The ethanol was removed *in vacuo*, the residue dissolved in water, and ether extracted. The extract was washed with 5% sodium hydroxide, dried, the ether removed, and the product fractionally distilled *in vacuo*. Fractionations were effected with a 1 foot vacuum-jacketed column packed with glass helices. The physical properties of the alkoxyalkoxy substituted benzaldehydes are shown in Table I.

Substituted Benzylideneaminoacetophenones. Two general methods were used. I: In this procedure equimolar amounts of the appropriately substituted benzaldehyde and p-aminoacetophenone were

TABLE I. Alkoxyalkoxy Substituted Benzaldehydes, RO— 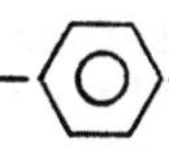 —CHO.

Compound	R-	B.P. (mm) °C	n_D^{20}
1	CH_3OCH_2-[a]	100-112 (1.0)	1.5445
2	$CH_3O(CH_2)_2-$[a]	144-6 (3.0)	1.5526
3	$CH_3CH_2O(CH_2)_2-$	150-4 (4.0)	1.5391
4	$CH_3O(CH_2)_3-$	142-7 (1.3)	1.5415
5	$CH_3(CH_2)_2O(CH_2)_2-$	139-40 (1.0)	1.5316
6	$CH_3CH_2O(CH_2)_3-$	140-50 (1.0)	1.5290
7	$CH_3(CH_2)_5O(CH_2)_2-$	140-50 (0.1)	1.5176

Note a: Reported by Weygand, et al., Ref. 8.

dissolved in benzene, about 0.1 g of benzenesulfonic acid added, and the solution refluxed until the calculated amount of water was collected in a Dean-Stark trap. The benzene was removed in vacuo, and the oily residue was triturated with hexane to induce crystallization. The solid was recrystallized from the solvent shown in Tables II and III. II: Equimolar amounts of the substituted benzaldehyde and p-aminoacetophenone were dissolved in methanol and refluxed on a steam bath for 1/2 to 1 hour. The condensation product generally crystallized on cooling the reaction mixture or, if not, the solvent was removed in vacuo and it was recrystallized from the solvent shown in Tables II or III.

The compounds were recrystallized until thin layer chromatography indicated a single compound. A 0.025 mm layer of alkaline silica gel G was used for the T.L.C's. with a developing agent containing benzene and tetrahydrofuran in a ratio 40:5. Visualization was accomplished with ultraviolet light.

Satisfactory analyses were obtained for all reported compounds.[9] The structures were confirmed by NMR and infrared analyses.

Thermal Measurements. The mesophases were identified and the transition temperatures measured by standard techniques[4] on a Leitz polarizing microscope. The hot stage temperatures were calibrated at the melting points of high purity standards; the thermometric accuracy is estimated as better than 0.5°C.

The heats of fusion and phase transitions were determined by differential scanning calorimetry on a Perkin-Elmer DSC-1B instrument. Hermetically sealed aluminum pans were used as sample containers. The instrument constant was determined by calibration with NBS Benzoic Acid Standard. The reported heats are believed to be accurate to 5%.

RESULTS AND DISCUSSION

Liquid crystalline phases were exhibited by all but one of the compounds synthesized. They are listed together with the melting points and transitions temperatures in Tables II and III for series

TABLE II. p-Alkoxybenzylidene-p-aminoacetophenones and their Thermal Properties.

Serial #	End-Chain	Synthesis & Recryst. Solvent	Transition	T °C	ΔH kcal/mole	ΔS e.u.
A1	$CH_3(CH_2)_2$-	II, m	Cr - N	92.0	7.35	20.1
			N - I	101.7	0.117	0.31
			S_1 - N	90.9	0.731	2.01
			S_2 - S_1	67.6	0.361	1.06
A2	$CH_3(CH_2)_3$-[a]	I, i	Cr - S	82.4	7.16	20.0
			S - N	96.4	0.429	1.16
			N - I	108.8	0.143	0.37
A3	$CH_3(CH_2)_4$-	II, m	Cr - S	80.2	7.52	21.3
			S - N	104.4	0.732	1.94
			N - I	106.3	0.123	0.33
A4	$CH_3(CH_2)_5$-	II, m	Cr - S	70.9	7.72	22.4
			S - N	112.1	0.714	1.85
			N - I	113.3	0.174	0.45
A5	$CH_3(CH_2)_6$-	II, m	Cr - S	80.7	9.57	27.1
			S - I	115.5	1.093	2.81
A6	$CH_3(CH_2)_7$-	II, m	Cr - S_1	70.0	8.99	26.2
			S_1 - I	118.3	1.159	2.96
			S_2 - S_1	51.8	0.118	0.36
A7	$CH_3(CH_2)_9$-	II, m	Cr - S	81.3	10.95	30.9
			S - I	119.8	1.31	3.34
A8	$CH_3(CH_2)_{11}$-	II, m	Cr - S	87.9	13.2	36.6
			S - I	120.4	1.44	3.83

Notes: a - Reported by Castellano et al., Ref. 10; m - methanol; i - isopropanol.

TABLE III. p-Alkoxyalkoxybenzylidene-p-aminoacetophenones and their Thermal Properties.

Serial #	End-Chain	Synthesis & Recryst. Solvent	Transition	T °C	ΔH k/cal mole	ΔS e.u.
B1	CH_3OCH_2-	II, m	Cr - I	94.0	9.56	26.0
B2	$CH_3O(CH_2)_2$-	I, h	Cr - I	93.7	8.14	22.2
			N - I	90.2	0.203	0.56
			S_1 - N	83.1	0.547	1.54
			S_2 - S_1	70.7	0.334	0.97
B3	$CH_3CH_2O(CH_2)_2$-	I, h	Cr - S_1	73.2	7.28	21.0
			S_1 - I	76.6	1.018	2.91
			S_2 - S_1	55.6	0.310	0.94
B4	$CH_3O(CH_2)_3$-	I, h	Cr - S	71.6	6.41	18.6
			S - I	73.5	0.918	2.65
B5	$CH_3(CH_2)_2O(CH_2)_2$-	I, h	Cr - S	47.5	6.41	19.9
			S - I	71.9	0.953	2.77
B6	$CH_3CH_2O(CH_2)_3$-	I, h	Cr - S	58.6	4.73	14.3
			S - I	72.8	0.957	2.77
B7	$CH_3(CH_2)_5O(CH_2)_2$-	I, e	Cr - S	53.2	8.26	25.3
			S - I	72.3	1.106	3.20

Notes: m-methanol; h-hexane; e-hexane, followed by ethanol/water.

A and B, respectively. Homeotropic texture was characteristic of the nematic phases of these compounds, while the smectic phases showed the usual focal-conic morphology.

Also included in the Tables are the measured heats of fusion and heats of transitions, and the derived entropy changes.

In fusion, the variations in the observed thermal quantities, as expected, show little regularity with molecular structure. The melting points of the compounds in series B tend to be lower than the melting points in the alkoxy series, but exceptions do occur. The entropies of fusion vs chain-length are plotted in Fig. 1. In series A, a monotonic increase is observed with increasing chain length - a regularity not shown by series B. A noteworthy point is that compounds in series B, differing only in the position of the ether oxygen, have widely different entropies of fusion, as well as melting points. Interpretation of the thermal quantities related to

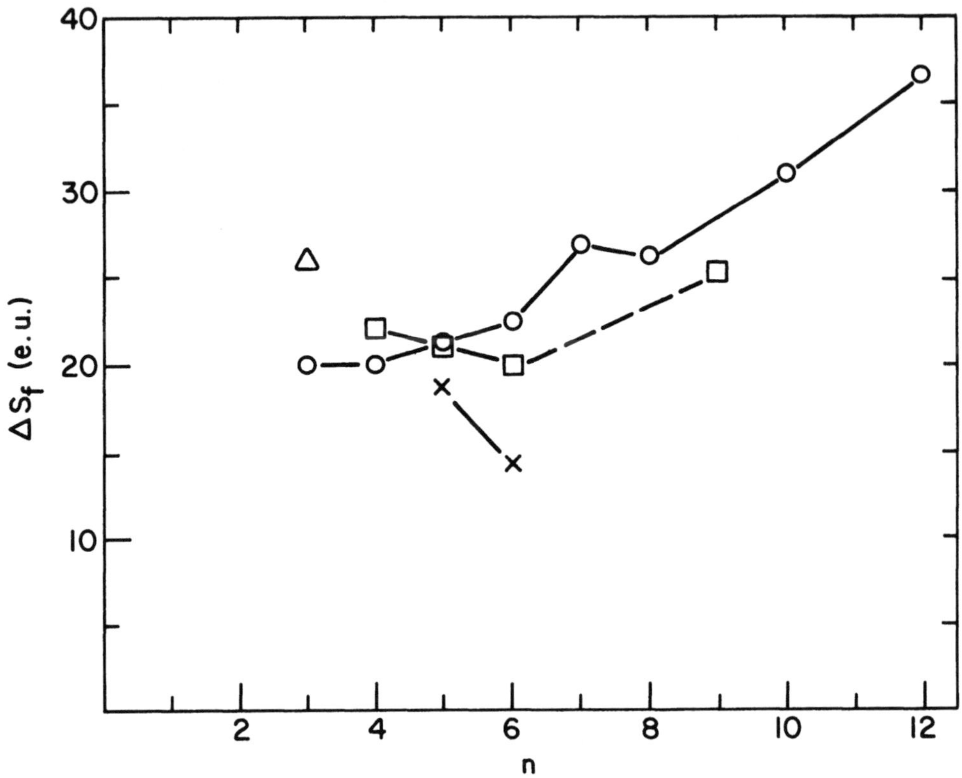

Fig. 1. Entropy of fusion vs chain length, n. For R=$CH_3(CH_2)_m$- the symbol is ○ ,and n=m + 1. For R=$CH_3(CH_2)_pO(CH_2)_q$- : △-- q=1, □--q=2, X -- q=3, and n=p + q + 2.

fusion are hindered by solid-to-solid phase transitions occurring in the investigated temperature range in some of the compounds. Most of the compounds in both series had a strong tendency to supercool from the melt, allowing also the study of monotropic liquid crystalline phases.

The upper transition temperatures for the observed smectic and nematic phases are plotted against chain length in Fig. 2. It is immediately obvious that all transition temperatures of the alkoxy-alkoxy compounds are well below the transition temperatures pertaining to the alkoxy compounds of equal chain length. This is in line with available data[8] on a limited number of alkoxyalkoxy substituted liquid crystals containing different functional groups. If we accept, following Gray,[4] the transition temperature for disappearance as a measure of the stability of the mesophase, the first conclusion is that oxygen substitution destabilizes both the nematic and the smectic liquid crystalline phases.

The transition temperatures for the disappearance of the S_1 phases of the compounds of series A fall on a smooth line with increasing chain length. The plot of the nematic clearing points in the same series shows the usual even-odd alternation and a less rapid overall rise with increasing chain length; it is intersected by the smectic line at n = 7. In contrast to series A, the transition temperatures of the disappearance of the smectic states of the alkoxyalkoxy substituted compounds decrease with increasing chain length. Furthermore, all compounds in series B having an end-chain longer than $CH_3O(CH_2)_2$- exhibit smectic phases only, indicating that the oxygen substitution destabilizes nematic phases to a greater extent than the smectic ones.

Perhaps the most significant fact is the insensitivity of the smectic transition temperatures to the location of the ether oxygen atom in the end-chain. As seen from the results, the smectic-isotropic transition temperatures of compounds B3 and B4, and of compounds B5 and B6, differ only by 3.1 and 0.9°C, respectively. The transition temperatures for the disappearance of the smectic phases of the alkoxy compounds of identical chain length, namely A3 and A4 are, on the other hand, approximately 30 and 40°C higher, respectively. This observation allows the assessment of the relative importance of permanent dipole interactions and dispersion forces in the maintenance of liquid crystalline order.

Assuming for the moment that the conformation of the end-chain is fixed in a zig-zag form independent of oxygen substitution (formulae II, III, and IV), the vector sum of the permanent dipole moments associated with the oxygen atoms in II is smaller than the dipole moment of the single oxygen atom in IV, whereas the vector sum in III is larger than the dipole moment in IV. Similar conclusions are obtained if a fixed cogwheel-type conformation is assumed,

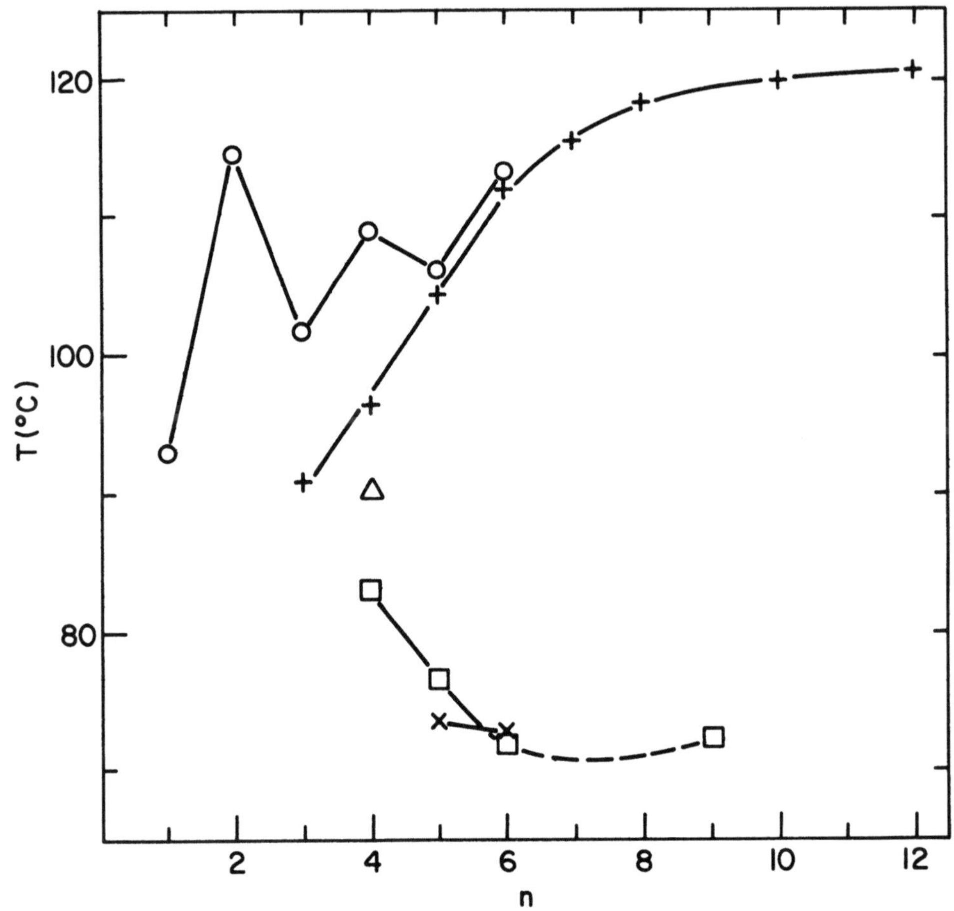

Fig. 2. Upper transition temperature of the mesophases vs chain-length. For $R=CH_3(CH_2)_m-$: ○ -- nematic, + -- smectic. For $R=CH_3(CH_2)_pO(CH_2)_q-$: △-- nematic, □-- smectic (q=2), x -- smectic (q=3). Data for the methoxy and ethoxy compounds are from Ref. 11.

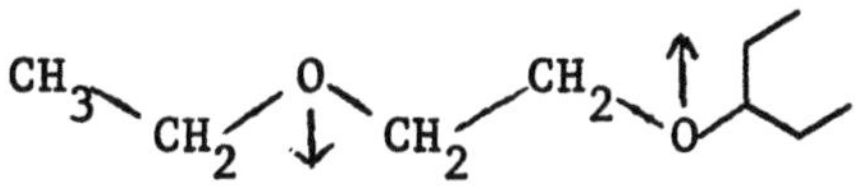

(II)

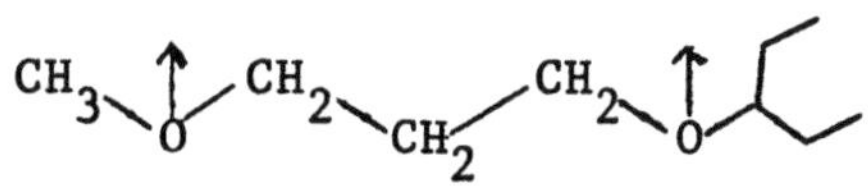

(III)

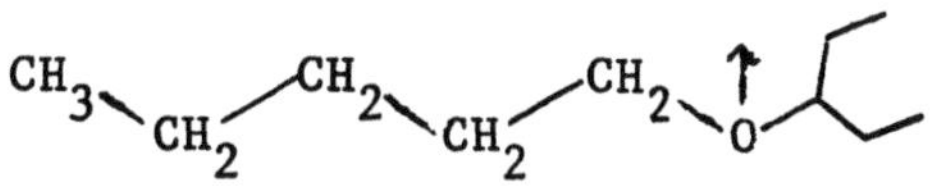

(IV)

and for compounds with longer end-chains. If the assumption of the fixed conformation is relaxed, i.e., the end-chain is allowed to adopt a conformation with the largest intramolecular interaction of the permanent dipoles, then an inspection of models indicates a widely different geometry (and repulsion envelope) for end-chains with oxygen atoms separated by two or three methylene groups, respectively. Under any of the above assumptions, if the permanent dipole - permanent dipole interactions provided a major contribution to the forces responsible for mesophase order, the location of the ether oxygen atom should greatly influence the transition temperature. The experimental results are clearly to the contrary.

The polarizability of the molecule, on the other hand, is reduced uniformly by approximately 1.2 x 10^{-24} cm^3 on substitution of a methylene group by an ether oxygen,[12] independent of location. Dispersion forces, which originate in the polarizability, should be reduced similarly independently of the location of the oxygen substitution. The experimental data, then, lead to the conclusion that dispersion forces play a dominant role in determining liquid crystalline order.

The trends in the entropy difference betwen the isotropic and smectic phases on one hand, and the isotropic and nematic phases on the other, are examined both for series A and B in Fig. 3. To facilitate comparisons in the cases of the smectic phases of those compounds also having a nematic phase, the plotted values represent the sum of the entropy changes observed at the smectic-nematic and nematic-isotropic transitions. No corrections were applied, however, to include specific heat contributions during passage through the nematic state. A general upward trend in entropies is observed with increasing chain length and there are signs of even-odd alternations, although in the case of smectic phases the latter is hardly discernible from the anomalies. As with the transition temperatures, the entropy changes in the alkoxyalkoxy compounds are not sensitive to the location of the ether oxygen atom. It is noteworthy that the entropies of transition in series B are invariably above the entropy changes of the corresponding compounds in series A. This we interpret tentatively as due to the higher entropy of the isotropic liquid in

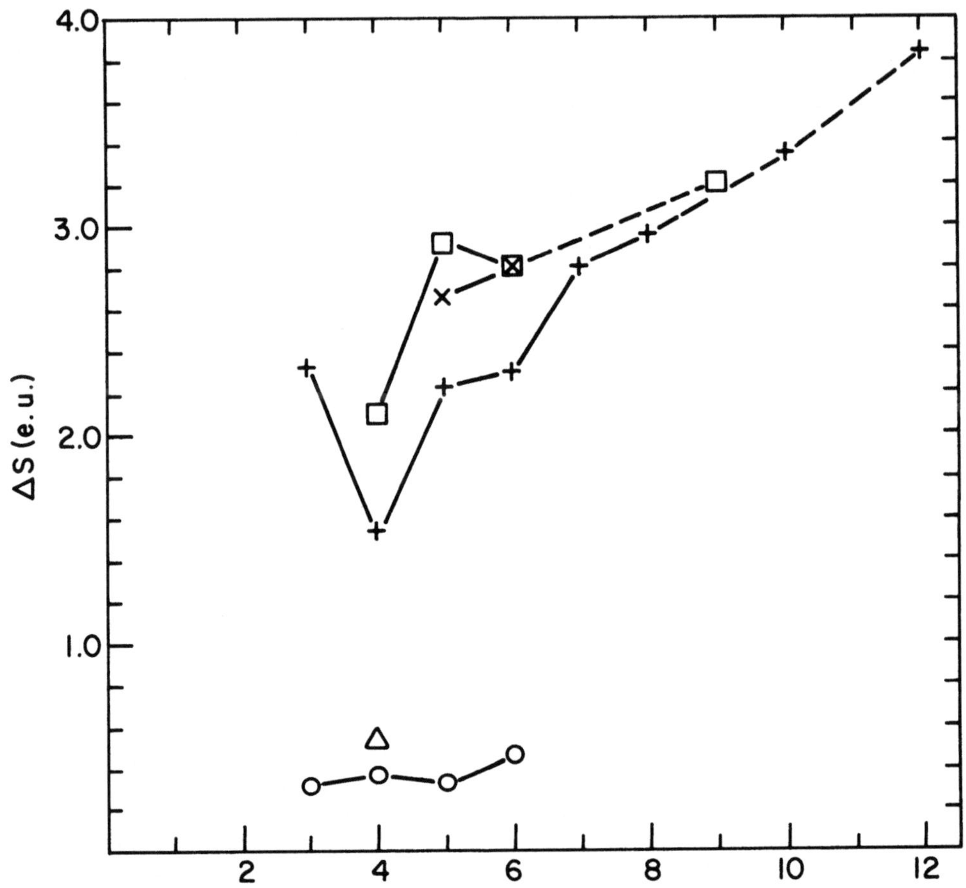

Fig. 3. Entropy difference between the liquid crystalline and the isotropic liquid phase vs chain length. For symbols see caption of Fig. 2.

the case of the alkoxyalkoxy compounds rather than an increased order in the mesophase. This interpretation depends on the notion that internal rotations in the mesophase are prevented by the requirement of packing, but in the normal liquid are hindered by the intramolecular torsional potential only. It is compatible with the fact that the barrier to internal rotations about carbon-oxygen single bonds are lower than about carbon-carbon single bonds.[13,14]

ACKNOWLEDGMENTS

We acknowledge gratefully the assistance of Miss Alice A. Ouye in the syntheses, and of Mr. Harold A. Huggins in the thermal and optical measurements.

REFERENCES

1. M. Born, Ann. Phys. (Leipzig) 55, 221 (1918).
2. W. Maier and A. Saupe, Z. Naturforsch. 14a, 882 (1959); 15a, 287 (1960).
3. A. De Rocco, 2nd Int. Liquid Crystal Conf., Kent, Ohio (1968), Paper 38.
4. G. W. Gray, Molecular Structure and the Properties of Liquid Crystals (Academic Press, New York, 1962).
5. W. R. Young, I. Haller, and L. Williams, 2nd Symp. Ordered Fluids and Liquid Cryst., preceeding paper.
6. Tables of Interatomic Distances, L. E. Sutton, Editor (The Chemical Society, London, 1958).
7. Estimated from data in Landolt-Bornstein, Zahlenwerte und Funktionen (Springer Verlag, Berlin, 1951), 6th Ed., Vol. I, Part 3, p. 415.
8. C. Weygand, R. Gabler, and N. Bircan, J. prakt. Chemie [2] 158, 266 (1941).
9. Analyses were performed by Berkeley Analytical Laboratories, Berkeley, California.
10. J. A. Castellano, J. E. Goldmacher, L. A. Barton, and J. S. Kane, J. Org. Chem. 33, 3501 (1968).
11. R. Walter, Ber. 58, 2304 (1925).
12. Estimated from increments tabulated in A. Bondi, Physical Properties of Molecular Crystals, Liquids, and Glasses (Wiley, New York, 1967), p. 461.
13. L. H. Scharpen and V. W. Laurie, Symp. Mol. Struct. Spectry., Columbus, Ohio (1965), Paper S9.
14. P. H. Kasai and R. J. Myers, J. Chem. Phys. 30, 1096 (1959).

CAPILLARY VISCOMETRY OF CHOLESTERIC LIQUID CRYSTALS

Wolfgang Helfrich

RCA Laboratories, David Sarnoff Research Center

Princeton, New Jersey

ABSTRACT

A recently proposed model of pluglike capillary flow of liquid crystals is studied in some detail for the case of cholesterics. The emphasis is on outlining the conditions under which pluglike flow may be found in circular capillaries. Numerical examples are given, and comparison with the few existing experimental data is made.

I. INTRODUCTION

A few years ago, it was shown by Porter, Barrall, and Johnson[1] that cholesteric and smectic liquid crystals may appear extraordinarily viscous in capillary flow. Viscosities up to 10^6 or 10^3 times larger than in the isotropic state were calculated at low flow rates with Poiseuille's law. At higher speeds of flow there was no marked difference between the mesophase and the isotropic liquid. Viscosities of normal magnitude may be expected if the flow velocity is perpendicular to the helical axis of the cholesteric state or to the unique axis of the smectic state. The excessive values were tentatively attributed to yield stress.[1] Recently[2] we pointed out that capillary flow of cholesteric and smectic mesophases may be pluglike, the velocity profile being flat rather than parabolic, and that this mechanism possibly explains high viscosity values. In the following we intend to give a more detailed treatment of pluglike flow of cholesteric liquid crystals, outlining some of the conditions which must be met for its observation.

II. BASIC FEATURES OF PLUGLIKE FLOW

We consider a cholesteric liquid crystal in a circular capillary whose helical axis is parallel to the capillary axis. To begin with, the orientation pattern is thought to be strictly immobile and rigid. Accordingly, any flow through the capillary requires molecular rotation which produces the torque per unit volume

$$-(\kappa_1 + \kappa_2)\, d\varphi_u/dt \tag{1}$$

The angle φ_u denotes the local direction of the long axes of the molecules corresponding to the unique axis of nematics and d/dt is the material derivative with respect to time. The proportionality constant $(\kappa_1 + \kappa_2)$ is discussed below. If there is only flow in the direction of the capillary axis, denoted as z axis, one has

$$d\varphi_u/dt = (2\pi/P_o)v_z$$

where v_z is the flow velocity and P_o the pitch of the helix. The energy ϵ dissipated cm^3 and sec is

$$\epsilon = (\kappa_1 + \kappa_2)\left(\frac{2\pi v_z}{P_o}\right)^2$$

It must be equal to the work done by the force p', where p' = -dp/dz is the negative pressure gradient. Therefore,

$$v_z = \frac{1}{(\kappa_1 + \kappa_2)}\left(\frac{P_o}{2\pi}\right)^2 p' \tag{2}$$

The ratio v_z/p' has been called "permeability".[2] The total flux per sec in z direction is

$$Q = \left(\frac{P_o}{2\pi}\right)^2 \frac{R^2\pi}{(\kappa_1 + \kappa_2)}\, p' \tag{3}$$

where R is the radius of the capillary. For comparison, Poiseuille's law reads

$$Q = \frac{\pi R^4}{8\eta}\, p', \tag{4}$$

with η being the viscosity. Eq. (3) can be written in the form of Eq. (4) if we introduce the apparent viscosity

$$\eta_{app} = (\kappa_1 + \kappa_2)\frac{R^2\pi^2}{2P_o^2} \tag{5}$$

This simple description of pluglike flow was given before.[2] We will now discuss some of its aspects in greater detail.

III. CIRCULAR FLOW IN CIRCULAR CAPILLARIES

Unless a capillary is rectangular, with one side much longer than the other, the shear stress produced by the molecular rotation will produce shear flow. In the circular capillary the forward motion is accompanied by a circular flow. We describe it in a right-handed cylindrical coordinate system z, r, φ. The z axis coincides with the center line of the capillary and r is the distance from it. For symmetry reasons, the velocity $\underline{v}$ of stationary flow must obey the relations

$$v_z(z + \Delta z, r, \varphi + \Delta z/P_o) = v_z(z, r, \varphi)$$

$$v_r(z + \Delta z, r, \varphi + \Delta z/P_o) = v_r(z, r, \varphi)$$

$$v_\varphi(z + \Delta z, r, \varphi + \Delta z/P_o) = v_r(z, r, \varphi)$$

Note that the pitch is regarded as positive for right-handed screws. Similarly any periodic part $\hat{p}$ of the hydrostatic pressure will satisfy

$$\hat{p}(z + \Delta z, r, \varphi + \Delta z/P_o) = \hat{p}(z, r, \varphi)$$

The total pressure may be written as

$$p = \hat{p} - p'z$$

All these variables are periodic in z, the period being $P_o/2$ if there is no polarity. The nonperiodic part of the velocity consists of a uniform forward flow v_z and a circular flow v_φ which depends on r. Considering an imaginary cylinder of radius r around the capillary axis one finds for the average tangential stress $\bar{t}_{r\varphi}$ exerted on the cylinder wall by the molecular rotation

$$\bar{t}_{r\varphi} = -\left[(\eta_1 - \eta_2 + \kappa_2)\overline{\cos^2\psi} + (\eta_2 - \eta_1 + \kappa_1)\overline{\sin^2\psi}\right]\frac{2\pi v_z}{P_o} = -\frac{1}{2}(\kappa_1 + \kappa_2)\frac{2\pi v_z}{P_o}$$

provided the periodic part of the flow can be neglected. Here ψ is the angle between the local alignment of the long axes of the molecules and the radius vector. The quantities $-(\eta_1 - \eta_2 + \kappa_2)(2\pi v_z/P_o)$ and $-(\eta_2 - \eta_1 + \kappa_1)(2\pi v_z/P_o)$ are the stresses where the alignment is perpendicular and parallel to the cylinder wall, respectively. They contain the viscosity coefficients (η) and the shear-torque coefficients (κ) for alignment parallel to the velocity gradient (subscript 1) and to the velocity itself (subscript 2).[3,4,5] The stress results in circular shear flow, the shear rate being

$$S = \frac{\kappa_1 + \kappa_2}{\eta_1 + \eta_{12}/2 + \eta_2} \frac{2\pi v_z}{P_o}$$

We use the fact that the average viscosity is $\eta_1\overline{\cos^2\psi} + \eta_{12}\overline{\cos^2\psi}\ \overline{\sin^2\psi} + \eta_2\overline{\sin^2\psi}$ where η_{12} is another viscosity coefficient. Consequently, the velocity of circular flow is

$$v_\varphi = -\frac{\kappa_1 + \kappa_2}{\eta_1 + \eta_{12}/2 + \eta_2}\, r \log\left(\frac{R}{r}\right) \frac{2\pi v_z}{P_o} \qquad \text{(in cm sec}^{-1}\text{)}$$

Note that for $R \gg P_o$ one has $v_\varphi \gg v_z$ for most r. The local energy dissipation by the combined molecular rotation and shear flow is

$$\begin{aligned} \epsilon(\psi) = {} & (\eta_1\cos^2\psi + \eta_{12}\cos^2\psi \sin^2\psi + \eta_2\sin^2\psi)(\omega - S)^2 \\ & + (\eta_2\cos^2\psi + \eta_{12}\cos^2\psi \sin^2\psi + \eta_1\sin^2\psi)\, \omega^2 \\ & - (\eta_1 + 2\eta_{12}\cos^2\psi \sin^2\psi + \eta_2 - \kappa_1 - \kappa_2)(\omega - S)\omega \end{aligned}$$

where $\omega = 2\pi v_z/P_o$ is the angular velocity of the molecules. The average is easily found to be

$$\overline{\epsilon(\psi)} = (\kappa_1 + \kappa_2)\left(1 - \frac{1}{2} \frac{\kappa_1 + \kappa_2}{\eta_1 + \eta_{12}/2 + \eta_2}\right)\left(\frac{2\pi v_z}{P_o}\right)^2$$

Thus one obtains for the total flux

$$Q = \left[(\kappa_1 + \kappa_2)\left(1 - \frac{1}{2} \frac{\kappa_1 + \kappa_2}{\eta_1 + \eta_{12}/2 + \eta_2}\right)\right]^{-1} \frac{P_o^2 R^2}{4\pi}\, p' \tag{6}$$

and the apparent viscosity

$$\eta_{app} = (\kappa_1 + \kappa_2)\left(1 - \frac{1}{2} \frac{\kappa_1 + \kappa_2}{\eta_1 + \eta_{12}/2 + \eta_2}\right) \frac{R^2\pi^2}{2P_o^2} \tag{7}$$

An extra pressure $\hat{p}(\psi)$ can be chosen such as to satisfy the requirement of local energy conservation. It will be seen that $\hat{p}(\psi)$ as well as the periodic part of the shear produce radial and tangential forces that are not balanced in the idealized flow pattern here considered. Inspection shows, however, that their effect is negligible for $r \gg P_o/2$. In other words, except for the core of the capillary ($r \lesssim P_o$) the flow pattern and the permeability are adequately described by the simplified model. Eqs. (6) and (7) should be applicable whenever the radius of the capillary is much larger than the pitch of the cholesteric helix.

IV. BALANCE OF TORQUES AT THE CAPILLARY WALL

In stationary flow the frictional torque has to be matched by an elastic torque in order to conserve angular momentum. This holds for any volume element of the liquid crystal as well as for the capillary wall. The distortion of the orientation pattern necessary to balance the elastic torque will be discussed in the next section. Here we deal with the balance of torques at the wall and some related problems such as wall alignment. We will continue to assume a rigid orientation pattern but explore the conditions for its immobility.

The question of mobility raises difficult problems, practically as well as mathematically. Ideally, a fixed wall alignment matching the undistorted cholesteric structure would be desirable. Since the wall alignment is sensitive to surface treatment, there is perhaps a chance to approach this situation in the case of flat capillaries by providing the inner sides of the broad faces with a suitable grating. The orientation pattern should also be held in place by an irregular wall alignment because a certain minimum energy will be required to break it loose and put it in motion. The maximum distortional couple stress τ_{max} sustainable by a cm^2 of the wall may be estimated to be

$$\tau_{max} \approx na \approx 10^2 \text{ g sec}^{-2} \qquad (8)$$

where $n(\approx 10^{14}\ cm^{-2})$ is the molecular surface density and $a(\approx 1$ eV) the difference in adhesional energy between "right" and "wrong" molecular alignment.

The couple stress actually sustainable by the wall is likely to be smaller in most cases. However, as will be seen below, the stability of the orientation pattern imposes another limit on the couple stress which is more stringent.

In pluglike flow there is always a boundary layer in which the velocity v_z is determined by viscous shear since on the capillary wall $\underline{v}$ should be zero as in isotropic liquids. The approximate thickness ℓ of the layer is given by the condition that at $r = R - \ell$ Poiseuille flow has the same velocity as pluglike flow. Disregarding the anisotropy of the viscosity one obtains

$$\ell \approx \frac{2\eta}{R} \frac{v_z}{p'}$$

Of course, it is always assumed that $\ell \ll R$ so that the effect of the layer on the total flux is negligible with ideal and fixed wall alignment. However, the boundary layer of Poiseuille flow is important if the molecular alignment at the walls is unrestrained. In this case it prevents the plug from falling through the tube without any internal or rotary motion. The kind of pluglike flow which should be encountered in the absence of restraints on the alignment is considered next.

As before we make the assumption that the variation of the flow within a half pitch is negligible. Then the forward velocity $v_z(r)$ in a capillary of circular cross section obeys the differential equation

$$-\frac{\eta_1 + \eta_3}{2}\left(\frac{d^2 v_z}{dr^2} + \frac{1}{r}\frac{dv_z}{dr}\right)$$

$$+ (v_z - V_z)(\kappa_1 + \kappa_2)\left(1 - \frac{1}{2}\frac{\kappa_1 + \kappa_2}{\eta_1 + \eta_{12}/2 + \eta_2}\right)\left(\frac{2\pi}{P_o}\right)^2 = p'$$

Here V_z is the velocity of the orientation pattern and η_3 the viscosity if the local alignment of the molecules is perpendicular to the plane of shear. The boundary conditions are $dv_z(0)/dr = 0$ and $v_z(R) = 0$. Apart from additive constants the solution is a Bessel function. Taking account of the first boundary condition, one obtains

$$v_z = V_z + v_{z,o} - C\, I_o(ar), \qquad C > 0 \tag{9}$$

where

$$v_{z,o} = \left[(\kappa_1 + \kappa_2)\left(1 - \frac{1}{2}\frac{\kappa_1 + \kappa_2}{\eta_1 + \eta_{12}/2 + \eta_2}\right)\right]^{-1}\left(\frac{P_o}{2\pi}\right)^2 p'$$

and

$$a^2 = \frac{2(\kappa_1 + \kappa_2)}{\eta_1 + \eta_3}\left(1 - \frac{1}{2}\frac{\kappa_1 + \kappa_2}{\eta_1 + \eta_{12}/2 + \eta_2}\right)\left(\frac{2\pi}{P_o}\right)^2, \qquad a > o$$

For $r \gg P_o$ the argument of I_o will be much larger than unity, allowing use of the asymptotic formula

$$I_o(ar) \approx e^{ar}(2\pi ar)^{-1/2} \tag{10}$$

We also note

$$I_o(0) = 1, \quad dI_o(ar)/dr > 0 \quad \text{for all } r$$

The second boundary condition, $v_z(R) = 0$, yields with (9) and (10)

$$0 \approx V_z + v_{z,o} - C \exp(aR)(2\pi aR)^{-1/2} \tag{11}$$

Before calculating V_z, one has to determine C from the relation

$$\int_o^R 2\pi r v_z(r)\,dr - V_z \pi R^2 = 0 \tag{12}$$

which reflects the fact that the total torque exerted on the orientation pattern must be zero if no torque is sustained by the capillary wall. Because of (10), one has for $R \gg P_o$,

$$\int_o^R I_o(ar)\, 2\pi r\, dr \approx (2\pi R)^{1/2}\, a^{-3/2}\, e^{aR}$$

Accordingly, Eq. (12) leads to

$$C \approx v_{z,o} \left(\frac{\pi}{2}\right)^{1/2} (Ra)^{3/2}\, e^{-aR}$$

With (11), one obtains for the velocity fo the orientation pattern

$$V_z \approx v_{z,o}\left(\frac{aR}{2} - 1\right) \approx v_{z,o}\,\frac{aR}{2}$$

How good is this result? Accuracy may be expected only if $a \ll 1/P_o$ which should be rare since $a \approx 1/P_o$ is likely to be the rule. If this condition is not met the original assumption of a flow varying negligibly over the distance $P_o/2$ is not valid. However, qualitatively we can draw the important conclusion that the flow is pluglike except for a boundary layer of the approximate thickness 1/a. The velocity over most of the capillary cross section is roughly equal to that of the orientation pattern. It is larger than with a fixed orientation pattern by the approximate factor R/P_o and smaller by about the same factor than in Poiseuille flow (4), provided $\kappa_1 + \kappa_2, \eta_1 + \eta_3, \eta_1 + \eta_{12}/2 + \eta_2$ are all of the same order of magnitude and, of course, $R >> P_o$.

It is questionable whether there are interfaces which do not influence the alignment. However any interface will sustain only a limited distortional torque. Larger torques should loosen the orientation pattern. Pluglike flow with fixed orientation pattern ($v_z = v_{z,o}$) may therefore abruptly turn into pluglike flow with slippage of the orientation pattern ($v_z \approx v_{z,o} aR/2$) as p' rises above a critical value. A special case which we only mention is uniform wall alignment. It would be associated with disclinations, i.e. linear alignment discontinuities, spiraling along the wall with the pitch of the cholesteric helix. If the disclinations are directly on the wall[6] the orientation pattern can be shown to be practically immobile under most circumstances because of the large energy required to move the disclinations through the fluid at rest. More probably, the disclinations are at a distance of about $P_o/2$ from the interface. This should result in pluglike flow with slippage of the orientation pattern, similar to the case without restraints.

Another problem in this context is leakage. By this we mean an irregular Poiseuille type flow in a layer at the capillary wall despite a fixed orientation pattern. It will occur if irregular wall alignment causes the orientation pattern near the wall to differ from that in the bulk so that flow with little or no molecular rotation is possible. Since the orientation pattern is fixed, such leakage may approximately be described as Poiseuille flow between two concentric cylinders. The total flux would be proportional to the third power of the layer thickness d. An estimate shows that it should be negligible for $d^3R << P_o^2R^2$. This may be simplified to give $d << P_o^{2/3}R^{1/3}$. With $P_o << R$ it is therefore more than sufficient to have $d \approx P_o$.

V. DEFORMATION OF THE ORIENTATION PATTERN

The orientation pattern is of course not quite rigid under flow. Instead, it is likely to be rather pliable in view of the smallness of the moduli of curvature elasticity.[7] In order to withstand the pressure force $p' = -dp/dz > 0$, the orientation pattern deforms in such a way as to exert on the material the force per cm^3

$$\bar{f}_z = - p'$$

acting in the negative z direction. The forward flow in the stationary twisted orientation pattern implies a screwing motion in z direction. Consequently there is a torque per unit volume

$$\bar{m}_h = -\frac{P_z}{2\pi}\, p'$$

around the helical axis $\underline{h}$. With deformation the helical axis need not be parallel to the capillary axis. However, the second formula is applicable only if there is no conical deformation of the helix (see below). P_z is the thickness of the cholesteric double layer as measured in z direction. It may differ from P_o but for geometrical reasons it must be uniform all over the cross section of the capillary. With circular flow both force and torque are averages over $P_z/2$. The pressure force will result in a depression of the cholesteric layers in the direction of flow. We assume that in the undeformed state $\underline{h} \parallel z$, so the depression is deepest in the center of the capillary. The deformation of the orientation pattern may also be regarded as a distortion in the planes normal to the capillary axis.

Assuming that there is no conical deformation of the helix, we distinguish between the local tilt γ of the cholesteric layers, i.e. of the layers of constant alignment, and the local tilt γ_h of the helical axis. (Both tilts are taken positive.) The angle γ_h as a function of γ is calculated for the special case $k_{11} = k_{33}$ by minimizing the free distortional energy

$$g = \frac{1}{4}(k_{11} + k_{33})\left(\frac{2\pi \sin(\gamma - \gamma_h)}{P_z \cos\gamma}\right)^2$$

$$+ \frac{1}{2}k_{22}\left(\frac{2\pi \cos(\gamma - \gamma_h)}{P_z \cos\gamma} - \frac{2\pi}{P_o}\right)^2$$

Here k_{11}, k_{22}, and k_{33} are Frank's[7] elastic moduli for splay, twist, and bend, respectively. Variations of the pitch within $P_z/2$ and terms arising from the variation of γ with r are neglected, which should be permissible for $P_o \ll R$. In an immobile orientation pattern one has $P_z = P_o$ even with flow, so γ_h may be obtained from minimizing

$$-\left[k_{22} - \frac{1}{2}(k_{11} + k_{33})\right]\sin^2(\gamma - \gamma_h) - 2\,k_{22}\cos(\gamma - \gamma_h)\cos\gamma$$

by varying $(\gamma - \gamma_h)$. It is easily seen that for $(1/2)(k_{11} + k_{33}) \geq k_{22}$ the solution is always $\gamma_h = \gamma$, implying that the helical axis is generally normal to the layers. For $(1/2)(k_{11} + k_{33}) < k_{22}$ the solution may be

$$\cos(\gamma - \gamma_h) = \left[1 - (k_{11} + k_{33})/2k_{22}\right]^{-1} \cos\gamma\,,$$

particularly if k_{22} is much larger than the other elastic moduli, but at sufficiently small γ one will again have $\gamma_h = \gamma$. On the other hand, the helical axis can be kept in the direction of the capillary axis ($\gamma_h = 0$) by means of electric or magnetic fields, if the polarizability is of the proper anisotropy. The distortional free energies are

$$g = \tfrac{1}{2} k_{22} \left(\frac{2\pi}{P_o}\right) \left(\frac{1}{\cos\gamma} - 1\right)^2 \quad \text{for } \gamma_h = \gamma \tag{13}$$

and, again for the special case $k_{11} = k_{33}$,

$$g = \frac{1}{4} (k_{11} + k_{33}) \left(\frac{2\pi}{P_o}\right)^2 \tan^2\gamma \quad \text{for } \gamma_h = 0 \tag{14}$$

Eqs. (13) and (14) permit the calculation of γ as a function of r and p′. The stress $t_{rz} = -dg/d\tan\gamma$ linked with the distortion is

$$t_{rz} = -k_{22} \left(\frac{2\pi}{P_o}\right)^2 \left(\frac{1}{\cos\gamma} - 1\right) \sin\gamma \quad \text{for } \gamma_h = \gamma \tag{15}$$

and

$$t_{rz} = -\tfrac{1}{2} (k_{11} + k_{33}) \left(\frac{2\pi}{P_o}\right)^2 \tan\gamma \quad \text{for } \gamma_h = 0 \tag{16}$$

The integrated stress exerted on a cylinder of constant r must balance the force acting on the material in the cylinder. Therefore,

$$2\pi r t_{rz} + \pi r^2 p' = 0 \tag{17}$$

which together with (15) or (16) gives γ.

It can be shown that the distorted orientation pattern of the type $\gamma = \gamma_h$ becomes unstable, tending to collapse by a common re-alignment in the direction of the helical axis, i.e., through a conical deformation of the helix, if the twist is about twice its natural value. The limit is reached at exactly $\cos\gamma = \frac{1}{2}$ for $P_z = P_o$, $k_{11} = k_{22} = k_{33}$. The stability of a distorted orientation pattern of the type $\gamma_h = 0$ should rise with the strength of the supporting field, so it can possibly be made quite large. Confining ourselves to the first case (no fields) we derive from (15) and (17) and $\cos\gamma = \frac{1}{2}$ the maximum pressure force p'_{max} sustainable by the orientation pattern

$$p'_{max} \approx k_{22} \left(\frac{2\pi}{P_o}\right)^2 \sqrt{\frac{3}{4}}\, \frac{2}{R} \tag{18}$$

Insertion of $k_{22} = 1\cdot10^{-6}$ dyne, $P_o = 3000$ Å, and $R = 0.01$ cm yields $p'_{max} = 8\cdot10^6$ dyne cm^{-3}. The corresponding couple stress exerted on the capillary wall (or on the layer 1/a) is roughly[8]

$$\tau \approx \frac{P_o}{2\pi} p'_{max} \frac{R}{2} \approx 2\cdot 10^{-1} \text{ g sec}^{-2}$$

which is much more restrictive than the maximum τ_{max} derived above in (8). Of course, the flux may start deviating from its ideal behavior (6) before this considerable deformation is reached. Combining Eqs. (3) and (18) one obtains a simplified formula for the maximum possible total flux Q_{max} in the regime of pluglike flow:

$$Q_{max} \approx \sqrt{\frac{3}{4}}\, 2\pi \frac{k_{22} R}{\kappa_1 + \kappa_2}$$

This important limit does not depend on the pitch.

In order to compare with the results of Porter et al.[1] measured on cholesteryl acetate and with R = 0.0375 cm we take $k_{22} = 1\cdot 10^6$ dyne and $\kappa_1 + \kappa_2 = 3\cdot 10^{-2}$ g cm^{-1} sec^{-1}, both from p-azoxyanisole. The result is $Q_{max} = 7\cdot 10^{-6}$ cm^3 sec^{-1}. Use of the more exact Eq. (6) instead of (3) should raise the result, perhaps to $1\cdot 10^{-5}$ cm^{3}sec^{-1}. This is about an order of magnitude less than the smallest flux in these experiments which was $2\cdot 10^{-4}$ cm^3 sec^{-1}, as deduced from the given shear rate of 10 sec^{-1}. The difference is rather large, but the estimate is necessarily crude and Porter et al. pointed out that even at this low speed of flow (ca. 0.4 cm sec^{-1}) the apparent viscosity had not yet reached a maximum. It has been shown earlier[2] that with the above values for $\kappa_1 + \kappa_2$ and R and with P_o = 3000 Å an apparent viscosity of $2\cdot 10^5$ poise is calculated for Eq. (5). This is in good agreement with the experimental result.[9]

VI. ALIGNMENT OF THE HELICAL AXIS

Up to now it has been supposed that in the center of the capillary the helical axis is always parallel to the capillary axis. This "symmetrical" configuration may be rare in freshly filled capillaries; instead the direction of the helical axis is likely to vary from place to place. If the variation is such that no particular direction predominates one would expect the flow to be controlled by permeation almost as much as in the symmetrical case. However, the flow itself will influence the orientation pattern. Some arguments are given in Appendix A which indicate that the orientation pattern, if broken loose from the walls, may tend to take the symmetrical configuration. Accordingly, there is a possibility that the helical axis turns into the direction parallel to the capillary axis if the flow is just strong enough to make the orientation pattern slip. When the flow stops the newly formed orientation pattern may become fixed in its instantaneous position by the action of a nonuniform wall alignment. It would then be possible, at not too large pressure gradients, to measure permeation in the symmetrical configuration and with an immobile orientation pattern. A more reliable way to establish the symmetrical configuration is, with suitable materials, the use of an electric or magnetic field.

As mentioned above, the cholesteric twisted structure cannot be distorted indefinitely. Above a certain pressure gradient it breaks down, at first at the capillary walls, permitting the orientation pattern to slip. The maximum cross section of a cylinder of "symmetrical" alignment becomes smaller as the pressure is increased further. One may speculate that twisted but strongly misaligned regions form between such a cylinder and the walls. Eventually, the helical axis may be mostly parallel to the radius vector, or flow alignment as known from nematic liquid crystals might dominate.

VII. CONCLUDING REMARKS

The mechanism of pluglike flow was proposed in order to explain the curious experimental results of Porter, Barrall, and Johnson[1] and of some other workers.[1] The approximate agreement of theory and experiment indicates that the model may be useful in the understanding of capillary flow, at least in the case of cholesteric liquid crystals. However, the experimental data are too few and too incomplete to allow firm conclusions. Likewise, the theory is sketchy and does not take account of all possibilities. For instance, we did not consider disclinations as a potential means to relax a strained orientation pattern which is not strained enough to transform without their intervention. (The formation of disclinations at the wall is possibly a transient phenomenon because a disclination source, as represented by some alignment irregularity on the wall, may be expected to be exhausted after emitting one disclination loop.) The measurement of the dependence of total flux on capillary radius or the optical observation of the velocity profile appear to be attractive ways to check the theory.

APPENDIX A

Here we wish to argue that an orientation pattern that is not attached to the walls but slips on all sides tends to assume the symmetric configuration which is characterized by the helical axis being parallel to the capillary in the center of the tube. We consider a cholesteric sample whose helical axis is uniformly at an angle $\gamma_o < \pi/2$ from the capillary axis. The cholesteric layers which now are tilted will be depressed if a pressure is applied. Taking $k_{22} \leq k_{11} = k_{33}$ and denoting the local tilt by γ one has for the distortional energy density [see Eq. (13)].

$$g = \frac{1}{2} k_{22} \left(\frac{2\pi}{P_z} \frac{1}{\cos\gamma} - \frac{2\pi}{P_o} \right)^2$$

The problem under consideration is essentially static and P_z may be expected to adjust in such a way as to minimize the total distortional energy. However, the adjustment takes time and for a certain period after turning on the flow, which increases with the length of the capillary, the original value $P_z = P_o/\cos\gamma_o$ should be a good approximation. One derives for the stress t_{tz} on the axis of maximum downstream slope (the axis normal to and cutting the z axis is considered)

$$t_{tz} = \frac{dg}{d\tan\gamma} = k_{22} \left(\frac{2\pi}{P_z} \right)^2 \left(\frac{1}{\cos\gamma} - \frac{1}{\cos\gamma_o} \right) \sin\gamma$$

The derivative

$$\frac{d\,t_{tz}}{d\tan\gamma} = -k_{22} \left(\frac{2\pi}{P_z} \right)^2 \left(1 - \frac{\cos^3\gamma}{\cos\gamma_o} \right)$$

which may be called an effective local force constant rises with γ, at least for small $|\gamma - \gamma_o|$. Greater stiffness of the upstream than the downstream half of the cholesteric layer means that the edge of the former sustains the larger part of the total force exerted by p' on the layer. The resulting faster slippage of the upstream half will reduce the tilt, provided it does not lead to oscillations. The effect is possibly enhanced by an anisotropic viscosity since η_3 may be expected to be smaller than η_1, as is the case in p-azoxyanisole.

A complete treatment of the stability of symmetric alignment would be much more complicated, involving two dimensions (r and φ) and consideration of the inflexion of $t_{tz}(\gamma)$ at $\gamma = 0$. However, there is little doubt that in the case $k_{22} \leq k_{11}, k_{33}$ slip flow should tend to produce the symmetric configuration. Some tentative explorations suggest that for very large k_{22}, as compared to k_{11} and k_{33}, the situation may be reverse.

REFERENCES AND FOOTNOTES

1. R. S. Porter, E. M. Barrall II, and J. F. Johnson, J. Chem. Phys., 45, 1452 (1966).
2. W. Helfrich, Phys. Rev. Letters, 23, 372 (1969).
3. J. L. Ericksen, Arch. Rational Mech. Anal., 4, 231 (1960).
4. F. M. Leslie, Proc. Roy. Soc., A307, 359 (1968).

5. Ericksen and Leslie use six viscosity coefficients (α or μ). We prefer to distinguish between four viscosity (η) and two shear-torque coefficients (κ). (They have already been used in a forthcoming theory of conduction-induced alignment.)

   The two sets of coefficients are related as follows:

$$\eta_1 = \tfrac{1}{2}(\mu_4 + \mu_5 - \mu_2) \qquad \eta_2 = \tfrac{1}{2}(\mu_4 + \mu_6 + \mu_3)$$

$$\eta_{12} = \mu_1 \qquad \eta_3 = \tfrac{1}{2}\mu_4$$

$$\kappa_1 = \tfrac{1}{2}(\mu_5 + \mu_3 - \mu_2 - \mu_6) \qquad \kappa_2 = \tfrac{1}{2}(\mu_6 + \mu_3 - \mu_5 - \mu_2)$$

   The sum $\kappa_1 + \kappa_2$ equals the coefficient $-\lambda_1$ of Ericksen and Leslie.

6. There are some indications in the older literature of disclinations bound to glass walls.
7. F. C. Frank, Disc. Faraday, 25, 19 (1958) or Ref. 4. A fourth constant occurring in curvature elasticity, k_{24}, is neglected because it affects the distortional energy only through the surface alignment, as was first shown by J. L. Ericksen, Arch. Rational Mech. Anal., 10, 189 (1962).
8. In the considered deformation of the orientation pattern, the deformational torques contain components normal to the capillary wall. These components must also be balanced. With fixed wall alignment, no balancing by shear is required as their integral over $0 < \varphi < 2\pi$ vanishes. An analysis of the orientation pattern at the wall appears difficult. Shear is involved if the wall alignment is unrestrained, complicating the situation even more.
9. R. S. Porter informed us of recent experiments which indicate that pure cholesteryl acetate may have only a smectic but no cholesteric mesophase. However, he is convinced that the impure cholesteryl acetate used in the viscosity measurements of Porter et al. (Ref. 1) was cholesteric.

KINETIC STUDY OF THE ELECTRIC FIELD-INDUCED CHOLESTERIC-NEMATIC TRANSITION IN LIQUID CRYSTAL FILMS: 1. RELAXATION TO THE CHOLESTERIC STATE

J. J. Wysocki, J. E. Adams and D. J. Olechna

Xerox Research Laboratories

Rochester, New York

ABSTRACT

Relaxation of cholesteric liquid crystals from the electric-field-induced nematic state back to the cholesteric state was studied by measuring the transmission of polarized light through the sample. Variables included time, temperature, wavelength and bias. The relaxation transient was found to include optical rotation which corresponded to the optical sense of the material. The rotatory dispersion furthermore was correctly specified by deVries' formula. The results point out the significance of the basic cholesteric parameters of pitch and sense. An attempt is made to relate the relaxation phenomena to the molecular system.

I. INTRODUCTION

Application of an intense electric-field to a cholesteric liquid-crystal film causes a phase transformation in which the optically negative material is converted into an optically positive state.[1,2] The transformation is reversible; upon removal of field or its reduction below a threshold value, the material relaxes back to the cholesteric state. The optically positive state can be either smectic or nematic in nature. While it is not easy to determine experimentally which of these forms is the correct one, some evidence[3,4] and a consideration of mesophase morphology suggest that the transformed state is nematic.

The transformation is an interesting one, for it connects two of the three mesophases in a fashion previously unknown. Thus, new information on the internal nature of these materials can hopefully be obtained. For example, from the curvature-elasticity theory developed by Frank [5] it was shown [6,7] that the field dependence of the transformation is determined by the dielectric anisotropy, elastic moduli, and pitch of the helical arrangement in the material. Some of these predictions have already been confirmed by experimental results.[4,8]

We felt that additional information on the nature of cholesteric and nematic materials could be obtained from a kinetic study of the transformation and its subsequent relaxation. Accordingly, we began a study of the changes in polarized light transmission which occur when an electric field is applied or removed from the material. It is to be understood of course, that this study, while meaningful in terms of molecular properties and arrangements, does result in data which can not as yet be directly related to the molecular system. Changes in transmission may be due to changes in birefringence, optical activity, absorption, and scattering; each of these can be related in principle to the molecules and their arrangement. But it is not trivial to sort out each effect and to find its relationship to the molecular system. Thus, only inferences on the internal nature of the system can be made at present from the data. These inferences are, however, quite valid and useful. For example, the question arose as to the nature of the driving force for the transformation.[9] Data from a kinetic study of the transformation provided the following information:

(1) The threshold field was independent of polarity. Once achieved, the transformation was unaffected by reversals in field polarity.

(2) Fastest recovery to the cholesteric state was achieved when the field was removed (i.e., the sample was shorted) rather than reduced below threshold to a field of either polarity.

(3) Use of low-frequency a.c. fields also caused the material to transform. Superimposed on the transformation transient was a component of light transmission which varied at double the applied frequency. When the material was completely transformed, the alternating component decreased to a very low value.

These observations are consistent with a driving force resulting from induced rather than permanent dipoles, and it was so concluded.

We will cover here primarily the case of the relaxation transient; i.e., the transformation having been achieved, the field is then removed and the material relaxes back to the cholesteric state. Future publications will cover the transformation transient.

II. EXPERIMENTAL CONDITIONS

A. Experimental Procedure

All measurements were made on material in the sample configuration shown in the upper part of Fig. 1, using the schematic apparatus shown in the lower part. The sample consisted of two glass plates coated with conducting tin oxide and separated by a formed insulating spacer. Liquid crystal material was put into the spacer void. The assembly was then made rigid by epoxying the outer edges together. It was found in early work[3] that electrostatic squeezing due to the applied fields did not play a significant role in the material's behavior.

The spacer was 1-mil Mylar. Because of spread of liquid crystal during insertion, the final cell thickness was typically 50% greater; i.e., 1-1/2 mils. The total area of cell assembly was 3.6 cm while the active area containing only liquid crystal was $\approx 0.1 cm^2$.

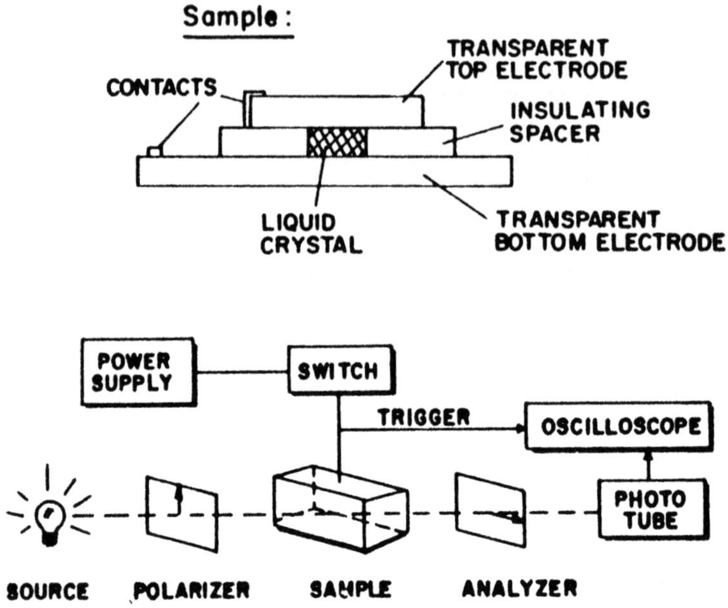

FIGURE 1. Sample configuration and schematic arrangement of apparatus.

The schematic arrangement of apparatus shows only essential parts, which are a light source, polars (generic name for a polarizer-analyzer combination), and phototube. A switch applied voltage to or removed it from the sample, activating the sweep of the oscilloscope at the same time. Removal of voltage involved simultaneous shorting of the sample unless otherwise noted. The oscilloscope display was photographed for subsequent analysis. In actual practice, a Leitz Ortholux polarizing microscope equipped with a hot stage and a 10-power objective was used to supply the polarizer-analyzer combination. The phototube was an RCA 7102 photomultiplier with an S-1 spectral response. A green band-pass filter was generally used to reduce the infrared content of the incandescent source to prevent saturating the phototube. This filter, referred to hereafter as the green filter, had peak transmission at 520 mμ and 10% transmission at 420 mμ and 580 mμ. The phototube output was checked with a variety of voltage and light conditions to make sure saturation or fatigue did not add significant error to the results.

Spectral data were obtained by use of interference filters with a half-peak band width of 10 mμ in place of the green filter.

B. Materials and Operating Conditions

All data were obtained with mixtures of cholesteryl chloride (CC) and cholesteryl nonanoate (CN). These ingredients were obtained from either Eastman Kodak, Aldrich Chemicals or K & K Laboratories. The ingredients were recrystallized one or more times from ethyl alcohol, and their transition temperatures were found to agree with literature values to within one degree.

Most of the study involved a 59 wt.% CC/41 wt.% CN mixture, which had an isotropic transition temperature of 68°C. Other mixtures studied were 55/45, 50/50, and 45/55 CC/CN.

The operating conditions for the study were established after determining the influence of the transformation voltage and its duration upon relaxation at 25°C. The relaxation transient was found to be independent of voltage polarity, of voltage magnitude from 125 to 500 volts (transformation threshold: 73 volts), and of voltage duration from 1 sec to 4 hours when the transformation voltage was 400V. In times less than 1 sec, complete transformation did not occur. Since the operating conditions were so uncritical, it was arbitrarily decided to operate with transformation voltages of 400 volts applied for 15 sec.

III. RESULTS

A. Transformation Threshold

Although this paper deals primarily with relaxation to the cholesteric state, it is necessary for our purposes to consider the threshold voltage for transformation to the nematic state. These data are shown in Fig. 2 as a function of temperature. In obtaining these data, we redefined the conditions corresponding to threshold. In our early work,[3] threshold was defined as that voltage at which the field of view with a 50x objective became transformed completely. The threshold is here taken to be the voltage at which the first perceptible sign of transformation is detected. Since the time for a change to occur becomes very long near threshold, each datum in Fig. 2 involved observation times up to hours.

The data in Fig. 2 are now rather well understood.[4,7,10,11] The threshold voltage is inversely proportional to the pitch of the mixture. The mixture is made up of right-(CC) and left-handed (CN) ingredients, the pitches of which vary with temperature at different rates. Since the reciprocal pitch of a mixture is related to the weighted sum of the reciprocal pitches of its ingredients,[10] with due regard to their chirality

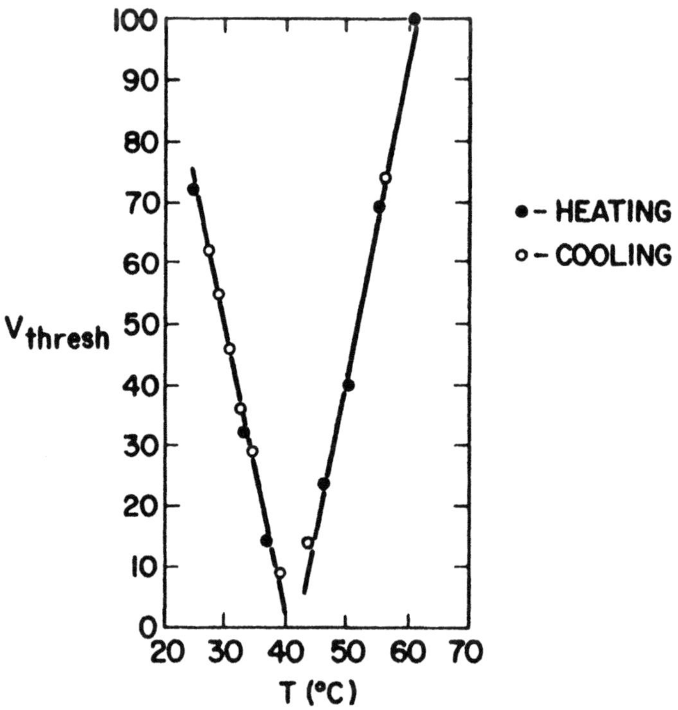

FIGURE 2. Voltage threshold for transformation vs. sample temperature. 59/41 wt.% CC/CN.

or optical sense, the chirality and pitch of this mixture will contain a discontinuity at that temperature at which the weighted averages of the ingredient pitches are equal in magnitude.

The mixture, for example, is left-handed at room temperature, being dominated by the nonanoate component. As the temperature is increased, the effectiveness of this component decreases while that of the chloride increases. Thus, the material pitch increases and the threshold field decreases. Near 42°C, the individual weight-averaged pitches become equal in magnitude, and the material pitch becomes large. The threshold field reaches a minimum since it is only required to align the molecules and not to "uncoil" the helix as well. At higher temperatures, the right-handed component dominates in an ever-increasing fashion. The material thus becomes right-handed; its pitch decreases, and the threshold field increases.

B. Visual Appearance of the Relaxation

After transformation, the material appears dark between crossed polars. With conoscopic viewing conditions, a well-defined uniaxial interference figure is obtained which gives a positive birefringence color pattern when a gypsum first-order-red plate is used.

Although the transformed material appears dark between crossed polars, it is not completely black because of the presence of a thin surface skin which remains untransformed even at the highest voltages used here. This skin has the multi-domained birefringent texture called "undisturbed" in the literature.[12,13]† The presence of the skin obviously signifies a surface interaction between the glass and liquid crystal strong enough to resist the applied field. As the field is increased, the skin becomes fainter, as if layers of it were gradually being removed.

Even though the skin thickness changes, its <u>pattern</u> does not change either with time or applied field; the pattern remains the same even after the material is

† In the undisturbed texture, the helical axis of the material is parallel to the substrate while in the planar or Granjean texture, the helical axis is perpendicular to the substrate.

cycled through the isotropic state. It thus seems to be established by unvarying factors in the glass surface.

When the applied field is removed, the surface skin intensifies and appears to grow according to its already present pattern. At the same time, the bulk material begins to change in a fascinating, difficult-to-describe fashion since its features do not correspond to any of the textures described for liquid crystals. Between crossed polars, the material assumes a puffy, cloud-like character which culminates in a bright flash. During the flash the transmitted light intensity is two or more times greater than in the steady-state with no voltage applied.

When the polars are uncrossed to another value of polar angle, the field of view is very bright initially. As relaxation proceeds, the transmission decreases and then an irregular honeycomb-like pattern develops in which a color play takes place. The sequence of colors is yellow, red and blue. The material then relaxes into the undisturbed texture.

The above sequence of events is independent of stage or tandem-polar rotation. Furthermore, no interference figure is observed during relaxation using conoscopic conditions.

These visual aspects are maintained as long as the material is cholesteric in nature. As the sample temperature is increased, no change in sequence occurs except that it becomes slower. Near 42°C, the sequence changes. A surface skin which grows when the field is removed is still present. However, there is no puffiness, no honeycomb pattern, and no color play. The undisturbed texture appears for a while and is then changed to an interference-like pattern such as seen with oil films of various thickness. Thread-like discontinuities are found in this texture which appears to be the steady-state texture; i.e., it persists for hours. Above 42°C, the relaxation sequence to the cholesteric state reappears, and the color play seems to be the same as below 42°C. The transient sequence becomes faster and faster as the temperature is increased until it is too rapid to detect visually. The appearance of the relaxation thus corresponds closely to the nature of the material just as the threshold voltage does.

C. Time Dependence of the Relaxation

The photomultiplier (PM) signal obtained when the sample was shorted (0 bias) is shown in Fig. 3. Unless noted hereafter, the data have been obtained with the 59/41 CC/CN mixture and the green filter described above.

The shape of the signal corresponds quite well to the visual description. The light transmission increases rapidly to a maximum at a time of ≈4 sec and then decreases to a steady-state value when the material has adopted the undisturbed texture. (In absolute terms, peak transmission amounts to 75%, and steady-state transmission to 26% of the light transmitted by parallel polars with no sample present.) The signal is seen to be independent of polar orientation with respect to the sample. An equivalent test also performed showed the signal to be independent of sample stage rotation, regardless of the polar angle. We thus conclude that the change in transmission is due either to a homogeneous non-birefringent texture or to a random distribution of birefringent areas. The steady-state signal is due without doubt to the latter case.

The relative PM signal obtained for three conditions of the polars is shown in Fig. 4. The crossed and parallel polar curves are essentially mirror images of each

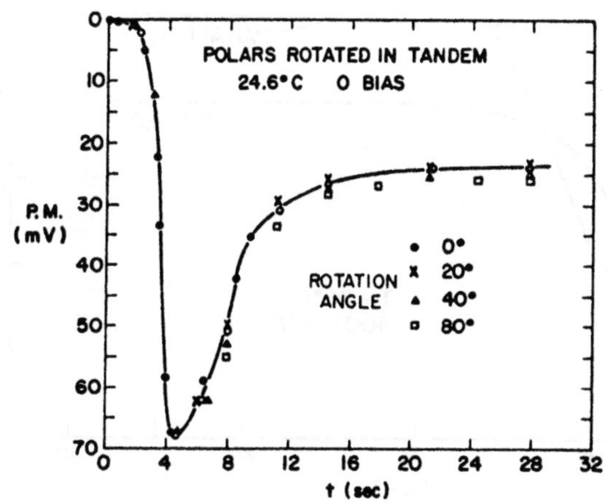

FIGURE 3. Time dependence of photomultiplier signal. Crossed polars rotated in tandem. Green filter. 59/41 wt.% CC/CN.

other at early times. When the analyzer was removed, the character of the curve radically changed; there was no sign of a light peak. Even visual examination of the transient in this case showed no unusual behavior, just a uniform decrease in light output. This decrease is attributed to scattering from the discontinuities which exist in the undisturbed texture. Since significant scattering occurs only after the light peak has subsided, we conclude that scattering does not play a major role in the early stages of the transient. Therefore, we can eliminate a random distribution of birefringent areas as a source of the light peak.

The PM signal was a sensitive function of polar angle as shown in Fig. 5. Features of these curves also correspond well with the visual observations. It is seen, for example, that minimum transmission occurs at a time which depends upon polar angle. The curves eventually coalesce, showing that the light output is independent of polar angle at long times. This is the behavior expected of the random distribution of birefringent areas known to be present at these times.

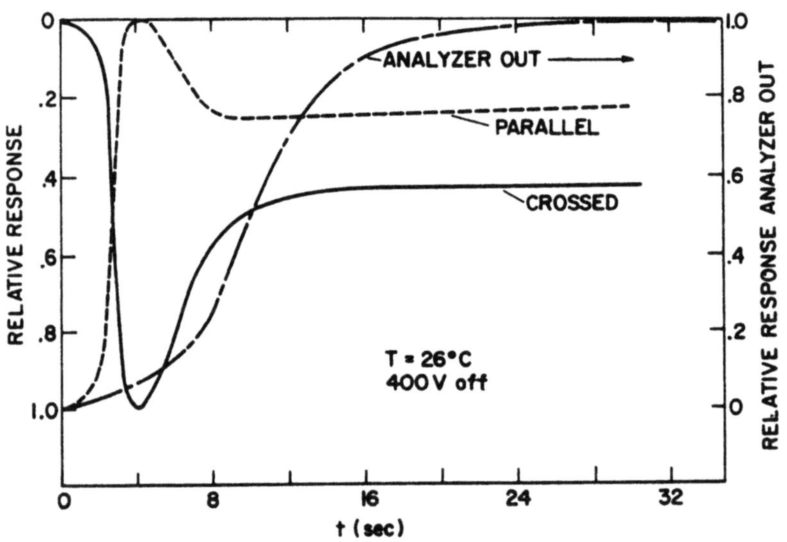

FIGURE 4. Time dependence of relative response for various polar conditions. Green filter. 59/41 wt.% CC/CN.

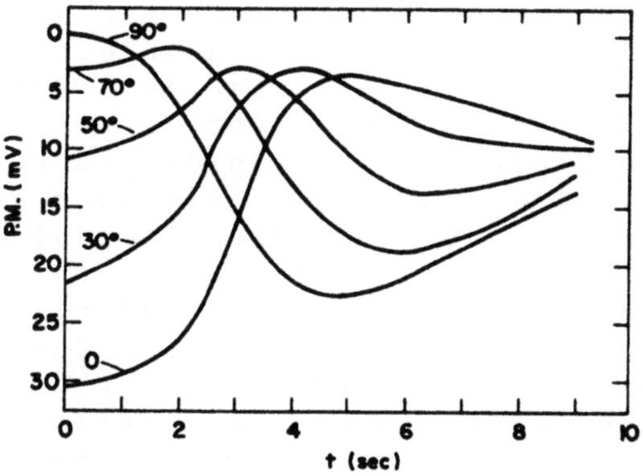

FIGURE 5. Time dependence of photomultiplier signal for various polar angles. 59/41 wt.% CC/CN. 24°C. Green filter.

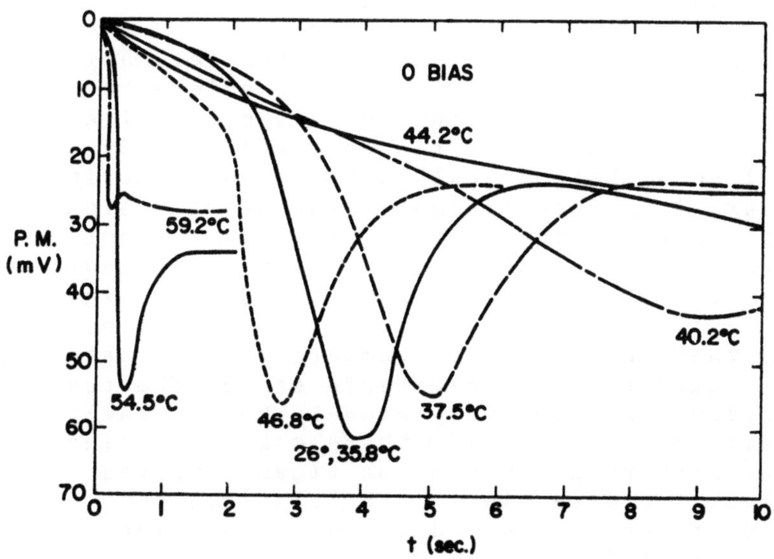

FIGURE 6. Time dependence of photomultiplier signal at various temperatures. Crossed polars. Green filter. 59/41% wt.% CC/CN.

The temperature behavior of the PM signal with crossed polars is shown in Fig. 6. A selection of data was made to show the trends more clearly. Between 26°C and 36°C, the signal changed very little. Above 36°C, the light peak occurred at rapidly increasing times; it disappeared around 42°C, only to reappear above 44°C. It subsequently occurred at rapidly decreasing times. The peak is seen even at 59°C. These data indicate the significance of the nature of the material at the various temperatures. In this instance, the nature of the material is related to the position in time of the light peak. When the material is cholesteric, the peak occurs at finite times; when nematic, at infinite time. While a change in viscosity with temperature may also play a role in the relaxation, the predominant effect seems to be determined by the difference in free energy between the cholesteric and nematic states.

D. Optical Rotation During Relaxation

Before steady state is achieved, the relaxation transient has been shown not to be due solely to scattering from domain discontinuities or to birefringence of a uniform birefringent area. Study of Fig. 5 will show that optical rotation is occurring during relaxation since minimum light transmission is observed at progressively longer times as the polar angle is reduced. To show the rotation more clearly, we have resorted to polar plots of the data. Figure 7 shows the PM signal as a function of polar angle. At zero time, when the field was removed and the material allowed to relax to the cholesteric state, the transmission pattern followed very closely Malus' law; i.e., the intensity varied as $\cos^2\phi$ where ϕ is the polar angle. The pattern at subsequent times revolved in a counter-clockwise direction. (For clarity, only the bottom lobe of the 2-second data is shown. The top lobe is symmetrically placed with respect to the bottom.) Before complete depolarization occurs, each of the curves has a cusp and minimum transmission at an angle 90° away from that for maximum transmission. The gradual increase in value of minimum transmission is attributed to depolarization as the undisturbed texture forms. After 20 seconds, the transmitted light was completely depolarized. Thus, these data show the transmitted light to consist of two components, one of which is presumed to be linearly polarized and the other,

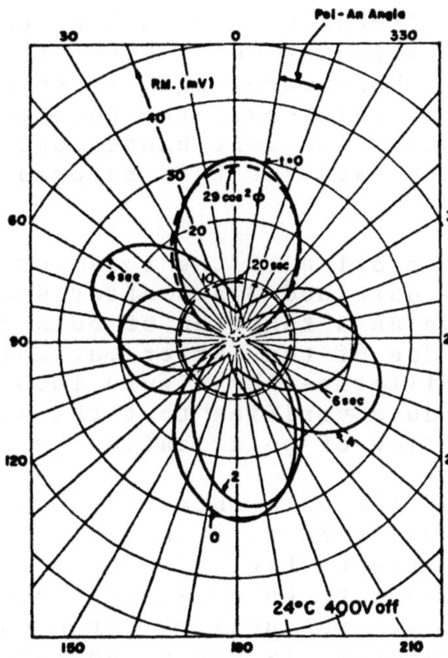

FIGURE 7. Polar plot of photomultiplier signal with time as parameter. Green filter. 59/41 wt.% CC/CN.

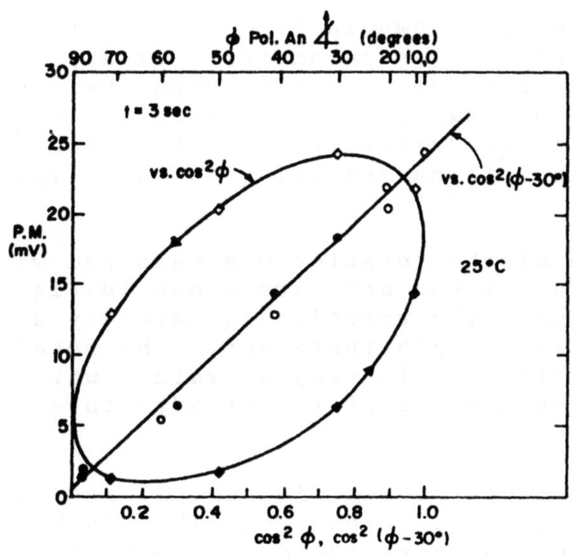

FIGURE 8. Photomultiplier signal vs. square of cosine of polar angle, 3 seconds after field removal. Green filter. 59/41 wt.% CC/CN.

unpolarized. The linearly polarized component revolves in a counter-clockwise fashion. Since the green filter used passes light whose wavelength is shorter than the pitch[13] of the material ($\approx 3\mu$), the counter-clockwise rotation corresponds to a left-handed optical sense, which is indeed the case for this material at room temperature.

Another demonstration of the rotation is shown in Fig. 8. The data obtained 3 sec after start of relaxation are shown as a function of both $\cos^2\phi$ and $\cos^2(\phi - 30°)$. Use of the latter variable results in collapse of the elliptic-like figure into a straight line. This procedure verifies that the sample had rotated the direction of polarization by 30°.

Polar plots of the data obtained at elevated temperatures are shown in Figs. 9 and 10. When the sample is in the nematic range (Fig. 9), very little rotation takes place, a result to be expected since unstressed nematic materials are not active. At higher temperatures where the material is again cholesteric (Fig. 10), rotation reappears but, in this instance, in a clockwise or right-handed sense. (Only the upper lobes are plotted in Fig. 10 for clarity.) This is precisely the sense expected at these temperatures.

The rotation is shown in Fig. 11 as a function of time with temperature as parameter. At any temperature, the rotation is zero at zero time when the field is removed; it then grows to a saturation value (not shown here). This figure summarizes the above results where the rotation is left-handed below 42° and right-handed above.

Dispersion of the rotation was measured with interference filters. These data are shown in Fig. 12 as a function of time. The rotation is markedly dispersive, decreasing as wavelength increases. The rotation saturates at long times (6-10 sec) at values which we believe are representative of the steady-state cholesteric material.

The shapes of the curves in Fig. 12 are similar and appear to be universal in nature. This is shown to be the case in Fig. 13 where the data are replotted in terms of a function of θ/θ_{max} where θ is the rotation. (The 900 mμ data were not included because they are the

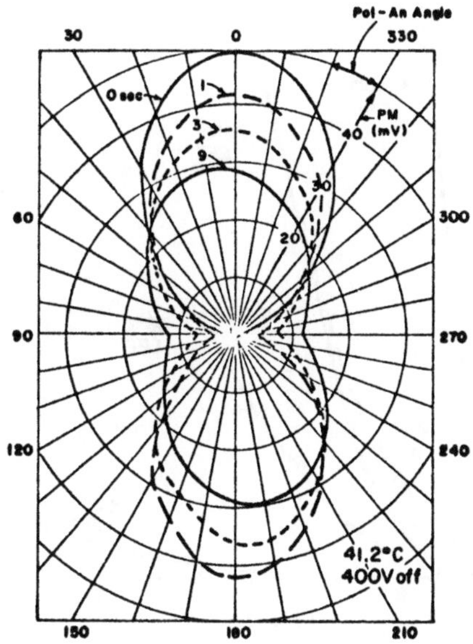

FIGURE 9. Polar plot of photomultiplier signal with time as parameter. Green filter. 59/41 wt.% CC/CN.

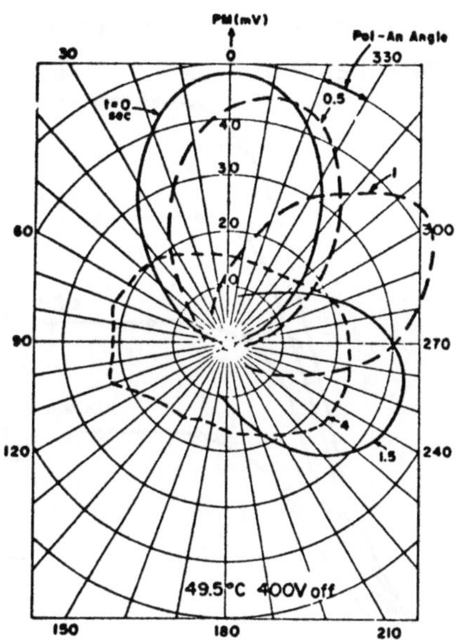

FIGURE 10. Polar plot of photomultiplier signal with time as parameter. Green filter. 59/41 wt.% CC/CN.

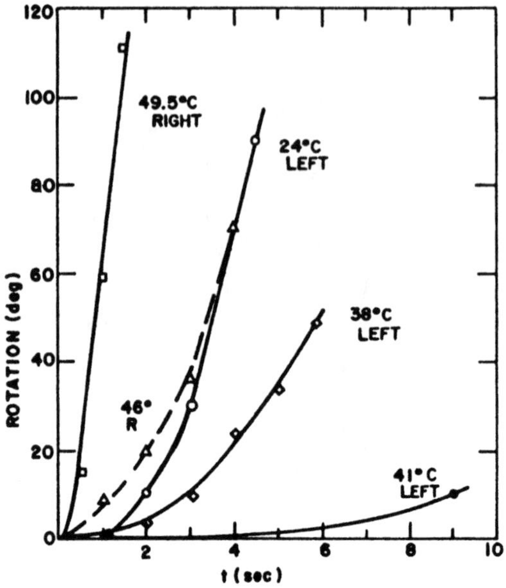

FIGURE 11. Time dependence of rotation with temperature as parameter. Green filter. 59/41 wt.% CC/CN.

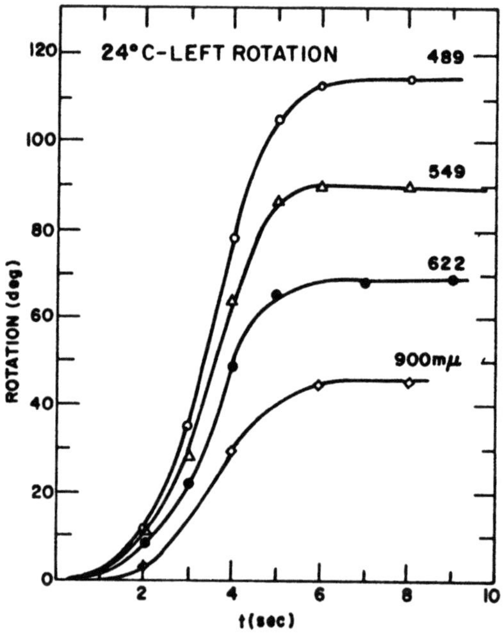

FIGURE 12. Time dependence of rotation with wavelength as parameter. 59/41 wt.% CC/CN.

most inaccurate of the group. The polars are no longer completely effective at this wavelength.) At any given temperature, the data plot into the same line, establishing the universal nature of the time dependence. The straight-line fit in Fig. 13 coupled with the behavior at zero time shows the time dependence of θ/θ_{max} is:

$$\frac{\theta}{\theta_{max}} = \frac{b\left(e^{at}-1\right)}{1 + b\left(e^{at}-1\right)} \tag{1}$$

The values of a and b are given in Fig. 13. Both of these coefficients appear to have an inverse-pitch dependence since they become small as the material becomes nematic-like.

The dispersive nature of the rotation is shown in Fig. 14 where the data are plotted as a function of reciprocal wavelength squared with time as parameter. At each time, the data appear to fit deVries' equation[14] for optical rotation in cholesteric materials;

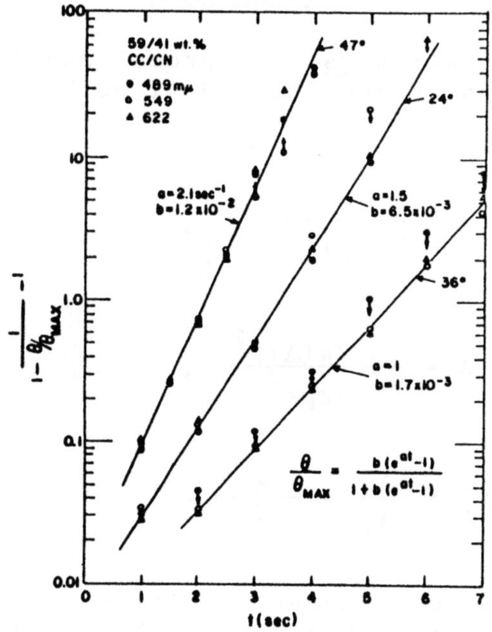

FIGURE 13. Time dependence of relative-rotation with temperature and wavelength as parameters. 59/41 wt.% CC/CN.

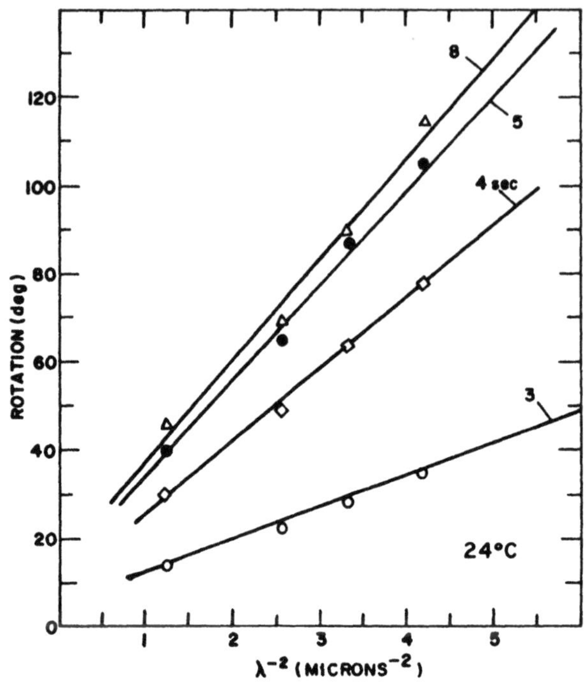

FIGURE 14. Rotation vs. reciprocal wavelength squared with time as parameter. 59/41 wt.% CC/CN.

i.e.,

$$\theta = \frac{A}{\lambda_r^2\left(1 - \lambda_r^2\right)}, \tag{2}$$

where
$$A = -\frac{2\pi(\Delta\epsilon)^2}{32p\epsilon^2}, \tag{3}$$

and
$$\lambda_r = \frac{\lambda}{\lambda_o} = \frac{\lambda}{p\epsilon^{1/2}}. \tag{4}$$

The symbol p denotes material pitch; $\Delta\epsilon$ is the anisotropy in dielectric constant parallel and perpendicular to the major molecular axis; ϵ is the average dielectric constant for these directions; and λ is the wavelength. To test the applicability of this equation, other mixtures whose pitch values correspond more closely to visible wavelengths were studied. Results of one study are shown in Fig. 15 where time-saturated values are compared to deVries' equation using the parameter values shown. The comparison shows deVries' equation to be an adequate representation of the experimental data.

Further confirmation of this was obtained by comparing the values of λ_o deduced from the best fit to saturated values of rotation with values[13] deduced independently in optical studies on equivalent mixtures. This comparison is shown in Table I together with the dielectric anisotropy $\Delta\epsilon$ and birefringence Δn computed from deVries' formula, assuming a dielectric constant of 3, for the sample whose thickness was known. It is seen that the optical and relaxation data are in close enough agreement to warrant use of deVries' equation to represent the data.

TABLE I

Comparison of Optical and Relaxation of λ_o

Mixture (wt.%)		λ_o [13]*, Optical (μ)	λ_o Relaxation (μ)	$\Delta\epsilon$†	Δn†
45/55	CC/CN	0.9	1.2	----	----
50/50		1.22	1.3	----	----
55/45		1.97	1.9	----	----
59/41	24°	3.5	2.8	0.38	0.11
	36°	----	3.0	0.45	0.13
	47°	----	3.0	0.44	0.13
	53°	----	1.6	0.38	0.11

* Pitch in ref. 13 is defined as $\frac{\lambda o}{2n} \approx \frac{\lambda o}{3}$.

†Computed from deVries' formula.

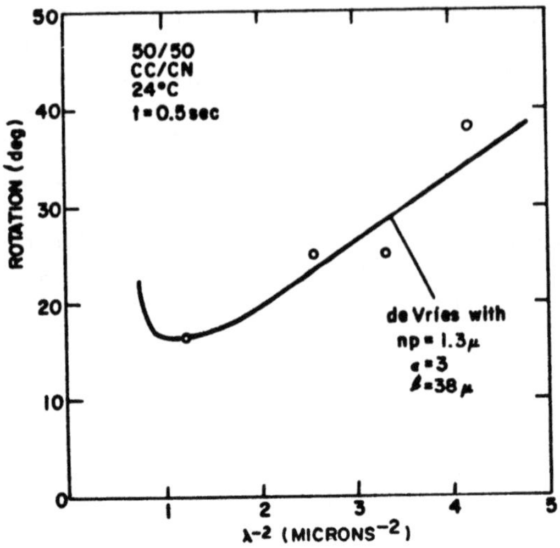

FIGURE 15. Rotation vs. reciprocal wavelength squared. Data (0) compared with deVries' equation.

E. Bias Effects

Besides the case where the field was removed from the sample, the study also included the case where the field was simply reduced to a value below threshold. Such bias fields were found to have an effect, as shown in Fig. 16, much larger than anticipated. The surprising aspect of these data is that voltages as low as 17 volts have a measurable effect upon relaxation while voltages of about 70 volts are needed for the transformation. Relaxation obviously has a larger sensitivity to voltage than transformation does. The effect of bias is to prolong relaxation. The peak transmission is pushed toward infinite time as the bias voltage approaches threshold. Comparable behavior is observed at other polar angles.

Although relaxation is strongly dependent on magnitude of bias, it is rather insensitive to bias polarity, a strong indication that induced rather than permanent dipoles are dominant in the transformation. These data are shown in Fig. 17.

A consequence of the bias dependence is that optical activity during relaxation is severely slowed down, as

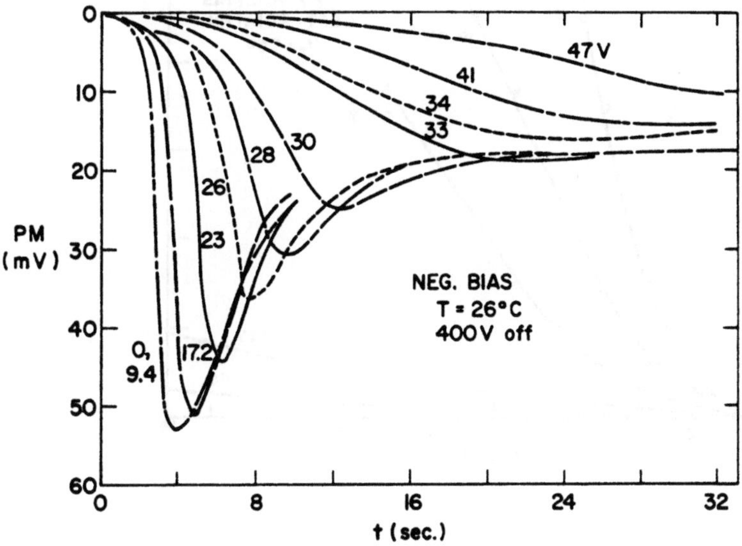

FIGURE 16. Time dependence of photomultiplier signal with bias voltage as parameter.

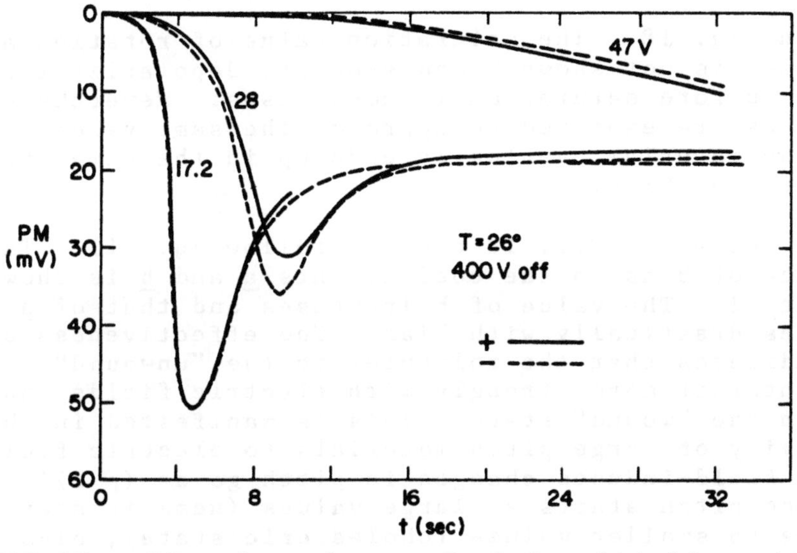

FIGURE 17. Time dependence of photomultiplier signal with bias voltage and polarity as parameters. Crossed polars. Green filter. 59/41 wt.% CC/CN.

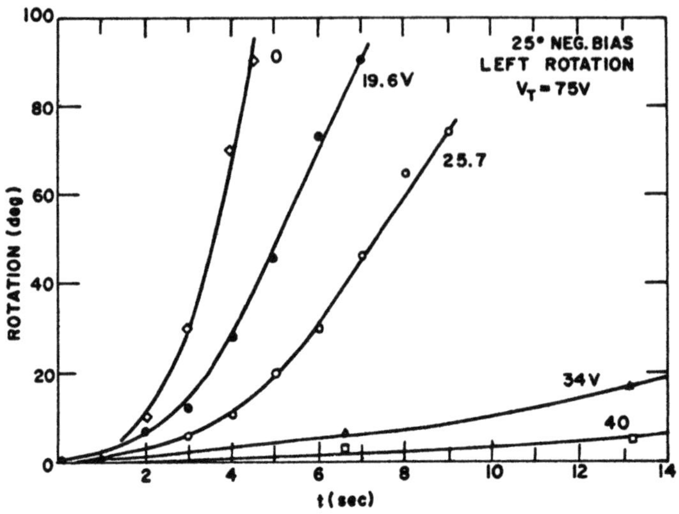

FIGURE 18. Time dependence of rotation with bias voltage as parameter. Green filter. 59/41 wt.% CC/CN.

shown in Fig. 18. The saturation value of rotation at long times is not shown because severe depolarization occurred before saturation in most cases. Nevertheless, the curves are expected to approach the same value because the material eventually ends up in the same state regardless of bias.

The curves in Fig. 18 closely follow Eq. (1). The influence of bias on the coefficients a and b is shown in Table II. The value of b increases and that of a decreases drastically with bias. The effectiveness of bias indicates that the molecules in the "unwound" state interact more strongly with electric fields than those in the "wound" state. This is manifested in the sensitivity of large pitch materials to electric fields[7] because field-induced changes in pitch go as $(p\ E)^4$. Since the pitch starts at large values (nematic state) and goes to smaller values (cholesteric state), bias fields have a larger effect during relaxation than in the steady state.

TABLE II

Effect of Bias on Coefficients of Eq. (1)*

Bias Voltage (volts)	a (Sec^{-1})	b ($X10^{-3}$)
0	1.47	7.1
-19.6	0.91	12
-25.7	0.66	9.8
-34	0.21	18

* Experimental Parameters
Green filter, 24°C
59/41 wt.% CC/CN

IV. DISCUSSION

Although the relaxation transient has not been unraveled in terms of an unequivocal picture of the internal molecular motions involved in going to a helicoidal cholesteric state from a linear nematic one, a number of suggestive clues have been obtained which will clarify the ecology of this system.

The following qualitative picture has developed. The cholesteric-nematic phase transition is due to an interaction of induced dipoles with the applied field. As a consequence, the helicoidal form is disrupted, and the resulting assembly of molecules is aligned into a linear structure consistent with the nematic form. When the field is removed or reduced below threshold, these molecules relax back to the cholesteric state. With large pitch, the material becomes nematic-like and the internal driving force from the transformed to the cholesteric state is reduced. Thus, the relaxation time increases.

The initial stage of relaxation displays optical activity but not birefringence; i.e., the polarization maintains its linear nature. The molecular structure is probably of a planar-like type with its optic axis parallel to the field direction. Equations formulated by deVries for a similar texture consequently apply here also.

Most of the data support this picture. The behavior of threshold voltage and optical rotation with temperature clearly show the cholesteric or nematic nature of the material and help specify the distinguishing features of the cholesteric state: pitch and sense. The bias dependences observed indicate that interactions which reduce the sensitivity of the cholesteric state to electric field are not complete in initial parts of the relaxation. One such interaction is that which determines pitch. The pitch goes from infinity at zero time to a smaller value in the steady-state. According to theory, the field-induced change in pitch depends directly upon pitch. Thus, large-pitch materials are influenced more by a given field than small-pitch materials are.

Three aspects of the data require further discussion. First, if the material were indeed in a planar-like state during relaxation, interference figures should be obtained with conoscopic observation. These figures are easily observed, for example, when the material is manipulated to achieve the Grandjean texture. Of uncertain significance in this matter is the presence of the surface skin which is birefringent and which relaxes more quickly than the bulk. This skin is optically in series with the bulk material, and its birefringence would tend to destroy the conditions required for interference figure formation. The question is whether or not it would also mask the influence of bulk phenomena on light transmission. If not, we would argue that the bulk relaxes initially in a different fashion than the skin, and while the skin has an undisturbed texture, the bulk has a planar-like one. Indeed, the natural relaxation may be to a planar-like texture, but the surface skin may force the bulk texture into the undisturbed one at long times, determining, as it were, the pattern for the bulk to assume.

Secondly, the time dependence of the pitch deduced from the data appears to be inconsistent with deVries' formula even though the saturation values are not. When the field is removed, rotation starts at zero and grows to a saturation value. In the same sequence, the pitch initially is infinite (nematic) at zero time and it drops to a finite value (cholesteric) in steady state. At a wavelength small in comparison to any of these values of pitch, deVries' formula predicts that rotation is directly proportional to $(\Delta\epsilon)^2 p$. Thus, for fixed $\Delta\epsilon$, large pitches would lead to large rotations. However, an inverse dependence seems to hold experimentally where large pitches correspond to small rotations. Since we believe deVries' formula correctly describes the observed rotation, we must suggest that the factor $(\Delta\epsilon)^2$ accounts for the discrepancy. At zero time, $\Delta\epsilon$ must be <u>effectively</u> zero (consistent with an aligned nematic state) and it must grow to the value for the material in question as the helicoidal state is resumed. This explanation is consistent with the idea that bias retards the molecules in moving from the aligned nematic state to the cholesteric state. Thus, $\Delta\epsilon$ would grow at a slower rate and, consequently, so would the rotation.

The third aspect concerns the general sequence of events which occur when a field is applied to a cholesteric liquid crystal. The texture just before the transformation has been invariably found to be the undisturbed one. Even if the material had the Grandjean texture initially, it quickly adopted the undisturbed state at fields much lower than the threshold field[3]. By symmetry and by argument based on excluded volume, one would expect that relaxation should lead immediately to the undisturbed state. But since this state is birefringent and does not exhibit optical activity, it cannot be the one assumed initially by the material. Instead, a planar-like texture must be invoked for consistency with experimental observations.

Obviously, the nature of this planar-like texture must be established. This and the other two items are some of the points which our future study will hopefully clarify.

Acknowledgements:

We acknowledge the discussions, assistance and general support of Dr. J. Becker and R. Madrid.

REFERENCES

1. J. J. Wysocki, J. E. Adams, and W. Haas, Phys. Rev. Letters 20, 1024, (1968).

2. A cholesteric-nematic phase transition can also be induced by magnetic fields. See, for example, E. Sackmann, S. Meiboom and L. Snyder, J. Am. Chem. Soc. 89, 5981 (1967).

3. J. J. Wysocki, J. E. Adams, and W. Haas, "Electric-Field Induced Phase Change In Cholesteric Liquid Crystals", in Proc. of 2nd International Conf. on Liquid Crystals; Kent, Ohio; 12-16 Aug. 1968 (to be published in Molecular Crystals).

4. H. Baessler and M. Labes, Phys. Rev. Letters 21, 1791 (1968).

5. F. Frank, Disc. Faraday Soc. 25, 19 (1958).

6. P. de Gennes, Solid State Comm. 6, 163 (1968).

7. R. Meyer, Appl. Phys. Letters 12, 281 (1968).

8. G. Durand, et.al., Phys. Rev. Letters 22, 227 (1969).

9. J. J. Wysocki and J. E. Adams, Bull. Am. Phys. Soc. 14, 739 (1969).

10. J. E. Adams, W. Haas and J. J. Wysocki, Bull. Am. Phys. Soc. 14, 739 (1969).

11. J. E. Adams, W. Haas and J. J. Wysocki, Phys. Rev. Letters 22, 921 (1969).

12. W. Haas and J. E. Adams, J. Appl. Opt. 7, 1203 (1968).

13. J. E. Adams, W. Haas, and J. J. Wysocki, J. Chem. Phys. 50, 2458 (1969).

14. H. deVries, Acta Cryst. 4, 219 (1951).

RECENT EXPERIMENTAL INVESTIGATIONS IN NEMATIC AND CHOLESTERIC MESOPHASES(+)

Orsay Liquid Crystal Group
Service de Physique des Solides
Faculté des Sciences
Orsay, France

ABSTRACT

The development of the Frank thermodynamical description of liquid crystals has recently allowed quantitative predictions on the static and dynamical behavior of the nematic and cholesteric mesophases. We report here the results of our experimental investigations to measure the parameters introduced by the theory, in two main directions : (a) the magnetically induced cholesteric to nematic phase transition, in low concentration mixtures of cholesteryl esters in nematic materials, appears as a general method for measuring the "twist" elastic constant of the nematic solvent ; (b) the study of the thermal fluctuations of anisotropy in nematic materials, observed by high resolution spectral analysis of the Rayleigh light scattered by the liquid crystal, allows the determination of the Leslie viscosity coefficients. Finally, we report on the observation of a new type of disclination line in cholesteric mesophases (double disclination line).

This paper is a short review of the experimental work performed during the current year by the Orsay Liquid Crystal Group on the physical properties of liquid crystal mesophases. The aim of this effort has

(+) Work partially supported by D.G.R.S.T. (under contract N° 6801 194).

been to test recent theoretical predictions concerning some elastic, magnetic and optical properties of liquid crystals, in the frame work of the continuum thermodynamical model. Our results have been partially published in form of letters, and experimental details can be found therin. We would like to emphasize here that the continuum model of liquid crystals is not a formal speculation ; it can indeed lead to an accurate characterization of the physical behavior of the mesophases, and it gives powerful methods for the measurement of the anisotropic physical constants involved in elastic and hydrodynamical properties of liquid crystals.

We first describe an experiment of magneto-elasticity, which allows the determination of the "twist" elastic constant(1) of a nematic material. We then report on results concerning the damping of thermal fluctuations of anisotropy in a nematic mesophase, which give at the present time 4 out of the 5 Leslie viscosity coefficients(2). Finally, we describe a new type of defect ("disclination") line recently observed in cholesteric mesophases.

I. MAGNETICALLY INDUCED CHOLESTERIC TO NEMATIC PHASE TRANSITION

The alignement of a cholesteric texture along an external magnetic field was first reported by Sackman and all(3), using an N.M.R. technique. De Gennes(4) and independently R.B.Meyer(5) have calculated this effect. We have optically observed this transition(6) using the Cano wedge method(7) for measuring the pitch of the helical cholesteric texture. We utilized high pitch samples composed of low concentration mixtures of active molecules (cholesteryl esters) in nematic solvents.

When a magnetic field H is applied perpendicular to the cholesteric screw axis, the pitch p of the texture increases with H and diverges logarithmically for a critical value H_c. Beyond H_c, the structure is nematic. The critical condition(4) is :

$$H_c\, p_o = \pi^2 (K_{22} / \chi)^{1/2}$$

where p_o is the zero field value for p, K_{22} the "twist" elastic constant of the material(1) and χ the anisotropic part of the diamagnetic susceptibility. Our experiment, which allows the simultaneous measurement of both

p_o and H_c, gives a determination of K_{22}, as χ is reasonably well-known[8]. In this high dilution regime ($\sim$ 1 % in mass of active molecules), χ and K_{22} characterize the nematic solvent alone.

Our present results, concerning p-azoxyanisol (PAA) and p-azoxyphenetol (PAP), are sumarized on the following table :

	t (°C)	$(K_{22} / \chi)^{1/2}_{(cgs)}$	K_{22} (cgs)
PAA	129	1.65 ± 0.15	3.1 ± 0.6 10^{-7}
PAP	147	2.8 ± 0.3	7.1 ± 1.4 10^{-7}

For PAA, this $(K_{22} / \chi)^{1/2}$ value is in good agreement with the old determination of Freedericks[9], using a different method. For PAP, to our knowledge, it is the first published value for K_{22}. The increase of the "twist" constant from PAA to PAP may be related to the larger length of the PAP molecule. A systematic measurement of K_{22} for a large class of nematic material is currently under way.

II. QUASI-ELASTIC RAYLEIGH SCATTERING IN A NEMATIC LIQUID CRYSTAL

In a nematic monocrystal, the mean molecular axis ("the director") undergoes thermal fluctuations of orientation, resulting in an intense depolarized scattering of light, via the modulation of the local refractive index. The effect has been observed by Chatelain[10] and recently quantitatively explained by De Gennes[11], by the existence of two modes of deformation for the spatial distribution of "directors", superpositions of "bending", "twist" and "splay" as introduced by Frank[1].

Using, in addition, the formalism of Eriksen[12] and Leslie[2] for the hydrodynamics of a nematic liquid crystal, where the angular motion of a director is frictionally coupled to the local velocity gradient of the fluid, the Orsay theoreticians have predicted the time dependent behavior of these thermal angular fluctuations[13]. The propagative character of these deformations depends of the dimensionless constant $K\rho / \eta^2$, where K and η are typical values for the Frank and Leslie elastic and viscosity coefficients, and ρ is the specific mass of the liquid crystal. In practice, the very small value of this parameter leads to non-propagative purely damped modes. Note that this analysis as yet ignores any possible piezoelectric effects as described by R.B.Meyer[14].

The dynamics of the director fluctuations can be observed as an intensity fluctuation of the light scattered from a nematic monocrystal. As the modes are non-propagating, the spectrum of the scattered light is not shifted but only broadened, corresponding to "Rayleigh" scattering.

Using a high resolution light beat laser spectrometer, we have measured the angular dependence of the width of the Rayleigh line[15], in a PAA sample. We have verified the dissipative character of the two modes, which can be isolated by a convenient geometry. From the knowledge of the elastic constants for this material[16] it is possible to fit the experimental data with the predicted analytical form for the angular dependence of the Rayleigh width. This fit is then a measurement of the Leslie viscosity coefficients. In fact, the number of these coefficients has recently been reduced from 6 to 5 by Parodi[17]. Our present results concern 4 out of these 5 coefficients. As they appear quadratically in the theoretical form, we obtain at the present time two possible sets of tentative data presented on the following table, using Leslie's notations :

$\alpha_{cos} \cdot 10^2$	α_2	α_3	α_4	α_5
set a	- 6,7	2,4	1,7	3,3
set b	- 14	-4,9	7,4	29,5

The two sets (a) and (b) can be physically characterized by the relative values of two friction coefficients : $\gamma_1 = \alpha_3 - \alpha_2$ measures the coupling between the angular motion of the "director" and the vorticity of the fluid ; $\gamma_2 = \alpha_6 - \alpha_5 = \alpha_2 + \alpha_3$ measures the coupling between the angular motion of the "director" and the shear rate of the fluid. The sets (a) and (b) correspond respectively to situations where $\gamma_1 > |\gamma_2|$ and $\gamma_1 < |\gamma_2|$.

At the present time, we have no decisive argument to arbitrate between (a) and (b), although the set (b) seems more compatible with old viscosity estimations[18]. An independent careful measurement of one of these viscosities is meaning for a definite choice.

III. DOUBLE DISCLINATION LINES IN CHOLESTERIC MATERIALS

Using the Cano-wedge method[7] for the pitch measurement of cholesteric materials, in the course of the magneto-elastic experiment previously described, we have observed a new type of disclination (defect) line ; in a wedge of variable thickness, Grandjean[19] and later Cano[7] have reported the appearance of regular striations, the Frank disclination lines[1], which allow discrete jumps of the helical torsion, in order to match the half-pitch p/2 periodicity of the helix and the increasing thickness of the wedge ; between two adjacent lines, the thickness increment of the wedge is then p/2. In fact, in addition to these well known lines, we have observed[20] double disclination lines, corresponding to thickness increment of 2. p/2. These double lines are stable in the deeper region of the wedge, and split, in the shallow region, into pairs of simple lines. Chatelain and all have independently observed these double lines.[25] These two types of lines have distinct optical and magnetic properties, as previously reported[21]. For instance, a magnetic field, orthogonal to the line, induces a zig-zag "buckling" of the double-lines, beyond a critical value roughly equal to half the critical field H_c introduced in section I, while the simple lines never zig-zag. These observed properties are compatible with a new model proposed by Kleman and Friedel[22] for the structure of the lines, involving the folding of cholesteric layers of quantized thickness around the line ; these defect lines would be the liquid crystal analogous of edge dislocations in solid crystals.

IV. CONCLUSION

It appears clearly that the macroscopic thermodynamical description of the liquid crystals, introduced in the thirties by Zocher[23] and Oseen[24], and presently developped by the various theoreticians already quoted, is able to quantitatively describe the properties of the liquid crystal mesophases on a scale larger than the wave length of light. Beyond testing the validity of these models, our experiments give new methods to measure the fondamental physical constant involved in the description of these anisotropic materials, the knowledge of which is necessary for the technological applications of liquid crystals.

REFERENCES

1. F.C.Frank, Discuss.Faraday Soc. 25, 19 (1958)
2. F.M.Leslie, Quant.Journ.Mech. and Appl.Math. 19, 337 (1966)
3. E.S.Sackmann, S.Meiboom, L.C.Snyder, J.Am.Chem.Soc. 89, 5891 (1967)
4. P.G.de Gennes, Sol.State Com. 6, 163 (1968)
5. R.B.Meyer, Appl.Phys.Let. 12, 281 (1968)
   " " " 14, 208 (1969)
6. G.Durand, L.Léger, F.Rondelez, M.Veyssié, Phys.Rev. Let. 22, 227 (1969)
7. R.Cano, Bull.Soc.Franç.Mineral. 91, 20 (1968)
8. V.Zwetkoff, Acta Physicochim. U.R.S.S. 18, 358 (1943)
9. V.Freedericks, V.Zwetkoff, Sov.Phys. 6, 490 (1934)
10. P.Chatelain, Acta Cryst. 1, 315 (1948)
11. P.G.de Gennes, Compt.Rendu 266 B 15 (1968)
12. J.L.Eriksen, Arch.Ratl.Mech.Anal. 4, 231 (1960)
13. Groupe d'étude des cristaux liquides, J.Chem.Phys. (to be published)
14. R.B.Meyer, Phys.Rev.Let. 22, 918 (1969). However, in the geometrical conditions used for the determination of the 4 viscosity coefficients given in this paper, a recent computation has shown that the damping of the modes is not sensitive to piezoelectric effects (O.Parodi, to be published)
15. Orsay Liquid Crystal Group, Phys.Rev.Let. 22, 1361 (1969)
16. V.Zwetkoff, Acta Physicochim. URSS, 6, 865 (1937), corrected, for the magnetic anisotropy values, following ref. 9
17. O.Parodi, to be published
18. M.Miesowicz, Nature, 158, 27 (1946)

19. F.Grandjean, Compt.Rend. 172, 71 (1921)

20. Orsay Liquid Crystal Group, Phys.Let. 28 A, 687 (1969)

21. Groupe Expérimental d'Etudes des Cristaux Liquides, Col.Soc.Franç.Phys. Montpellier (1969), to be published in J.Physique

22. M.Kleman, J.Friedel, Col.Soc.Franç.Phys. Montpellier (1969), to be published in J.Physique

23. H.Zocher, Phys.Z. 28, 790 (1927)

24. C.W.Oseen, Trans.Faraday Soc. 29, 883 (1933)

25. P.Chatelain, M.Brunet-Germain, Comptes Rendus, 266 571 (1968).

SMALL ANGLE X-RAY STUDIES OF LIQUID CRYSTAL PHASE TRANSITIONS

II. SURFACE, IMPURITY AND ELECTRIC FIELD EFFECTS

C. C. Gravatt* and G. W. Brady

Bell Telephone Laboratories, Inc., Murray Hill, N. J.

ABSTRACT

Phase transitions in nematic and cholesteric liquid forming materials have been investigated by small angle X-ray scattering. Measurements were performed on p-azoxyanisole in the isotropic and nematic liquid regions and on cholesteryl bromide in the isotropic and cholesteric liquid regions. For p-azoxyanisole premonitory phenomena have been observed in the isotropic liquid at temperatures considerably above the isotropic-nematic transition. These phenomena, which are indicative of ordering in the isotropic liquid, are strongly dependent upon sample purity, and are not evident in the purest materials. D.C. electric fields have also been found to induce ordering in the isotropic liquid. It was not possible to observe any similar ordering effects, or to detect the isotropic-cholesteric transition, in cholesteryl bromide by small angle X-ray scattering.

We have previously reported on results of a small angle X-ray study of the isotropic-nematic phase transition in p-azoxyanisole.(1) Before going into a discussion of the present experiments it is well to summarize our earlier findings and indicate some of the questions that required further investigation.

* Present address: National Bureau of Standards, Washington, D.C. 20234

The theory of small angle scattering as applied to liquid crystals was given in the previous publication.[1] In general, the angular dependence of the scattered intensity is measured as a function of temperature. If there exists scattering above that of the background, then this excess scattering is used to determine a length L which is a direct measure of the extent of spatial correlation in the liquid. It is a spherically averaged length over which any two volume elements are correlated as to electron density fluctuations. Figure 1 shows the temperature dependence of L as previously reported.[1] As the isotropic-nematic transition temperature ($T_c = 136°C$) was approached from above, the isotropic liquid developed regions of correlation which started at $\sim T_c+20°C$ with L $\sim$ 300 A, and as the temperature was lowered, increased to a value of 2000 A at T_c. Below T_c, L was essentially constant throughout the nematic region. Analysis of the data showed that the phase transition was first order.

Some questions remained unanswered at this point. Two of these, perhaps the most important, were surface effects and the role of impurities. It is well recognized that liquid crystal materials tend to orient on mica surfaces, with the orientation extending 0.1 mm into the liquid.[2] Since our measurements were made using a transmission technique with a 1.0 mm thick cell, about 20% of the scattering volume could be surface affected material, and thus be a strong contributing factor to the observed results.

With respect to impurity effects, research grade British Supply House p-azoxyanisole, of quoted purity 99%+ was used. As we noted, a slight discoloration of the material was evident after the experiments were completed, but since the measurements were made over a period of about six weeks of continuous scanning it is doubtful that the decomposition affected the results. However, the fact that the material changed color indicated that the impurity level was significant and that measurements on much purer materials were called for.

One further question with regard to the initial work was the asymmetrical character of the correlation effects around T_c. Frenkel heterophase fluctuation theory[3] predicts that they should be symmetrical. The constancy of L in the nematic region, as shown in Figure 1, raises the question as to what ordering scheme was responsible for the small-angle scattering. In particular, the question posed itself as to whether the observations were characteristic of a nematic-forming system and thus related to the threads (nematos) which form domain boundaries, or disclination lines in these compounds. We had contented ourselves in the preceding publication with estimating the number of correlated molecules, since we had as yet made no measurements on non-nematic liquid crystals. Evidently such experiments were also necessary.

In this communication we present the results of an extension of the previous study into the effects noted above, that is,

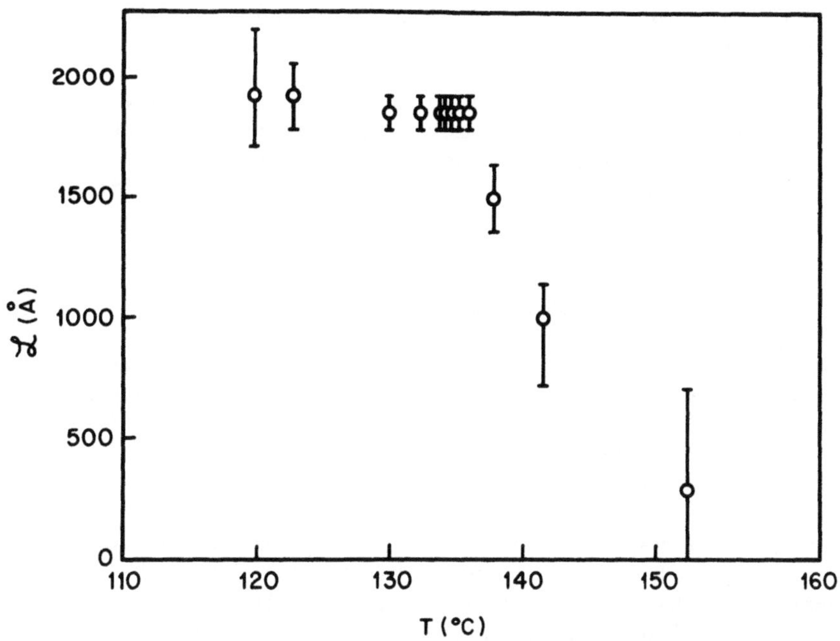

Figure 1 Temperature dependence of the correlation length

surface effects and sample purity. We also report briefly on an attempt to observe an isotropic-cholesteric phase transition in cholesteryl bromide, both pure and impure. Finally, electric field studies were made, both to measure their intrinsic effect, and also by their application to perturb any surface orientation that may have been present.

EXPERIMENTAL

The experimental arrangement used in this study was the same as that previously described,[4] except for the introduction of electrodes for the electric field studies. The modified Bonse-Hart diffractometer was again used in conjunction with the constant temperature cell capable of temperature control to 0.003°C. Slit height desmearing corrections were applied where necessary.[4] Platinum electrodes were inserted in the cell in a geometrical arrangement such that the electric field was perpendicular to the incident X-ray beam, and thus lay in the scattering plane. With this arrangement $\vec{E}$ and $\vec{S}$, the electric field and scattering vectors were parallel. Stabilized D.C. power of 400 volts/cm and 1000 volts/cm were applied to the liquid. The current was monitored during these measurements; the pure material (see below) had a conductivity of 10^{-9} (ohms-cm)$^{-1}$; the conductivity of the "impure" material was a factor of 10 higher.

The p-azoxyanisole samples used in the present study were supplied by the James Hinton Company, zone refined to a purity of 99.99%. The cholesteryl bromide was custom made Frinton Laboratories product, and was recrystallized four times from acetone before use.

RESULTS AND DISCUSSION

An extensive set of measurements similar to those previously reported were performed on the zone refined material. Repeat runs were also done on the 99% material. Great care was taken to avoid contamination. The interior of the cell was gold-plated and the filling was done in a He atmosphere.

The results were striking. For three separate runs on different samples of the Hinton material it was found that there was no change in the scattering patterns over the temperature range from 160°C to 120°C and in particular no change near T_c. Visual observation showed a phase transition at T_c = 136°C and no supercooling was observed. Thus no excess scattering over that of the background, (the 160°C curve[1]), was present, indicating that no correlation effects were present, or detectable, in the purest material. The repeat runs on the less pure material reproduced the results shown in Figure 1. In several cases, when working on a sample that showed no excess scattering in numerous temperature scans, an intensity increase would develop. This excess scattering always had a time dependence such that 24 hours might be required to obtain a stable intensity pattern. L values calculated from this stable scattering data would agree very well with those shown in Figure 1; in fact the sample would now duplicate exactly the results shown in the Figure. Visual observation of these samples always showed a red coloration. No chemical analysis was made to determine the degree of decomposition.

Having shown the marked effect of impurities, and before discussing them further, we can quickly dispose of two of the other points raised in the introduction. First, surface effects appeared to play no role in contributing to the observed results. Several grades of different quality mica surfaces were used, having in some cases no evident step dislocations or scratches, in others a large amount of them. A rubbed or scratched surface has been shown to induce a preferred orientation in nematic liquids.[6] Thin Be sheets were also used as windows. None of these experiments showed any difference on the presence or absence of scattering described above. The Be experiments were not as exact as those for mica because of the pronounced small angle scattering from the Be itself, but no surface effect was detected. The conclusion then is that such effects were absent in all these experiments, or alternatively were always present, but did not influence the results. It should be noted however, that when an impure sample was solidified and then cut open the red color was more pronounced near the mica windows.

Secondly, the scattering from cholesteryl bromide was measured above and below the isotropic-cholesteric transition temperature. All experimental details were the same as those for the p-azoxyanisole measurements. The presence of the Br atom resulted in a large amount of scattering. No temperature dependence was seen. Further, decomposition of the material did not result in measurable excess scattering. Thus the scattering behavior was identical to that of the pure nematic compound and did not show the marked impurity effects of the latter system.

With these observations we can attempt a rationalization of the results. We must emphasize that further experiments on different systems will be necessary before a more definite interpretation can be given. A salient feature of the impurity effect, probably the most important is that it manifests itself in the nematic material only. Whether it is a nucleation phenomenon, which promotes molecular ordering of a higher degree than in the pure material is a possibility, although why this does not manifest itself in the cholesteric material is not clear. Also, if the impurities ordered in a liquid crystal type of arrangement above T_c to produce a difference in electron density, this could give rise to the observed scattering, but this again does not seem a plausible explanation of the different behavior of the two materials. Our current thinking is that the nematic threads which mark the domain boundaries are probably segregation sites for the impurities, producing either an excess or deficit in electron density and that the observed correlation lengths are a measure of this. In the earlier measurements(1) we had hypothesized that L was a measure of the number of correlated molecules in a domain. The discovery of the impurity effect does not necessarily alter these estimates since L would now be a distance related to the boundary separation. Such an interpretation would explain why such small amounts of impurity produce such marked effects because the threads make up a smal part of the total value, probably of the same order of magnitude as that of the impurities. It would also explain the constancy of L throughout the nematic region since once it has formed, the physical properties are essentially temperature independent.

The set of measurements of the effect of electric fields on the scattering from p-azoxyanisole in both the isotropic and liquid crystal phases showed small but significant changes. The thermal conductivity of the electrodes decreased the stability

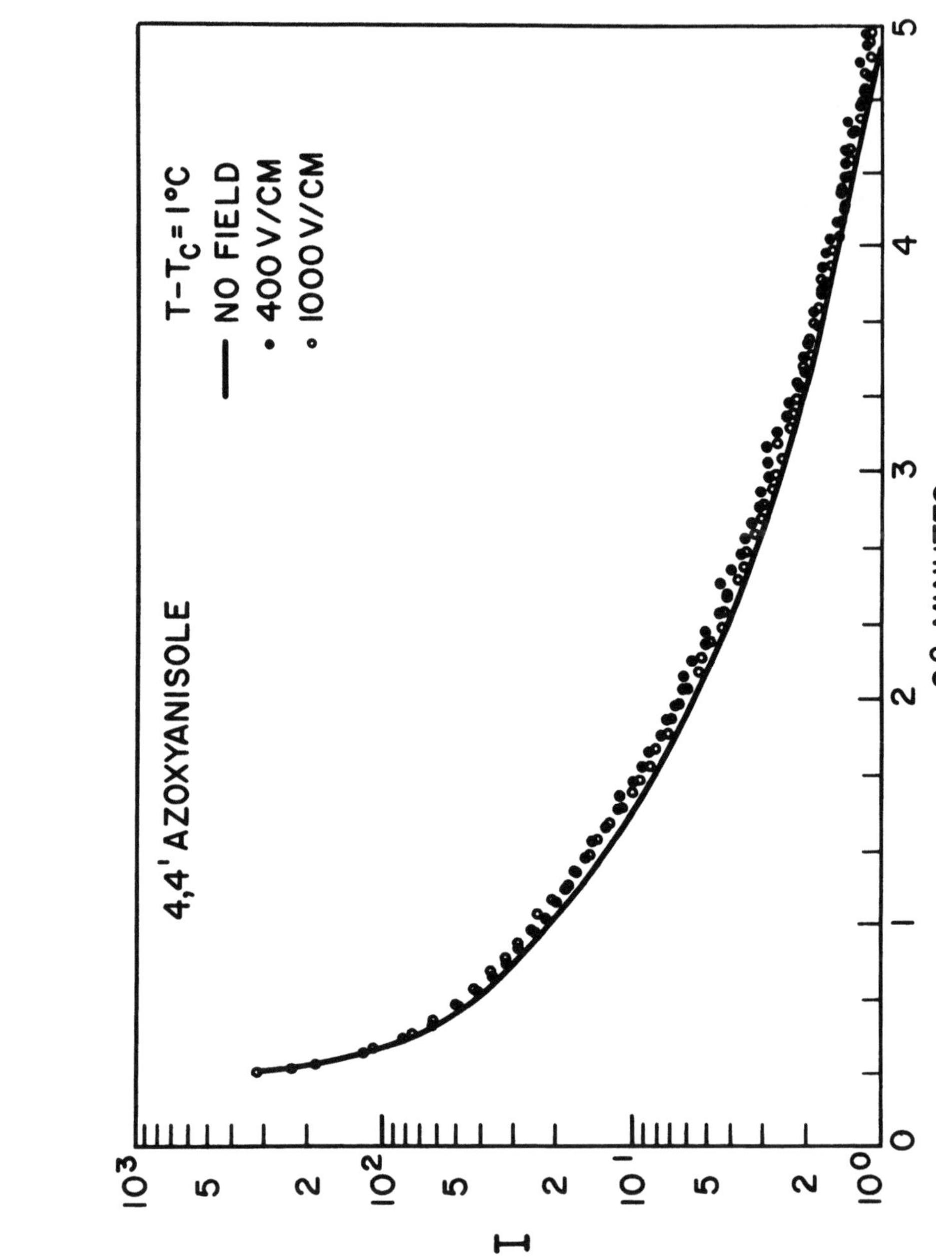

Figure 2 Plot of intensity versus 2 as a function of the applied electric field

of the temperature control to some extent and accordingly measurements were made at temperatures sufficiently removed from T_c so that the variation in $\Delta T = | T-T_c|$ was minor. Visual examination of the liquid showed considerable flow in the nematic region, but none was evident in the isotropic region The flow appeared to develop at field strengths slightly larger than 400 v/cm. Below this field strength disclination lines were visible but no macroscopic flow could be detected. At 1000 v/cm there was considerable flow.

The excess scattering as a function of field strength for $T-T_c = 1°C$ is shown in Figure 2. The solid line is the background scattering (T_c+20°C) for a pure sample and the points are unsmoothed data at the two field strengths listed. There appeared to be a very slight decrease in excess scattering for the higher field but the difference between the results for the two fields was within experimental error. Field effects of almost identical magnitude were observed at temperatures of 8.1 and 4.7°C above T_c. At T_c-1°C, the field effect was present but smaller than in the isotropic region, and also independent of field strength.

Correlation length analysis of the data was difficult and subject to large degree of error due to the small magnitude of the excess scattering, but the following results seem to be indicated. At temperatures above T_c, application of fields produced ordering characterized by a value of $L \approx 2000$ A ± 500 A independent of T, about the same magnitude as that induced by impurities at or near T_c, although as noted, the scattering was much less. At one temperature $\Delta T = -1°C$ the value of L was 1500 A ± 1000 A. All field effects were reproducible and no hysteresis was observed. A field-off run was made before and after each field-on run and the agreement was good.

Summarizing the above findings, it has been found that p-azoxyanisole does not show any excess scattering as a function of temperature for pure samples. Therefore the change in long range molecular ordering that occurs at T_c either does not result in electron density changes sufficient to produce measurable X-ray intensities or is masked by other effects. Impurities and D.C. electric fields were found to produce measurable excess scattering, but with different temperature dependence. At any temperature above T_c electric fields produce a correlation of extent seen very near T_c in impure materials. The magnitude of the field effect is considerably smaller than that produced by impurities. No change in scattering could be produced by using different types and qualities of surfaces for cell windows, nor was any observed in cholesteryl bromide.

REFERENCES

1. C. C. Gravatt and G. W. Brady, Molecular Crystals and Liquid Crystals, 7, 355 (1969).
2. J. G. Chistyakov, Soviet Physics Uspetki, 9, 551 (1967).
3. J. Frenkel, Kinetic Theory of Liquids, Dover Publications (New York, 1956).
4. C. C. Gravatt and G. W. Brady, J. Appl. Cryst., in press.
5. C. C. Gravatt, G. W. Brady and J. L. Lundberg, Rev. Sci. Inst., 39, 1701 (1968).
6. G. W. Gray, Molecular Structure and the Properties of Liquid Crystals, (Academic Press, New York, 1962), Chapter IV.

THE EFFECTIVE ROTARY POWER OF THE FATTY ESTERS OF CHOLESTEROL

J. E. Adams, W. Haas, J. J. Wysocki

Xerox Corporation, Xerox Square

W-114, Rochester, New York 14603

ABSTRACT

The pitch of two component mixtures of certain cholesteric liquid crystals is a strong function of chemical composition. We have found that over a wide range of materials the pitch of a mixture can be accurately represented by a weighted average of ingredients. If an effective rotary power, which we define as the inverse of pitch, is assigned to each constituent, the resultant pitch is just the inverse of the net effective rotary power of the mixture. In particular, if components with opposite intrinsic screw sense are mixed, there will exist one composition corresponding to no net rotation or infinite pitch. This technique was used to measure the effective rotary power of the fatty esters of cholesterol. Each ester was studied in mixtures with cholesteryl chloride. The results indicate a strong dependence on composition, with effective rotary power decreasing with decreasing aliphatic chain length. This technique provides a direct comparison of all esters at a common temperature. This is not possible in single component systems since there is no common cholesteric temperature and the strong pitch temperature dependence complicates studies. Although neither ingredient is liquid crystal at room temperature, the mixtures are liquid crystals over a wide compositional range and a modest extrapolation provides the single component data.

INTRODUCTION

The unusual optical properties associated with the Grandjean plane texture of the cholesteric mesophase are consistent with a molecular distribution consisting of layers of molecules stacked in a helical fashion. For example, dispersive reflection is explained in terms of internal Bragg-like scattering; and anomalous optical activity is related to the system's ability to distinguish right from left. Many of the optical properties of cholesterics can be derived from a knowledge of only the indices of refraction of the birefringent layers, and the pitch and sense of the helix. In many respects, these films act like a one-dimensional single crystal, with pitch playing the role of a lattice parameter. Dispersive reflection, for example, is, aside from the complications of refraction, equivalent to x-ray scattering in solid crystals. The homogeneity of solid single crystals is manifested in many experiments such as line broadening in NMR and X rays; equivalently, for cholesteric liquid crystals, the quality of the one-dimensional helical arrangement can be determined by line-width studies in optical-scattering experiments.[1] In fact, the homogeneity of cholesterics is much less than in solid single crystals. However, there is sufficient long-range order in these materials to provide the well-known spectacular dispersive reflection that has stimulated so much interest in these films. Furthermore, the sensitivity of the lattice parameter to certain stimuli in liquid crystals is orders of magnitude higher than in solid single crystals. For example, in certain cholesterics the pitch doubles with a temperature change of less than one degree centigrade.[2] There is another distinction between liquid crystals and solid crystals: The range of the liquid-crystal-lattice parameter is enormous, extending at least over three orders of magnitude. Samples can be conveniently prepared anywhere in this range, as opposed to the tedious and often impossible task of making systematic changes in solid lattice parameters. The technique for changing the pitch of a liquid crystal is to mix it with another liquid crystal. The mixture has a new pitch which can be predicted in terms of the pitches of the components.

The new helix has a homogeniety comparable to a single-component helix and, in fact, is not noticeably distinguishable optically from a single-component helix. This experimental fact suggests a new course of investigation. Both right-handed and left-handed cholesterics exist. Friedel[3] observed that by mixing two components of opposite sense, compensation occurred and that if the concentrations were adjusted properly, complete cancellation would result and the mixture would behave like a nematic. Much later, Cano[4] showed that in mixtures of cholesterics and nematics the nematic molecules affected pitch only by decreasing the density of active molecules. Recently, we have reported[5] that at least in certain compositional ranges the pitch of a mixture of two cholesterics can be expressed in terms of effective constituent pitches, using a simple linear additive approach. The word effective is stressed because in most cases neither ingredient is mesomorphic at the temperature of the experiment, while the mixture is. The important point here is that the role a molecule plays in a mixed helix can be characterized by a single parameter. Furthermore, it is possible to experimentally measure this parameter by an extrapolation of data from mixtures. Therefore, a comparison of helical-forming tendencies can be made for a wide class of cholesterics at a common temperature.

Helical-forming strength is conveniently expressed in terms of an effective rotary power which has dimensions of inverse pitch. The effective rotary power of a mixture is the direct weighted average of the effective rotary powers of the constituents. It is a measure of the structural role of an ingredient in the formation of a helical structure and should not be confused with the optical activity of the liquid-crystal film. Optical activity relates to the interaction of light with the helix. We report in this paper an experimental determination of the effective rotary powers of the fatty esters of cholesterol.

EXPERIMENTAL PROCEDURE

All chemicals were obtained from either Eastman Kodak or Aldrich Corporation. The esters studied were all of the form $C_{27}H_{45}OCO(CH_2)_LCH_3$. Ingredients were recrystallized from ethanol, and a C-H analysis performed by Galbraith Laboratories* yielded results which

---

* Galbraith Laboratories, Inc., P. O. Box 4187 - Lonsdale, 2323 Sycamore Drive, Knoxville, Tennessee 37921

TABLE I

RESULTS OF C-H ANALYSES[1]

	%C		%H	
L	Calc.	Obs.	Calc.	Obs.
0	81.24	81.55	11.29	11.25
1	81.34	81.35	11.39	11.25
2	81.51	81.37	11.48	11.46
3	81.63	81.59	11.57	11.55
4	81.74	81.77	11.65	11.64
5	81.85	82.02	11.73	11.60
6	81.96	82.26	11.80	11.60
7	82.05	81.96	11.87	11.82
8	82.15	82.09	11.94	11.98
10	82.32	82.28	12.06	12.18
14	82.61	82.89	12.26	12.26
16	82.74	83.00	12.36	12.36

1. Performed by Galbraith Laboratories.

are shown in Table I. The ingredients, all of which are crystalline at room temperature, were weighed and then dissolved in petroleum ether to promote better mixing. Samples were cast from solvent onto a black substrate. After solvent evaporation the liquid-crystal film was heated above the isotropic transition temperature and allowed to cool to drive off as much solvent as possible. To determine the influence of solvent on pitch, several samples were made by mixing dry ingredients directly in a heated crucible. No significant pitch discrepancies were observed between identical samples prepared by these two different procedures, although line-width studies suggest the mixing is more complete if a solvent step is included. All data reported were taken using this procedure. The films were caused to adopt the Grandjean plane texture by a cover-slip displacement. All measurements were taken at room temperature and with a free liquid-crystal surface. The film and substrate were mounted vertically on a spectrometer stage. The geometry for dispersive reflection is shown in Fig. 1. ϕ_i is the angle of incidence, and ϕ_s is the angle of reflection for a particular wavelength. The pitch, p, can be found using an expression derived by Fergason given below:

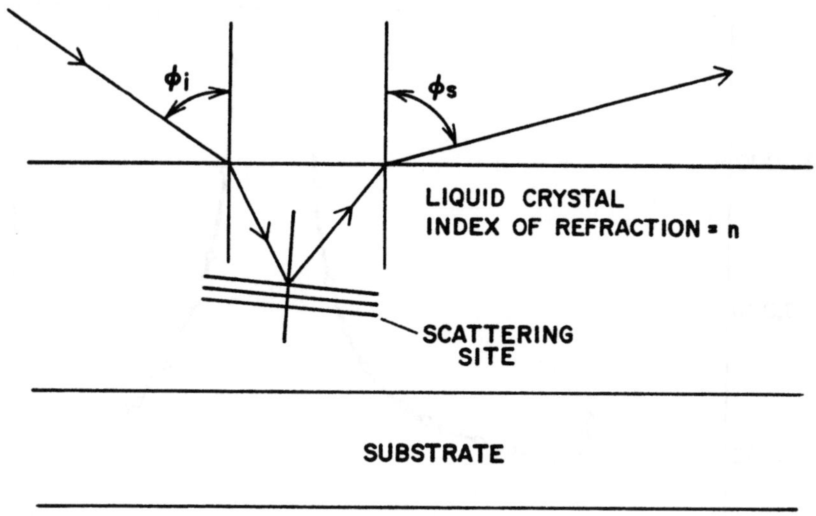

FIGURE 1. Geometry for dispersive reflection

$$\lambda = 2np\cos\frac{1}{2}\left\{\mathrm{Sin}^{-1}\left(\frac{\mathrm{Sin}\phi_i}{n}\right) + \mathrm{Sin}^{-1}\left(\frac{\mathrm{Sin}\phi_s}{n}\right)\right\} \quad (1)$$

Here, n is the index of refraction of the film. In practice, pitches were measured by fixing angle of incidence and observation and adjusting a monochromator for maximum signal. A photodiode was used as the detector. The wavelength range of this study was from 0.4 to 1.6 microns and was instrument-limited. However, in many cases sample stability also imposed a boundary on compositional range.

In all systems studied, the qualitative features of Friedel's observations were found to hold at least in certain compositional regions. Typical results for a mixture of an ester, all of which are left-handed, and cholesteryl chloride, which is right-handed, are shown in Fig. 2. In this case, cholesteryl chloride was mixed with cholesteryl nonanoate, and the per cent by weight of cholesteryl chloride in the mixture, we define as α. Although it is possible to separate the index of refraction from pitch experimentally by a fit to dispersion data, only the product is shown in this paper. Previous work[1] has indicated no particular trend in index

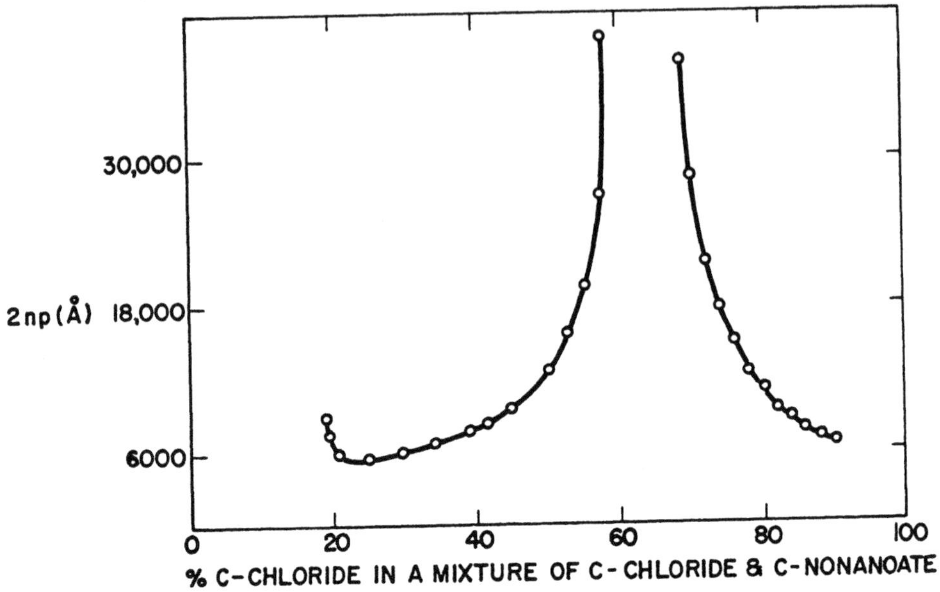

FIGURE 2. Pitch vs. composition in mixtures of cholesteryl chloride and cholesteryl nonanoate.

of refraction, all films having an index around 1.5. The most conspicuous feature of this data is the singularity at a composition we will call α^*. For this mixture the right and left-handed components essentially compensate and the result is a large pitch, nearly nematic film. Moving away from α^*, the sense of the film is determined by the dominant component. It is found experimentally that for $\alpha < \alpha^*$ the films are left-handed, and for $\alpha > \alpha^*$ the films are right-handed. There is a compositional region in which the data completely fail to follow an additive picture, in particular, in the region $\alpha \approx 20$. We believe that, in this nonanoate rich region, the film is tending toward the smectic mesophase. This conclusion is based on the observation[6] that only those esters which have a smectic mesophase, that is, $L > 6$, exhibit this anomalous region.

A linear additive argument would predict that

$$p = \frac{100}{\Sigma \alpha_i \theta_i} \tag{2}$$

where p is the pitch of the composite, θ_i represents the effective rotary power of the i th ingredient, and α_i, its percent by weight in the mixture. In the case of a two-component system, the pitch can be expressed in terms of the per cent by weight of one ingredient and the effective rotary powers of the two constituents. The reduced expression is given by

$$p = \frac{1}{\theta_A a/100 + \theta_B(1-a/100)} \tag{3}$$

where a is the per cent by weight of ingredient A. To determine the degree of fit, it is convenient to plot the inverse of pitch vs. a. Figure 3 shows the results of this plot in the cholesteryl chloride cholesteryl nonanoate system. The deviation in the $a \approx 20$ region is again a manifestation of the tendency toward the smectic state of these films. Because of the definition of pitch, it cannot assume negative values and a conventional representation of experimental results would appear as a V in this plot. We prefer to invert the ordinate scale for data for mixtures greater than $a = 60$ to demonstrate the fit more clearly, and to facilitate the most accurate extrapolation of data. A straight line extrapolation to $a = 0$ and $a = 100$ gives the effective rotary powers of the constituents. Typically, ten

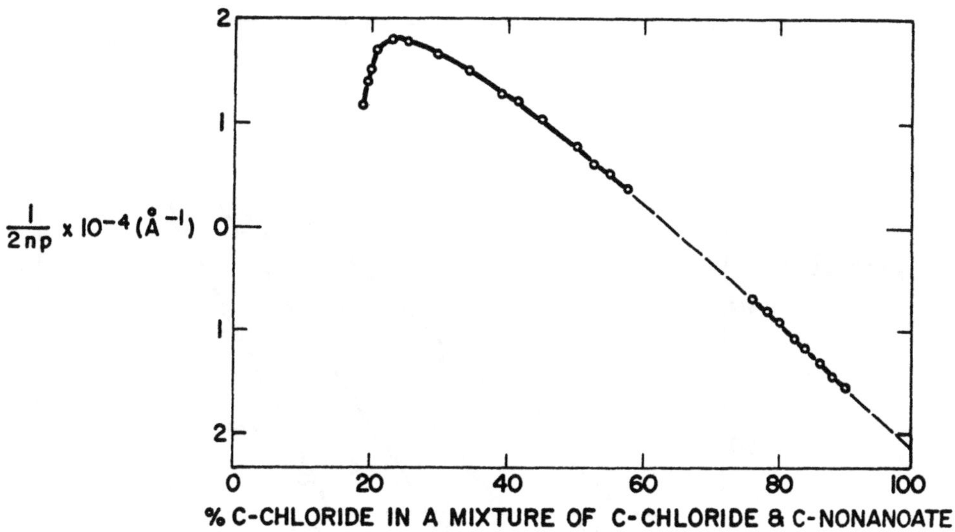

FIGURE 3. Inverse pitch vs. composition in mixtures of cholesteryl chloride and cholesteryl nonanoate.

points were taken for each ester; Fig. 4 shows some characteristic results. The practical experimental compositional range varied considerably from ester to ester and was limited principally by sample crystallization. Extrapolation of the data yields results shown in Fig. 5. There is a strong trend toward larger effective rotary powers in heavier esters. This is consistent with observations reported by Gray[6] in which he compares the range of color play in these esters as each is cooled through its own mesomorphic temperature range.

Mixtures of two components of the same sense were also studied, and good agreement with the general equation was found. A typical example of results in a system comprising two right-handed components is shown in Fig. 6. This particular system is a mixture of cholesteryl chloride and cholesteryl bromide and, as can be seen, the pitch varies monotonically from that of one pure component to that of the other. Figure 7 shows a similar plot in a system consisting of two left-handed ingredients. Here, a mixture of cholesteryl formate and cholesteryl nonanoate was chosen, and the shape of this curve is a manifestation of the trend toward larger effective rotary powers in longer chain esters.

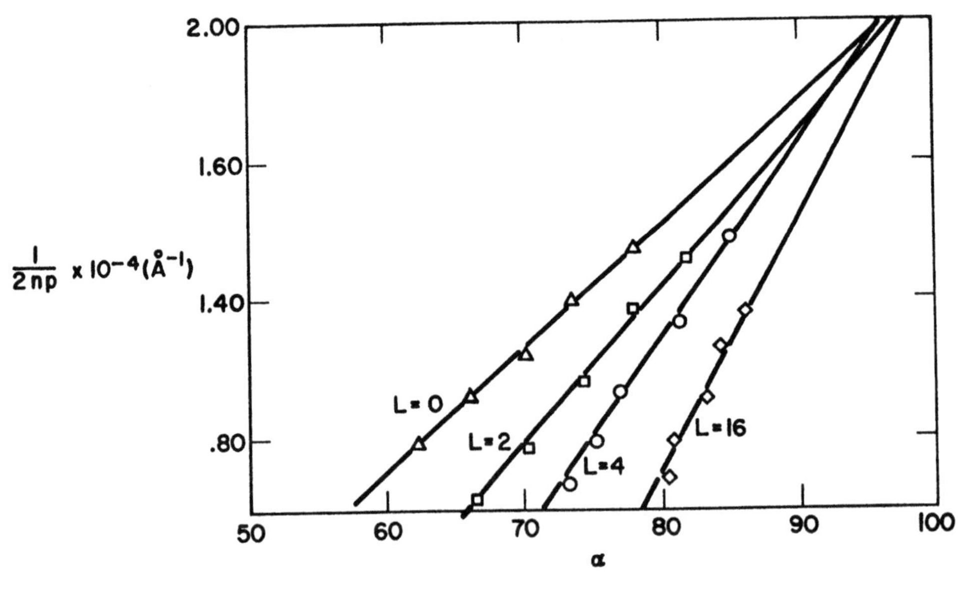

FIGURE 4. Inverse pitch vs. composition in four ester systems.

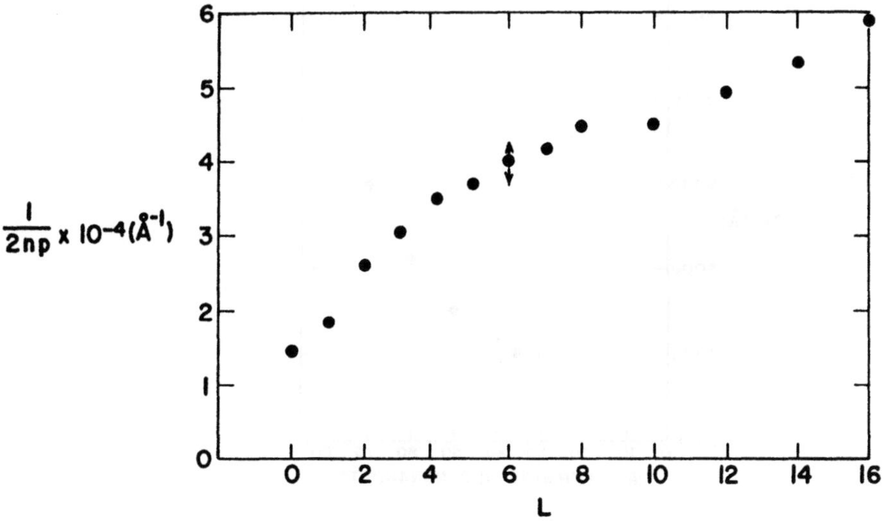

FIGURE 5. Effective rotary power vs. chain length

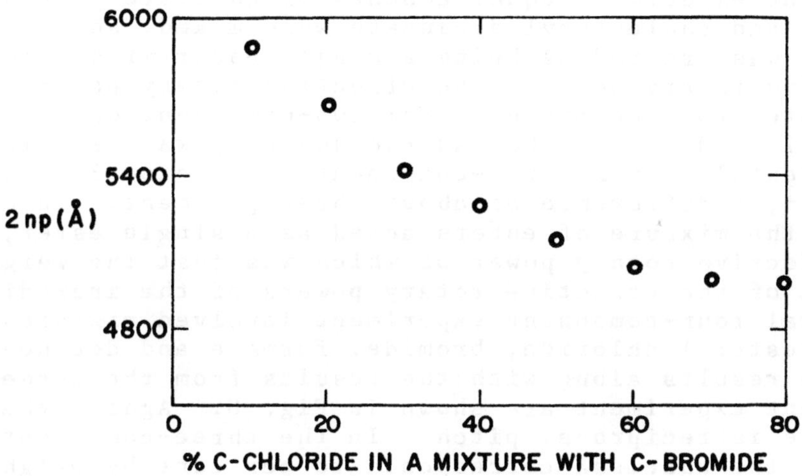

FIGURE 6. Pitch vs. composition in mixtures of cholesteryl chloride and cholesteryl bromide.

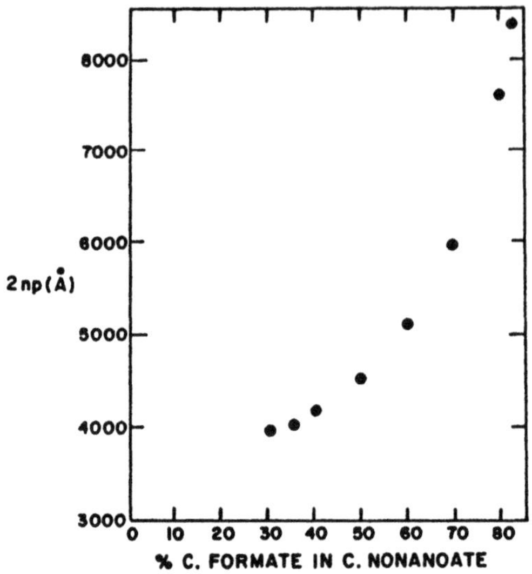

FIGURE 7. Pitch vs. composition in mixtures of cholesteryl formate and cholesteryl nonanoate.

We have also investigated the behavior of three- and four-component mixtures and found close agreement with the above prediction. In one particular three-component experiment equal amounts of cholesteryl propionate and cholesteryl decanoate were mixed, and this mixture was treated as being a single ester with some effective rotary power. The effective rotary power calculated for the mixture from two-component data was 3.17×10^{-4} Å^{-1}, whereas the quantity was determined experimentally from three-component data to be 3.29×10^{-4} Å^{-1}, a difference of about three per cent. In other words, the mixture of esters acted as a single ester, the effective rotary power of which was just the weighted average of the effective rotary powers of the ingredients. A typical four-component experiment involved mixtures of cholesteryl chloride, bromide, formate and decanoate, and the results along with the results from the three-component experiment are shown in Fig. 8. Again, the ordinate is reciprocal pitch. In the three-component results the abscissa corresponds to per cent by weight of cholesteryl chloride in the system. In the four-component results the abscissa corresponds to per cent by weight of a mixture of equal amounts of cholesteryl chloride and cholesteryl bromide in the system.

Again, it was found that any number of the ingredients could be considered as a single ingredient, and

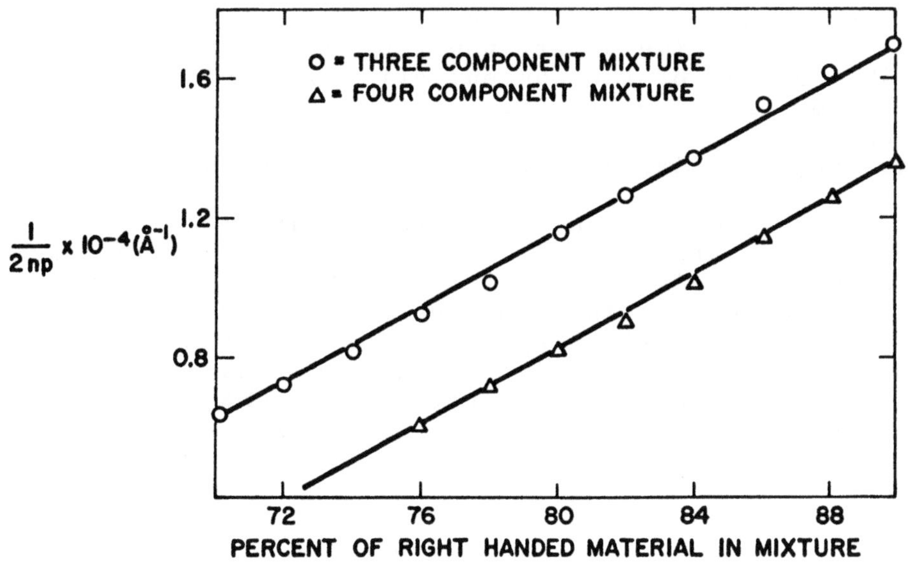

FIGURE 8. Inverse pitch vs. composition in three and four-component mixtures.

its role predicted from two-component data. These data are shown to indicate that no noticeable decrease in the quality of fit occurs as the number of ingredients is increased.

DISCUSSION

The anomalous optical properties of cholesterics involve the interaction of light with the helical structure formed by the molecules. This interaction includes interference effects caused by periodicity and is more complicated than just the sum of interaction with all the same molecules in a spatially random arrangement. In this sense the effects are cooperative even as Bragg reflection and energy gaps in solids involve cooperative effects. This feature arouses our interest in the degree to which the additive theory is followed in mixtures. The implication here is that each molecule plays a building block role and that this role is substantially unaffected by its environment. If this were not the case, the characterization of molecular role could not be as simple as a single intrinsic parameter, such as effective rotary power, but would have to include a statement about the modulation effects of neighbors. Of course, environment has some influence. We have found cases where the effective rotary power of one ingredient is a weak function of the concentration and effective rotary power of the other ingredient. The degree of usefulness of the concept of effective rotary power depends upon the size of environmental effects.

To our knowledge such effects are small and represent a perturbation on the helical-forming tendencies of individual molecules. If these effects were large, then some other intrinsic molecular property would serve as a better starting point for a systematic study of mixtures. The effective rotary power appears to us to be a useful molecular property and connecting it functionally to other known molecular properties is one of our aims.

CONCLUSION

In summary, mixtures of cholesterics produce helical arrangements which can be characterized by a single pitch parameter. The mixtures are homogeneous over dimensions much larger than the wavelength of visible light. By measuring pitch in a series of mixtures which are stable at some common temperature, it is possible to determine the helical-forming tendencies of the components. We have measured this property in the fatty esters of cholesterol and have found that aliphatic chain length plays the central role for these molecules. We intend to pursue this line of study in other systems in hopes of building up a reservoir of correlations which will shed light on the fundamental question of helix formation. Finally, we emphasize that the behavior of a molecule, in regard to its role in a helix, is a weak function of its environment. Studies of these effects, although they represent a considerable complication over the present work, may eventually provide a new level of insight into the cholesteric structure.

ACKNOWLEDGMENTS

The authors acknowledge helpful discussions with James Becker, Doris Olechna, Louis Leder and Frank Saeva and the technical assistance of Bela Mechlowitz and David Trost.

REFERENCES

1. J. E. Adams, W. Haas and J. Wysocki, J. Chem. Phys. 50, 2458 (1969).

2. J. L. Fergason, J. Appl. Opt. 7, 1733 (1968).

3. G. Friedel, Ann. Phys. 18, 273, (1922).

4. R. Cano, Bull. Soc. Franc. Min. Crist. XC, 333 (1967).

5. J. Adams, W. Haas and J. Wysocki, Bull. Am. Phys. Soc. 14, 6 (1969).

6. G. W. Gray, Molecular Structure and the Properties of Liquid Crystals (Academic Press Inc., New York, 1962).

MESOMORPHIC BEHAVIOUR OF THE CHOLESTERYL ESTERS-I : p-n-ALKOXYBENZOATES OF CHOLESTEROL

J.S.Dave and R.A.Vora

Chemistry Department, Faculty of Science
and Faculty of Technology and Engineering
M.S.University of Baroda, Baroda

ABSTRACT

A homologous series of Cholesteryl p-n-alkoxybenzoates has been synthesised and its mesomorphic behaviour studied. All the members of the series are enantiotropic mesomorphic. The higher members starting with cholesteryl p-n-heptyloxybenzoate exhibit enantiotropic polymesomorphism-smectic and cholesteric mesophases; the smectic mesophase increases with the increase in the alkyl chain length in the series at the cost of the cholesteric mesophase.

All the members of the series exhibit iridescent colours with the change in temperature, the colour intensity decreasing as the alkyl chain length increases and the last two members of the series do not show any colour change. At the smectic-cholesteric phase transition, a colour spectrum is observed both while heating as well as cooling. The higher members in the series show homoeotropic smectic texture on cooling.

The mesomorphic (1,2) state is a state of matter intermediate between a true crystal and a true liquid. It has the optical properties of a crystal and the mechanical properties of a liquid, Liquid crystalline substances are divided into three main classes ; (I) Nematic, (II) Smectic and (III) Cholesteric. (I) A nematic liquid crystal constitutes long rod shaped molecules which preserve parallel or nearly parallel orientation. (II) A smectic phase constitutes a stratified structure in which the molecules are arranged in layers with their long axis approximately normal to the plane of the layers. (III) A cholesteric phase has certain characteristics of its own which are markedly different from the smectic and nematic mesophases. It is optically active and shows extremely high rotatory power of the Grandjean plane texture. When illuminated with white light the most striking property of the cholesteric structure is that of scattering the light to give vivid colours. The colour of the scattered light at a particular angle to the surface of the film is dependant on (a) the substance (b) the temperature and (c) the angle of the incident beam.

Extensive work has been done on nematic and smectic mesophases and their structures are definitely known but comparatively very little is known about the cholesteric phase. Cholesteric phase is mainly exhibited by cholesterol derivatives from which it derives the name. The optical properties described above depend on a delicately balanced molecular arrangement. A change in shape or dipole-moment or any other disturbance which interferes with the weak forces between the molecules, results in a dramatic change ; reflection, transmission, birefringence, optical rotation and colour-all undergo marked transformations.

Daniel Berg (3) is of the opinion that in cholesteric liquid crystals the molecules are aligned to each other within the planes but with respect to the next plane there is a slight twist ; so, as one goes up the axis from one cholesteric plane to the next, the cholesteric molecules are spiral. The repeat distance is of the order of 1000 A°. This repeat distance gives the cholesteric phase its peculiar optical properties of scattering the light and appearing highly iridescent.

It is interesting to note that first liquid crystalline substance observed by Reinitzer (4) was

Table 1

Cholesteryl p-n-alkoxybenzoates.

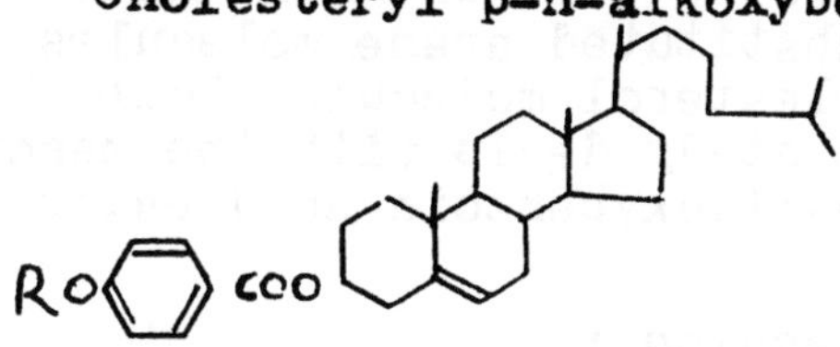

R =	Transition temperatures in °C: Smectic	Cholesteric	Isotropic
CH_3	-	180	268
C_2H_5	-	149.5	265
C_3H_7	-	141	253
C_4H_9	-	134	248
C_5H_{11}	-	148.5	236.5
C_6H_{13}	-	150	234.5
C_7H_{15}	138.5	160.5	222
C_8H_{17}	138	171.5	220.5
C_9H_{19}	128	176	213
$C_{10}H_{21}$	110	177.5	209
$C_{12}H_{25}$	128.5	179.5	200.5
$C_{16}H_{33}$	92	170.5	179.5
$C_{18}H_{37}$	47	161	163

cholesteryl benzoate. Friedel (5), Lehmann (6) and Jaeger (7) have studied the mesomorphic behaviour of the fatty acid esters of cholesterol. There is some disagreement about the transition temperatures and the number of mesophases in their results. Gray (8) studied homologous series of fatty acid esters of cholesterol ; he has reported modified transition temperatures for these compounds and some of the mesophases are identified which were missed in the previous work. Little work seems to have been done on the mesomorphic behaviour of the aromatic esters of cholesterol. Majority of the cholesteryl derivatives so far synthesised are with aliphatic side chains. The effect of substituted arene molecules as side chain in the cholesterol molecule should be interesting. The present study deals with the mesomorphic behaviour of p-n-alkoxybenzoic acid esters of cholesterol.

Results and discussion :

Following thirteen esters of cholesterol with p-n-alkoxybenzoic acids are synthesised and their mesomorphic behaviour studied. The melting point and transition temperatures of the mesophases are given in the table 1.

The cholesteryl p-n-alkoxybenzoic acid esters studied here give enantiotropic mesophases. Wiegand(9) has shown that cholesteryl anisoate melts at 162.5 - 163°C to cholesteric phase and clears to isotropic liquid at 236°C. Gray (1b) reports that it melts at 175°C to cholesteric phase and clears to isotropic liquid at 258°C with decomposition. In the present investigation the corresponding melting point and transition temperatures observed for this compound are respectively 180°C (m.p.) and 268°C (c.p.). In this study the slide was prepared by evaporating the solvent from solution of the compound. This on heating gives focal conic cholesteric texture which changes to plane cholesteric texture. The change from the plane cholesteric texture to isotropic transition is clearly observed under the microscope.

In this series the first six members show only enantiotropic cholesteric phase. The higher members starting with cholesteryl p-n-heptyloxy benzoate show enantiotropic smectic and cholesteric mesophases. In the corresponding fatty acid esters of cholesterol smectic phase commences with the heptanoate as a monotropic phase. All the aliphatic esters (8)

Table 2

p-n-Alkoxybenzoates (The temperatures are not corrected).

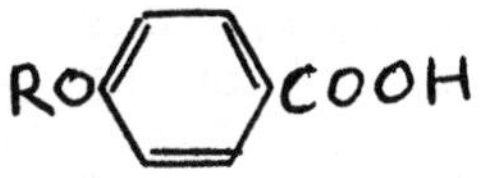

R =	Transition Temperatures in °C		
	Smectic	Nematic	Isotropic
Methyl	-	-	184
Ethyl	-	-	197
Propyl	-	146	156
Butyl	-	147	160
Pentyl	-	124	151
Hexyl	-	105	153
Heptyl	92	98	145
Octyl	100	107.5	146
Nonyl	94	117	144
Decyl	97	121.5	146.5
Dodecyl	95	129	137
Hexadecyl	84	-	131.5
Octadecyl	101.5	-	131

exhibit monotropic smectic phase except myristate which is enantiotropic smectic. The smectic-cholesteric transition temperatures of the fatty acid esters of cholesterol increase only upto cholesteryl laurate and then fall steadily through the myristate, palmitate and stearate. In this homologous series of cholesteryl p-n-alkoxybenzoates smectic-cholesteric and cholesteric-isotropic transition temperatures change in a regular manner when the series is ascended as is the case with other mesomorphic series. Cholesteric-isotropic transition temperatures fall smoothly with a regular alternation for the odd and even carbon-chain esters. Smectic-cholesteric transition temperatures rise to a maximum at the dodecyloxy benzoate and then fall off steadily as the chain length increases through the hexadecanoate and octadecanoate. This is graphically represented in the figure obtained by plotting the transition temperatures against the number of carbon atoms in the alkyl chain for the series. It can be seen from the graph that in consonance with other potentially mesomorphic series, the smectic phase length increases at the cost of the cholesteric phase, as the series is ascended. In both the series smectic-cholesteric transition temperature curve does not coincide with the falling cholesteric-isotropic curve and does not give rise to direct smectic-isotropic transitions for the longest chain esters which is the case in other potentially mesomorphic series. As a general rule the increase in alkyl chain length should increase the smectic tendencies of potentially mesomorphic compounds. Smectic behaviour is attributed to the relatively large lateral cohesive forces between molecules compared with the terminal cohesions. Since the lateral cohesions will increase as the terminal cohesions decrease with increasing alkyl chain length, it is possible to imagine a maximum in the smectic-cholesteric transition point curves, which will begin to fall only when the increasing lateral cohesions can no longer counterbalance the falling terminal cohesions(1c).

In order to compare the fatty acid esters of cholesterol with those of the p-substituted benzoic acid esters, the average smectic-cholesteric and cholesteric-isotropic transition temperatures for some corresponding esters have been calculated in each case. These values give an indication of the relative thermal stabilities of the smectic and cholesteric phases in the two series. In the calculation of the thermal stabilities, the monotropic transitions have been taken into consideration in the usual way.

Table 3

Cholesteryl p-n-Alkoxybenzoates.

R	Found % C	Found % H	Formula	Required % C	Required % H	Form
CH_3	80.48	10.30	$C_{35}H_{52}O_3$	80.769	10.0	White plates
C_2H_5	80.87	9.965	$C_{36}H_{54}O_3$	80.898	10.112	White plates
C_3H_7	81.25	10.27	$C_{37}H_{56}O_3$	81.02	10.218	White needles
C_4H_9	81.23	10.55	$C_{38}H_{58}O_3$	81.138	10.32	White needles
C_5H_{11}	80.79	10.21	$C_{39}H_{60}O_3$	81.25	10.416	White plates
C_6H_{13}	81.27	10.75	$C_{40}H_{62}O_3$	81.355	10.508	White needles
C_7H_{15}	81.28	10.48	$C_{41}H_{64}O_3$	81.456	10.596	White plates
C_8H_{17}	81.51	10.76	$C_{42}H_{66}O_3$	81.553	10.679	White needles
C_9H_{19}	81.60	10.83	$C_{43}H_{68}O_3$	81.645	10.759	White plates
$C_{10}H_{21}$	81.39	10.96	$C_{44}H_{70}O_3$	81.773	10.866	White plates
$C_{12}H_{25}$	81.94	10.93	$C_{46}H_{74}O_3$	81.899	10.979	White plates
$C_{16}H_{33}$	82.39	11.45	$C_{50}H_{82}O_3$	82.19	11.23	White plates
$C_{18}H_{37}$	81.91	11.14	$C_{52}H_{86}O_3$	82.321	11.345	White needles

Average transition temperature	Cholesteryl esters of Fatty acids	Cholesteryl esters of p-substituted benzoic acids	Increase
Smectic-cholesteric (C_7–C_{10})	78.0°C	171.0°C	93.0°C
Cholesteric-Isotropic (C_1–C_{10})	99.3°C	237.0°C	137.7°C

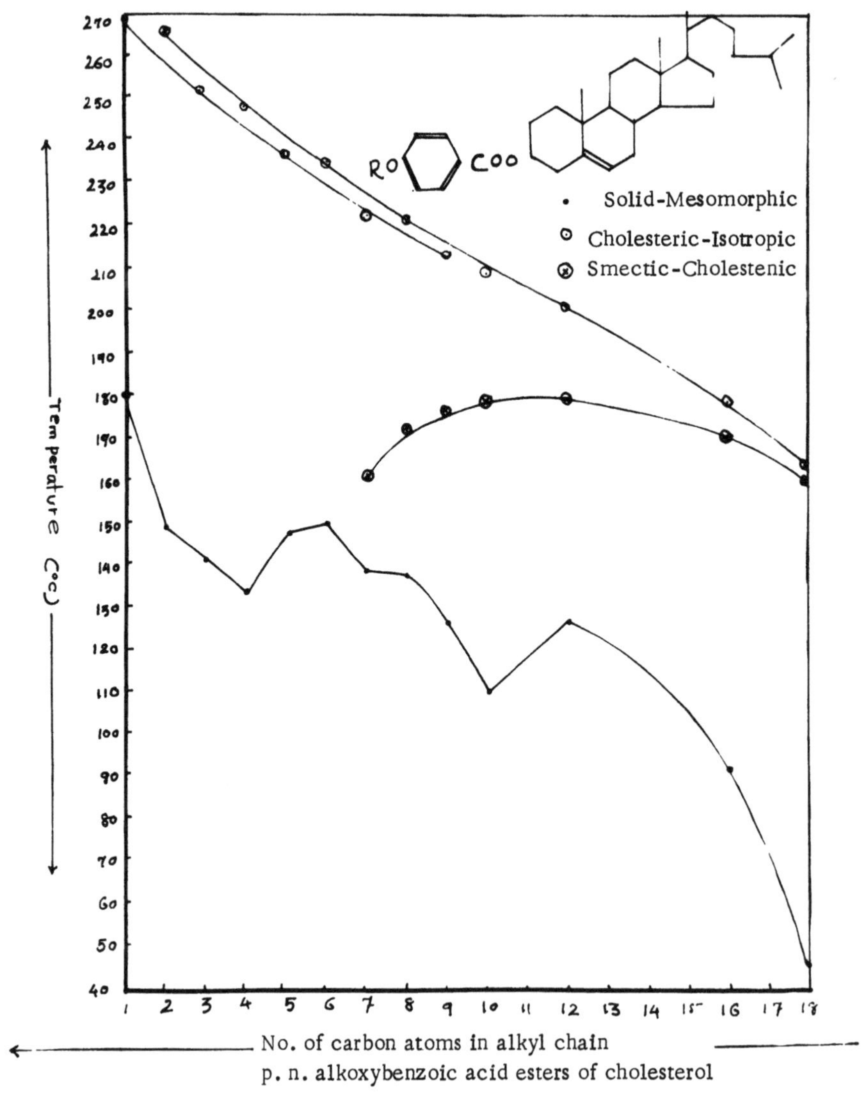

Figure 1.

The incorporation of the benzene ring in the cholesterylbenzoates increases the average relative stability of the cholesteric phase by 137.7°C and of the smectic phase by 93.0°C. The stability increases are appreciable in both cases, but the increase is greater in the cholesteric phase. This can be attributed to the conjugation arising out of the introduction of the benzene ring in the molecule.

In the polymesomorphic substances under investigation first the focal conic smectic phase is obtained which sometimes developes a homoeotropic texture. This on further heating gives cholesteric focal conic texture. Focal conic cholesteric phase texture changes to the plane cholesteric texture even with the slight disturbance in the melt. Plane texture gives high rotatory optical power and shows iridescent colours. On cooling first eleven members of the series give focal conic texture of cholesteric phase, just 0.2 to 0.5°C below the cholesteric-isotropic transition temperatures. Cholesteryl hexadecyloxy and cholesteryl octadecyloxy benzoates cool to give homoeotropic cholesteric phase 2°C below its cholesteric-isotropic transition temperatures. These cholesteric compounds on heating display red-green-greenish blue and blue colours. It is observed that the colour intensity decreases as one proceeds from member to member in the increasing order of the alkyl chain length. The phenomenon is reversible. At the juncture of the smectic-cholesteric transition, a rainbow spectrum is observed for a short time both while heating and cooling. The last two members of the series do not display any colour behaviour.

Experimental :

Determination of transition temperatures :

The melting points and transition temperatures of cholesteryl esters were determined with the help of Leitz Ortholux Polarising Microscope equipped with a heating stage. The slides were prepared by three different methods. In the first case the substance was heated on the slide upto its isotropic temperature ; a cover slip was placed over it and cooled. In the second method the substance was dissolved in a suitable solvent (chloroform). A few drops of the solution of the compound were placed on the slide and the solvent allowed to evaporate. A cover slip was then placed over it. In the third case the slides were prepared by heating the substance on the slide upto its mesomorphic temperature and putting the cover slip over it and cooling. The slides were then heated in the slot of the heating block of the microscope and the temperatures were noted. The temperature was raised gradually and within the critical region of the transition temperature to be noted the heating was regulated by three degrees per minute. In the first type of the slides, on heating, the solid changes to a focal conic cholesteric texture which in some cases transforms to homoeotropic cholesteric texture ; this sometimes makes the transition to isotropic form rather difficult to judge because both the homoeotropic and isotropic forms are non-birefringent. In the other two types of the slides, on heating generally the focal conic cholesteric texture gives the plane cholesteric texture. The change from plane cholesteric texture to isotropic transition can be clearly observed due to the high birefrigence and colourful plane texture of the cholesteric phase.

Preparation of compounds :

(1) p-n-alkoxybenzoic acids :

These acids were prepared by the method of Brynmor Jones (10). One mole of p-hydroxybenzoic acid was dissolved in concentrated aqueous solution of potassium hydroxide (2 moles). To this 1.1 mole of alkyl iodide or bromide was added and the mixture was refluxed for two to three hours for completion of the reaction. In the case of higher homologues the above method gives less yield and requires more heating period. So the following method which gives better yield in 1 to 3 hours refluxion period, was adopted. The aqueous potassium hydroxide was replaced by a solution of potassium hydroxide (2 moles) in 250 ml. methanol. The potassium salt of the alkoxy acid separates out which is filtered, dissolved in water and treated with acid (HCl). The alkoxy acid separates out. This is filtered and recrystallised from glacial acetic acid till it gives constant transition temperatures. The yield is 60 to 70 per cent. The melting point and the transition temperatures compare quite well with those given in the literature (11) and are listed in table 2.

(2) Preparation of acid chlorides (8) :

p-n-Alkoxybenzoyl chlorides were prepared by treating the corresponding p-n-alkoxybenzoic acids with thionyl chloride (20 per cent excess) and heating on a water bath till evolution of hydrochloric acid ceases. Excess of thionyl chloride is distilled off.

(3) Preparation of cholesteryl p-n-alkoxybenzoates (8):

Equimolecular proportions (0.01 mole) of cholesterol and acid chloride were taken in round bottom flask and N-N-dimethylaniline (20 ml.) was used as a solvent. The whole mixture was heated at 120°C in an oil-bath for two hours. The whole mass is added to ice-cold water when the oily ester formed separated out as solid. The solid is filtered and washed with 2N sulphuric acid to remove the dimethylaniline, followed by acetone. The solid is crystallised from chloroform-acetone (1:1) mixture to constant transition temperatures. All the benzoates crystallised as white plates or needles. The analysis of these compounds is given in table 3.

References :

1. Gray, G.W., "Molecular structure and the properties of liquid crystals". Academic Press, London, (1962), p-3.
1b. ibid p-192-196 (unpublished work).
1c. ibid p-189, J.Chem.Soc., 396, (1957).
2. Brown, G.H. and Shaw, W.G. The mesomorphic state, Chem.Rev. Vol.57 (6), (1957).
3. Daniel Berg, Natl.Acad.Sci.Natl.Res.Council, publ.No. 1356, 23-5, (1965), publ.1966.
4. Reinitzer, F., Monatsh, 9, 421, (1888).
5. Friedel, G., Ann.Phys. 18, 273, (1922).
6. Lehmann, O., Z.Phys.Chem., 56, 750, (1906).
7. Jaeger, Rec.Trav.Chim., 25, 334, (1906).
8. Gray, G.W., J.Chem.Soc., 3733, (1956).
9. Wiegand, C., Z.Naturforsch, 4B, 249, (1949).
10. Brynmor Jones, J.Chem.Soc., 1874, (1935).
11. Gray, G.W., J.Chem.Soc., 4179, (1953).

The authors take this opportunity to express their sincere thanks to Prof.S.M.Sethna for his keen interest in the work.

Baroda, 24-7-1969.

INDEX

T

U

V

W

X

Z

MIX
Papier aus verantwortungsvollen Quellen
Paper from responsible sources
FSC® C105338

If you have any concerns about our products,
you can contact us on
ProductSafety@springernature.com

In case Publisher is established outside the EU,
the EU authorized representative is:
Springer Nature Customer Service Center GmbH
Europaplatz 3, 69115 Heidelberg, Germany

Printed by Libri Plureos GmbH
in Hamburg, Germany